ANIMAL WELFARE SCIENCE

Animal welfare as a concept and practice is important to many disciplines and professions. From blue skies scientists to front line animal rescue, there is a need for this textbook, which distils and interprets the core concepts and language of animal welfare science.

Coverage includes the main theories and why we needed them; practical advice for how to assess animal welfare; illustrative examples of different species and the importance of their interactions; and the stresses and strains experienced across an increasingly diverse workforce. Foundational knowledge is combined with real-world examples and recent advances to communicate the content and spirit of this unique field.

This accessible text is designed to support tertiary animal welfare curricula but is also relevant to the study of human–animal interactions, anthropology, and related fields. It provides a jumping-off point for many areas of knowledge and application for anyone interested in how animal welfare and science intersect. *Animal Welfare Science: An Interdisciplinary Guide* was created to provide an informational grounding that fosters competency, credibility, and the confidence to engage with peers across the animal welfare space. It also provides an overview of the career landscape with practical advice about the challenges and pressure points that you may encounter. Along with copious color illustrations, bullet points, thought experiments, tables, and links to video and other online resources aid accessibility.

ANIMAL WELFARE SCIENCE

AN INTERDISCIPLINARY GUIDE

Emily Patterson-Kane
and Tina Rich

CRC Press
Taylor & Francis Group
Boca Raton London New York

CRC Press is an imprint of the
Taylor & Francis Group, an **informa** business

Cover Art: Air Element, a study from Joseph Sulkowski's allegorical mural APOKALUPSIS: An Uncovering.

First edition published 2026
by CRC Press
2385 NW Executive Center Drive, Suite 320, Boca Raton FL 33431

and by CRC Press
4 Park Square, Milton Park, Abingdon, Oxon, OX14 4RN

CRC Press is an imprint of Taylor & Francis Group, LLC

ISBN: 978-1-032-45112-1 (hbk)
ISBN: 978-1-032-44636-3 (pbk)
ISBN: 978-1-003-37545-6 (ebk)

DOI: 10.1201/9781003375456

Typeset in Times
by Deanta Global Publishing Services, Chennai, India

Contents

PART III EVALUATION

Preface

A need: This textbook arose from the realization that there was something missing in our teaching toolkit for animal welfare science. Where was the generalist text suitable for the student, professional, employee, enthusiast, or just advocate of animal welfare? This book aims to plug that gap with a text that provides a comprehensive and hopefully engaging overview to boost the confidence of any reader interested in animal welfare. But there is another imperative—the creation of a text that explicitly resists the urge to settle into a rigidly orthodox practice that differs by discipline or profession. Instead, we should hold space for the novel and perhaps revolutionary approach that might tangibly improve the lives of animals. The ethos of this book is to foster a mindset of openness, curiosity, and a generosity of spirit that is conducive to supporting the mission of animal welfare science.

The style: Together, we decided to deliver a picture of animal welfare science as it currently exists, rather than some simplified but distorted message. Both authors are classically trained scientists, fluent in the dry and dispassionate writing style of science. However, we opted for a more active voice for this text as we feel it is more inclusive. In this we're in good company, with at least one legendary scientist (Polly Matzinger) choosing to co-opt her canine companion as a co-author rather than be singularly responsible for her "active" voice (recounted in Ted Anton's *Bold Science*). As a nod to *Homo sapiens* the storyteller, we make liberal use of vignettes that illustrate themes, other viewpoints, or just the sheer breadth of the issues tackled. Animal welfare science is a broad space where the authors and the reader each have a distinct experience and point of view, and perfect agreement is not necessary to engage and learn.

Scientist as human: Now is a difficult time for science, and the public perception of scientists can be unflattering. "Robot-like and lacking emotions, as well as valuing knowledge over morality and being potentially dangerous" was the verdict published over eight years ago in *PlosOne* ("The Immoral Landscape? Scientists Are Associated with Violations of Morality", 2016). Reason and emotion are two sides of a cognitive heuristic that has kept *Homo sapiens* from extinction by combining energy-hungry brain power with a much "cheaper" gut reaction (intuition). In science we often find this pairing to be severed. One of René Descartes' many moments of genius gave us three-dimensional Cartesian coordinates—the space within which mathematical science would be explored. But despite his brilliance, Descartes was still emotionally deaf to the cries of the animals that he vivisected and condemned as soulless. The scientific study of life, and the living, presents a challenge to our abilities to abstract the world to study it—instead we must step outside of the box. Rather than negate emotion, we must recognize our own traits of empathy, sensitivity to social norms, and cognitive bias. We must be able to accurately interpret our own work and understand how we are emotionally influenced in, for example, designing theories, supporting dogmatic thinking, and deciding which animals have a culture.

And whose science? We also challenge the view of whose knowledge is authoritative. Some, with no classical training, painstakingly acquire a deep understanding of animal lives and behavior, only to have their efforts derided because of a lack of credentials. We acknowledge those contributions and refuse to delegitimize them because of a different language, practice, or knowledge system. To reintroduce the voices, passions, and personalities of some of the scientists we mention, we have provided links to online videos, films, books, and other resources that we would encourage you to access. And then there's the science itself. You might notice some humming and hawing in the text as to the nature of animal welfare science. Is it a discipline, is it a practice, a cross-disciplinary field? In truth, it doesn't matter. Animal welfare science is a shapeshifter—a newcomer whose practitioners have been audacious in challenging rigid definitions of animal welfare. Instead, aspects of animal life are mixed and matched to make bespoke models of welfare that make sense in their own context. The ideas of animal welfare science have skipped borders and continents, and changed mindsets on the way. In that spirit, we also had a desire to share some of the ideas and concepts that we wished we'd known earlier and have picked up along the way.

A manifesto for the mind: In writing this book the issue of consciousness became a bit of an obsession for one of us, who dragged the other into a deep dive of the great mind debate—what is the nature of mind? And again, it's no surprise that animal welfare scientists are in the thick of it, asking the question "Are we there yet?" The answer is a resounding no. The hard sciences have, in abstracting the thing to study it, created a conundrum. In turning an apple into a pear, scientists are aghast that they've collected data on a pear and not the apple, the apple being "mind" and the brain's machinations the pear. This anxiety is palpable, and heretical rumblings are manifesting as books published by those famous enough not to worry about ridicule from a scientific mainstream that has no idea of the right answer itself. The inclusion of this material is as much to demonstrate the uncertainty of our scientific worldview as it is to educate. This nuanced context provides an important background to even the most specialized or pragmatically applied activities that we carry out in service to or interaction with animals. Animal welfare is a space where some of the more important aspects of our understanding of consciousness and the consequences of ethical concern may play out over the next few decades.

Fighting the alternative fact: Those of us who are stateside will also know that we live in an era that has given us the phrase "alternative fact", and which provoked (according to the *New York Times*) a bump in the sale of George Orwell's dystopian novel *1984*. The extent of the suppression of scientific truth and expertise is a clear and present danger to multiple agencies and as a result, animal welfare. The tracking of anti-science activity at the federal, state, and local level is revealed by the *Silencing Science Tracker* run by the Climate Science Legal Defense Fund and the Columbia Law School's Sabin Center for Climate Change Law. Between November 8, 2016, and January 20, 2021, 346 such actions were identified. The war on science has become turbocharged in the Trump-Vance administration and risks becoming a national catastrophe. In short, our strange times have led us to touch on topics rarely covered in textbooks that warn and inform the unwary and provide pointers to support in an anxious contemporary workplace.

Acknowledgments

Whether we know them or not, the people who appear in this book are acknowledged for their contributions to the field—the inspiration, effort, and innovation they have provided.

Regarding the suggestions that led to the proposal of this book, we would like to acknowledge Dr Gail Golab and Dr Maya Gupta.

For their support and tolerance, we thank the members of our respective households and acknowledge their artwork and photography, which richly adorns this book. A huge thank you also to Janet Patterson-Kane, whose wisdom permeates its pages.

Thanks also to Mary and Dash, who had to hear far too much about consciousness. Cheers for not hanging up!

It was a joy to work with Joseph Sulkowski and Elizabeth Brandon; their artistry brought the cover of this book to life. We are also indebted to Janet Kessler for sharing her love and fascination of coyotes with us. All showed a remarkable generosity of spirit, and we are incredibly grateful for their contributions and suggestions.

Most of all, a thank you to our remarkable non-human animal educators, past and present. Many appear in the book, but one absentee whose influence is still greatly felt was named Lotte.

The last acknowledgment is to the process of investigation—the addictive sense of "foraging for pleasure" described by Jaak Panksepp. Multiple threads of investigation remain, linked to some extraordinary biology that we've yet to unravel. We're taking a deep breath now as the next dive will be so much deeper.

About the Authors

Emily Patterson-Kane trained as a behavioral psychologist in New Zealand at Waikato University and Victoria University of Wellington. She held the Killam Postdoctoral Fellowship to research the environmental needs of laboratory rodents at the University of British Columbia. Her published research investigates many aspects of animal cruelty, animal welfare assessment, and animal welfare history and education. Emily first took the role of Animal Welfare Scientist at the American Veterinary Medical Association and is currently employed as a director of research at the American Society for the Prevention of Cruelty to Animals®. She co-authored *The Sciences of Animal Welfare* and *Rethinking the American Animal Rights Movement* (Routledge, 2022). (Shown with Vienna, her canine collaborator.) Emily's website can be found at www.EmilyPattersonKane.com.

Tina Rich is a generalist scientist, which is a rare and adaptive thing in today's largely specialist world. After obtaining a PhD in Immunology from Birkbeck College (University of London), she took postdoctoral posts at the Brain Tumor Research Center (UCSF, San Francisco), the Laboratoire d'Immunogénétique Moléculaire (Institut des Cordeliers, Paris), and the Department of Pathology at the University of Cambridge. A Research Councils UK Fellowship supported her transition to tenure after being recruited to the Institute of Comparative Medicine (ICM) at Glasgow University's School of Veterinary Medicine. Tina's work at the ICM began an ongoing interest in non-human animal health and welfare. Since then, she has worked as a writer and editor, paralegal, and strategic contributor to non-profit organizations who have a focus on the welfare of the environment and animals. At this stage of her globe-trotting, Tina lives in New Jersey with two dogs and two cats, all of whom clearly communicate their welfare wants and needs.

Part I

Theory

The Houses of Parliament in the United Kingdom comprise the House of Commons and the House of Lords. Some of the earliest animal welfare laws were debated in these chambers and ultimately became law as Acts of Parliament. The social tensions that drove these developments and early theories of animal welfare are covered in Part I. Pencil drawing of the Houses of Parliament (#108068621) by Mike Demidov/Shutterstock.com.

DOI: 10.1201/9781003375456-1

Chapter 1
Introduction

1.1 THE UNAVOIDABILITY OF ANIMAL WELFARE

Animal welfare is defined as an ethically significant state of an animal ranging from negative to positive. As we become more sophisticated in understanding that pendulum of welfare, to what extent can we leverage the power of animal welfare science to support animals? As Marian Dawkins (2006) wrote: "Animal welfare science thinks big. It asks big questions about animal consciousness, animal health and animal emotions and it tackles big issues that affect millions of people and billions of animals".

While animal welfare is a particularly important and popular topic right now, it is neither a recent nor a passing fad. Historically, relatively few influential people have tried to argue that causing animals to suffer is ethically acceptable. However, a few researchers, most notably the French philosopher-scientist René Descartes (1596–1650), argued that animals are no more than complex machines or *automata* that are incapable of experiencing suffering. In an argument that descended to him from Aristotle and Thomas Aquinas, Descartes claimed that humans have no moral duty to protect animals from harm. Instead, Descartes viewed the animal as a spiritual void—a mindless, soulless creature. Proof positive of this, according to Descartes, was the inability of any animal to use language in a meaningful and expressive way.

Because he believed animals had no feelings, Descartes had no moral issues with his use of *vivisection* (surgery without *anesthesia*/induced unconsciousness) to learn about their anatomy and physiology (Cohen, 1936; Harrison, 1964; Cottingham, 1978). Nowadays, it is common to denigrate Descartes as an isolated historic aberration, or just a sadist, but his thinking was certainly not simplistic. Moreover, his position that animals are due little if any ethical consideration is not without its adherents up to the present day. Of course, we now know that animals have a multiplicity of ways to communicate with each other, and it's just that language happens to be a major form of *human* communication (Figure 1.1). Setting animals up for failure by comparisons with a human biology that they cannot replicate, but often exceed in other ways, is a feature

DOI: 10.1201/9781003375456-2

Figure 1.1 A meeting on the Jersey Shore. Two "chats" are going on here. One involves mostly human voice and body language; the other, canine body, olfactory, and acoustic systems. The English language dataset has ~600k words and the dog's nose has ~300 million olfactory receptors (vs. 6 million in humans), with a "stereo sniff" function. Which interaction is exchanging more data? Image reproduced with the permission of Janet Patterson-Kane.

of our shared history. Throughout this book, ample evidence will show that many non-human animals do possess welfare states, and so we should avoid harming them. These considerations then get caught up in an endless debate as to which animals have welfare, what kind of welfare, and what that means for our interactions with them.

Most people first develop the idea that animals have feelings in an intuitive way through empathy. But it soon becomes clear that animals experience the world in different ways from humans, and so intuition alone can get us into trouble. One obvious example is our predilection to biases such as *anthropomorphism* (inaccurately projecting human attributes onto non-human animals). Anthropomorphism literally means "shaped like a human". For example, when a chimpanzee seems to smile with their teeth gritted together, this is a fear grimace reflecting discomfort. The toothy *Dalmatian smile*, most practiced by its namesake, is a behavior that appears alarming but generally indicates joy (Figure 1.2). *Anthropodenialism*, the mirror bias of anthropomorphism, denies human attributes for non-human animals (de Waal, 1999). Walking the line between the two biases can be a very tricky exercise (Białek, 2023).

Figure 1.2 Meet Jupiter, a Dalmatian putting his best grin on. The smile/snarl combo is sometimes called a "smarl" and generally indicates a happy dog. After a weekend with Jupiter, the author's dog (not a Dalmatian) tried to contort his lips into a smarl as he noticed the attention that Jupiter was getting. Photo by Tina Rich.

Much of this book will cover how the sciences have helped us to develop a more complete understanding of animal welfare and will attempt to flag biases such as those we've just mentioned. The scientific method has helped us to conceptualize, quantify, and communicate our observations of animal welfare, and in doing so, help animals in a variety of settings. At the same time, persisting *anthropocentric* views of cognition (meaning human centered, see Goralnik & Nelson, 2012) have led us to misinterpret and underestimate the animal mind (Bräuer et al., 2020; Kamil, 1987).

The impetus and motivations for understanding and protecting animal welfare vary across the sciences, as they do across other domains such as ethics, religion, law, economies, and communities. Likewise, people's pervasive interests in animal wellbeing are many and varied. The key point is that the science of animal welfare truly succeeds when put into action. There will always be the argument of the pursuit of knowledge for its own sake, but animal welfare science has an immediacy that demands application.

Keywords: animal welfare, René Descartes

Drawing the line

This book will return repeatedly to the dilemma of deciding where to draw the line between species whose wellbeing is ethically important and those where it is not. The positioning of this line can be distressingly arbitrary. Some philosophers exclude most or all non-human animals from ethical consideration, just as we continue not to count them as *persons* under the law. This idea became especially important to philosophers during a period called the Enlightenment in the 17th and 18th centuries, a time when the scientific method was born in a society dominated by religious spiritualism. This revolution paved the way for a new scientific worldview of *materialism* (now often referred to as *physicalism*), which we will meet more fully in Part II. This was a truly radical moment and set in stone "how we think" about our practice of science. During this time, Descartes was just one example of a prominent animal welfare skeptic. Another was the German philosopher Immanuel Kant (1724–1804). Kant suggested that we only have duties toward rational beings and that all animals are without reason. However, Kant did fret that cruelty to animals might make a person cruel to people.

In the United States, animals still exist in an ethical twilight zone where there is some recognition that we have ethical duties toward them. However, animals are less likely to be protected if they are unowned, ugly, common, or invasive, have a poorly understood biology (i.e., we don't understand what hurts them), or are considered vermin.

Philosophers often hold quite diverse opinions of animal welfare. Some assert that we should treat animals more like insentient objects (animal welfare skeptics), while others embrace a moral equivalency between ourselves and non-human animals (animal liberation or animal rights). The result is that diametrically opposed recommendations for animal welfare might be made by researchers operating under different value systems (Fraser, 2008). In many Indigenous societies it is this winnowing of complexity to derive some hard and fast opinion that is seen as the enemy of wisdom (Yunkaporta, 2019). These clashing worldviews indicate another tension, the desire of many to study the world as we define it, rather than the world as it is.

Where do we draw the line that marks out those whose welfare should concern us? In 1989, in a memorable debate (*"Does the Animal Kingdom Need a Bill of Rights?"),* the prominent animal rights philosopher, Tom Regan, described non-human animals as "somebodies" rather than "somethings" (Figure 1.3). (Regan's powerful argument can be watched online at https://www.youtube.com/watch?v=xj-MJKFM0Zs.) Since then, science has even nudged some invertebrates into that class of animal considered to be a "somebody too intelligent to harm" (e.g., the octopus). Still, it is hard to avoid an arbitrary pecking order for animal welfare, which might be inverted in a different culture or tradition. On a pragmatic level, it would be difficult to do much of anything without harming some animals, like tiny eyelash mites, or the blobs of cells called placozoans, which bob around in water. Australian animal liberation philosopher Peter Singer was somewhat inconsistent on whether it was okay to eat oysters (although recent data for edible snails might provoke a rethink; Fossati

Figure 1.3 Is this crested pigeon a pest with a quiff or "a somebody"? If we pushed crested pigeons to extinction, would we try to recreate them genetically, as some are attempting for the passenger pigeon (Part II), or would they be terminally forgotten? Image reproduced with the permission of Janet Patterson-Kane.

et al., 2024), and Regan suggested that if one life had to be sacrificed for the sake of a group, it would be best if that life was non-human.

Regardless of what criteria you use to decide an individual's moral worth, it is likely that some members of the animalia do not have it. And when it comes to which animals to exclude from moral concern, is it better to draw a crooked intuitive and disputed line based on what makes us feel uncomfortable, or a clear and rational line with serious implications for our way of life no matter where we put it? One question raised by this book is whether we can or should draw this line at all, and what would happen if we did not.

Keywords: Peter Singer, Tom Regan, animal rights

1.2 ANIMAL WELFARE SCIENCE

Animal welfare science is the overlap between animal welfare and the sciences—a position that confers an extraordinary breadth of influence and reach. This definition might seem overly simplistic, but this overlap is a busy and sometimes contested space that does not belong entirely or even predominantly within any existing discipline. Despite its complexity, animal welfare science does have its own unique character and skills that require dedicated study to master. As an animal welfare scientist, regardless of your background or role, there are things you should know and things you should be able to do.

This book is designed for students contemplating or pursuing a career in the animal-related sciences or professions. It will equip them with the necessary understanding of animal welfare science that encompasses its theories and application. Coverage includes the species and settings that are most relevant to the animal-related professions, namely companion animals, livestock, wildlife, and animals used in research, testing, and teaching, with some inclusion of working animals and animals used in sport and entertainment.

As an interdisciplinary and introductory text, the goal is to outline the underlying principles and general scope of the field. Some topics may not be a focus of your own current or future professional activities. Nevertheless, an awareness and basic working knowledge of them is to be expected. In addition, for those wishing to specialize, these texts will supply a foundation for tackling more advanced topics in this emerging and evolving field. Each sub-unit includes a list of references for background reading or to pursue topics of interest.

Students of animal welfare science come from diverse backgrounds, take part in a range of training programs, and end up in many types of careers (and other activities). Most practicing animal welfare scientists have cobbled together their education from various sources, having been mentored by leaders in the area, or having graduated from one of the few dedicated university programs. Over recent years, courses and specializations in animal welfare have become more common, and professional opportunities are now more widely available.

In the process, it has become clear that several somewhat different traditions for animal welfare education have sprung up. These differences are associated with nationalities, charismatic theorists, professions, industries, and the approaches adopted by governmental or large non-governmental organizations (NGOs). While they overlap, they do provide noticeably different understandings and curricula. Textbooks also tend to have a scope that aligns with one or another trend. These different sets of values can be tricky to navigate, for example when jobs transition, and so a thorough grasp of the workplace mission and its culture is usually a rather urgent task on the to-do list. This becomes all the clearer as we realize that wellbeing and productive work depend on people aligning their labor with their ethical priorities, interests, and work styles.

This text is not seeking to establish an overly rigid orthodoxy of what animal welfare science is or should be. Instead, a descriptive approach has been taken to illuminate the developmental arc of animal welfare science with the goal of sharing expected proficiencies and opportunities. Organized animal welfare science has its immediate historical origins in the United Kingdom and is most developed in Europe, North America, Scandinavia, and Australasia. Only by continuing to adapt to the needs of diverse cultures will the field become more pervasive in other regions. Areas of study that are poorly represented in the English-language literature will tend to be under-represented. Given the breadth of the discipline, many subjects worthy of inclusion will be mentioned briefly or omitted due to limitations of space or the authors' familiarity with the subject.

Animal welfare science has several relevant dimensions. For this book they have been broken down into sections relating to theory, assessment, evaluation, interactions, and social norms. Across these sections, the field's history illustrates how practices evolved in a particular place and time. As such, animal welfare science is not static but responds to needs and challenges as they arise. Progress may be rapid, such as in nutrition and disease prevention, while other problems are more intractable, such as animal abuse and neglect. New challenges to animal welfare are posed by climate change, the threat of zoonoses, genetic editing, or the popularizing of new abusive practices such as animal crushing. Above all, the vision is to encourage the reader to join or collaborate with the cross-disciplinary community of animal welfare scientists in protecting and advancing the welfare of all animals.

Keyword: animal welfare science

Sciences and animal welfare

In this book there is a lot of discussion as to what animal welfare is. There is less about what science is, but this is a surprisingly hotly contested subject. In a general sense, science is a practice—an effort to carefully observe everything in the world to better understand and predict how it works. But you could pursue that goal without any scientific philosophy or method. For that reason, science is often described by its methodologies. For example, if you iteratively test hypotheses without knowing the outcomes, and then you corroborate or falsify those hypotheses using observable data (qualitative or quantitative), and you use a systematic approach and contribute to a shared body of knowledge, then you are participating in a form of empirical knowledge acquisition termed the *scientific method*.

Some people identify science very closely with experimentation, but many important phenomena can only be studied through passive observation (for example, planets). Others think that the scientifically valid method requires that we abstract the subject (i.e., separate it from ourselves) and study it as an impartial observer. This limitation was largely overthrown by the *Cognitive Revolution*, a movement born in the 1950s that revolutionized the scientific study of the *mind* (reportable only by the individual). To the present day, people strenuously disagree about what makes an activity scientific and whether their topic is a *hard science* (i.e., math, with greater precision and control over variables) or a *soft science* (i.e., sociology, where the results are more difficult to predict). Casting these arguments aside, what matters is that we participate in rigorous, imaginative, and compelling science. That we are alive to the fact that the scientific observer brings their own cultural baggage to the exercise, and so the scientific method is never culturally neutral. Every era has had its share of ridiculous mainstream scientific beliefs, and we will doubtless be no different.

Misunderstandings can be avoided by carefully trying to parse the context in which animal welfare is discussed. Always look at the time, place, speaker, audience, and the reason for the statement. For example, look at these seemingly contradictory statements. All are valid in the context of each author's personal understanding of what animal welfare is and what science is:

- "Animal welfare is a science" (https://www.worldanimalprotection.ca/education/what-is-animal-welfare)
- "Many professors and scientist [sic] think that animal welfare is not science" (Vučinić, 2015)
- "Animal welfare is not a science alone but is laden with values and common sense" (Fisher, 2007)
- "Animal welfare is science-based" (Batchelor, 2013)
- "The concept of animal welfare is not scientific but ideological" (Kim, 2005)
- "The study of animal welfare cannot be based entirely on science" (Lassen et al., 2006)

There are parts of animal welfare that are not science, just as there are parts of science that are not animal welfare—where to draw the line is where people disagree. While the field is described as animal welfare science, this author (EPK) thinks that it is best defined as the use of scientific approaches to address the issue of animal welfare, rather than a completely novel and separate discipline within science. Others will, of course, disagree, and this may change with time.

Keywords: the cognitive revolution, scientific method

1.3 ANIMAL WELFARE SCIENCE CAREERS

The importance of animal welfare in a career is sometimes direct and obvious, for example, as an animal trainer, researcher, animal control officer, or stockperson. However, other human activities may have unintended consequences for animals. For example, the scientists who developed the insecticide DDT in the 1940s certainly planned to kill a lot of insects. They presumably didn't set out to devastate the bald eagle population, but that's exactly what happened up until DDT was banned in the United States in 1972. Anyone working directly with animals should have a good understanding of animal welfare science, but this expertise (or an appreciation of its existence) is also crucial in all activities that have significant effects on animals, people, and the environment.

Many careers are grounded in academic disciplines—branches of knowledge that compile information and offer specialized training. In the field of animal

Table 1.1 Examples of animal welfare science disciplines

Biology • Conservation biology • Wildlife biology Ecology • Ecotechnology Ethology Neuroscience Physiology Zoology	Psychology • Comparative psychology Phenomenology • Interspecies phenomenology Sociology	Animal behavior • Clinical animal behavior Animal Studies • Critical animal studies Behavioral sciences Human Animal Interactions Ethnography • Multi-species ethnography • Etho-ethnology and Ethno-ethology	Anthropology • Anthrozoology • Critical anthropology Ethnozoology Social work • Veterinary social work Archeology • Zooarchaeology Ethnography Multi-species ethnography
Criminology • Green criminology Law enforcement • Animal law • Animal care and control • Humane law enforcement	Agriculture Biomedical research • Laboratory scientist • Laboratory technician Nutrition (Animal) Ecology	Economics Communications Science-science communication	Forensics • Veterinary forensics • Behavioral forensics Public Health • One Health Veterinary Medicine • Veterinary technician/ nursing

welfare science, these disciplines are continually evolving, seeking to refine their methods and expand their understanding. While it would be nearly impossible to compile a comprehensive list of all the disciplines relevant to animal welfare science, Table 1.1 highlights several well-established examples that offer tertiary-level education.

The relationship between animal welfare science and these disciplines is often complex. For instance, this book will illustrate how many influential animal welfare scientists have emerged from the field of psychology. Yet, in my experience (EPK), most psychologists remain primarily focused on human subjects. Although the disciplines listed vary widely in their scope and the extent to which they incorporate animal-centered perspectives, all have contributed prominent researchers and educators to the field of animal welfare science.

Keyword: discipline

Example: Veterinary medicine and animal welfare

Veterinary medicine is a prominent example of a profession based in a scientific discipline that explicitly includes animal welfare in its knowledge base and training methods. This placement of animal welfare within the discipline is not just a cultural practice but is a *codified* (regulated) aspect of training. Veterinary medical training uses a competency-based model, which sets expectations as to what a graduating veterinarian should know and be able to do. Welfare is integrated into the expected *day one competencies* that include: "incorporates animal welfare, client expectations, and economic considerations into the diagnostic or treatment plan" and "advises stakeholders on practices that promote animal welfare (The American Association of Veterinary Medical Colleges, AAVMC), https://www.aavmc.org/". Graduating veterinarians also take an oath that establishes practice "in keeping with the principles of veterinary medical ethics" and an "obligation [to] the continual improvement of my professional knowledge and competence". (You can read the veterinary oath online at the American Veterinary Medical Association, AVMA; https://www.avma.org/resources-tools/avma-policies/veterinarians-oath). The *Principles of Veterinary Medical Ethics* mentioned in the oath establish that a veterinarian must always consider the welfare of the patient, the needs of the client, the safety of the public, and the need to uphold the public trust vested in the veterinary profession. However, the primary consideration must always be to "the needs of the patient to prevent and relieve disease, suffering, or disability while minimizing pain or fear" (https://www.avma.org/resources-tools/avma-policies/principles-veterinary-medical-ethics-avma).

A veterinarian may also focus on welfare through a specialized job role. For example, most facilities that conduct research with animals have a committee that decides whether the research meets the necessary scientific and ethical standards (the Institutional Animal Care and Use Committee or IACUC). Other than the chair, the Animal Welfare Act specifies only two compulsory roles: 1) a person not affiliated with the institution to represent the community and 2) the veterinarian caring for the animals. This veterinarian is expected to safeguard the health and welfare of the animals employed in research, testing, or teaching activities at that facility.

Finally, a veterinarian may seek board certification as an animal welfare expert. There is a specialty college (American College of Animal Welfare, https://acaw.org/) where veterinarians can apply to be *boarded* (a board certification demonstrates additional specialist training) as an animal welfare specialist, equivalent to other specialty colleges for oncology, pathology, surgery, or shelter medicine.

Keywords: Institutional Animal Care and Use Committee (IACUC), Veterinarian's oath, boarded

1.4 CONVENTIONS AND FORMATTING

As animal welfare science is a broad and pluralistic field, this book aims to accommodate this diversity while covering the core history and curriculum. In doing so, some decisions needed to be made that would not be within the preferences of all readers. For clarity, some of these are shared below to make the authors' choices transparent.

Geography: The scope of this book is predominantly the United States. Material relating to other countries is supplied throughout because the field is international and informed by knowledge derived from other areas, particularly the United Kingdom.

Definitions: Extensive coverage is given to the diverse ways that animal welfare is defined within animal welfare science. After this section, animal welfare is broadly presumed to include considerations of physical, subjective, and natural states, and is most closely aligned with the *Three Circles* definition. This is not intended to assert that this definition is the *correct* one. It is, however, well known and not very prescriptive, and so its use tends not to exclude understanding from other points of view.

Animal-Based Research: Research is an important activity within science. However, a great deal of animal-based research can be challenged based on either a rejection of the use of animals for research, disagreements about how the treatment of animals in research may be justified or limited, or issues with the scientific methodology and rationale for animal use. In this book, research is referenced that many readers may find objectionable. The authors have chosen to include research important to shaping the field even when it caused suffering to animals.

Depictions of Suffering: No material is included in this book that would be generally regarded as explicit or gratuitous. However, it is the nature of the subject that situations of animal suffering will be described that you might find shocking and upsetting, which unfortunately comes with the territory.

Structure: Because this book provides introductory information on a wide variety of themes, the reader may wish to learn more about a topic that is only briefly mentioned. To assist further study, related literature is provided at the end of each chapter under the section titles and in alphabetical order. Where possible, open-source references have been used. Keywords are provided at the end of each section to indicate terms that students may wish to familiarize themselves with. These terms are only defined in the text when first discussed. A glossary clarifying the meaning of these terms and others is provided at the end of the book. Supplementary materials such as illustrative stories, examples, thought experiments and exercises, are not considered to be examinable material.

READINGS

The unavoidability of animal welfare

Białek, M. (2023). The new anthropomorphism debate and researching non-human animal emotions: A Kantian approach. *Roczniki Filozoficzne*, *71*(3), 205–229. https://doi.org/10.18290/rf23713.10

Bräuer, J., Hanus, D., Pika, S., Gray, R., & Uomini N. (2020). Old and new approaches to animal cognition: There is not “one cognition”. *Journal of Intelligence*, *8*(3), 28. https://doi.org/10.3390/jintelligence8030028. PMID: 32630788; PMCID: PMC7555673.

Cohen, L. D. (1936). Descartes and Henry More on the beast-machine—A translation of their correspondence pertaining to animal automatism. *Annals of Science*, *1*(1), 48–61.

Cottingham, J. (1978). ‘A Brute to the Brutes?’: Descartes’ treatment of animals. *Philosophy*, *53*(206), 551–559. https://doi.org/10.1017/s0031819100026371

Dawkins, M. S. (2006). A user’s guide to animal welfare science. *Trends in Ecology & Evolution*, *21*(2), 77–82. https://doi.org/10.1016/j.tree.2005.10.017. Epub 2005 November 10. PMID: 16701478.

de Waal, F. (1999). Anthropomorphism and anthropodenial: Consistency in our thinking about humans and other animals. *Philosophical Topics*, *27*(1), 255–280. https://doi.org/10.5840/PHILTOPICS199927122

Goralnik, L., & Nelson, M. P. (2012). Anthropocentrism. In R. Chadwick (Ed), *Encyclopedia of applied ethics* (2nd ed., pp. 145–155). Academic Press. ISBN: 9780123736321.

Harrison, R. (1964). *Animal machines: The new factory farming industry.* Vincent Stuart Ltd.

Kamil, A. C. (1987). A synthetic approach to the study of animal intelligence. *Nebraska Symposium on Motivation*, *35*, 257–308. PMID: 3332032.

Drawing the line

Fossati, P., Stefanini, F. M., Ravasio, G., & Coerezza, U. (2024). Welfare of invertebrates: A pilot study on a new land snail stunning technique. *Scientific Reports*, *14*(1), 8378. https://doi.org/10.1038/s41598-024-58133-4

Fraser, D. (2008). Understanding animal welfare. *Acta Veterinaria Scandinavica*, *50*(1), 1–7. https://doi.org/10.1186/1751-0147-50-S1-S1

Yunkaporta, T. (2019). *Sand talk: How indigenous thinking can save the world.* The Text Publishing Company. ISBN. 978-0062975645.

Sciences and animal welfare

Batchelor, C. E. (2013). *Ethical development in veterinary undergraduates: Investigating the value of a novel reflective exercise* [Unpublished doctoral dissertation]. University of Glasgow.

Fisher, M. W. (2007). Shelter and welfare of pastoral animals in New Zealand. *New Zealand Journal of Agricultural Research*, *50*(3), 347–359. https://doi.org/10.1080/00288230709510303

Kim, D. H., & Park, B. Y. (2005). In *Proceedings of the Korean Society for Food Science of Animal Resources Conference* (pp. 71–84). Korean Society for Food Science of Animal Resources.

Lassen, J., Sandøe, P., & Forkman, B. (2006). Happy pigs are dirty! Conflicting perspectives on animal welfare. *Livestock Science, 103*(3), 221–230. https://doi.org/10.1016/j.livsci.2006.05.008

Vučinić, M., & Radisavljević, K. (2015). Veterinarian attitudes towards animal welfare implementation into veterinary practice in Serbia. *Ochrana Zvířat a Welfare* (Proceedings of the 22nd International Conference of Animal Protection and Welfare 2015 ISBN 978-80-7305-759-6), 194.

Chapter 2

Theories—Foundational Texts

2.1 INTRODUCTION TO THEORIES

Animal welfare is a nuanced and hotly debated subject. To avoid confusion and ambiguity, scientists attempt to create clear descriptions of exactly what animal welfare means. These may show how animal welfare is scientific, aspects that might or might not constitute an animal's welfare, and how animal welfare can be measured and assessed. These descriptions usually take the form of a scientific *theory*. A scientific theory is a description of the structure of some aspect of the natural world. This description should be specific enough that an iterative process of testing can either support it or prove it to be wrong (a theory that can be disproven is termed *falsifiable*). Definitions of animal welfare have implications far beyond the pages of journals and textbooks. They tangibly influence outcomes such as:

- How laws and regulations govern animal use
- What products are attractive to consumers
- What kinds of research can attract funding and recruit students
- What careers are available to individuals
- What the community finds acceptable

These and other factors shape how people interact with animals and so which animals flourish, live well, experience suffering, or even exist at all.

Most animal welfare scientists will have developed a way of thinking about animal welfare that they find meaningful and useful. Likewise, they will also have ideas about science that reflect their personal philosophy of its role in society. Initially, these ideas might be difficult to articulate clearly, but with time they will be fleshed out and can provide structure for future professional activities and roles. A theoretical framework can be used to clarify goals and assumptions and to find work and other animal welfare activities that are fulfilling. As a suggestion, it is a good idea to start with well-established and popular frameworks that you can adapt or replace with new or more obscure options.

DOI: 10.1201/9781003375456-3

In addition to finding the best framework to use, it is important for a scholar in animal welfare to be conversant with the influential theoretical frameworks used in the field. Each theory addressed in this chapter is widely used, with variations between countries, cultures, industries, and settings. Moreover, each has some advantages and some limitations. Being familiar with a range of approaches helps a professional to demonstrate their expertise and to adapt to working in different settings while also keeping a healthy awareness of potential biases and blind spots. It is also essential to be able to accommodate a diversity of approaches when working with colleagues. Breakthroughs can be achieved by different groups working together even if their motivations and theoretical justifications are quite different.

Keywords: falsifiable, scientific theory

Henry Spira and reintegrative shaming

Journalist and social activist Henry Spira (1927–1998) organized no fewer than 409 institutional members to oppose the Draize test in cosmetics testing. (The *Draize Test* assesses how irritating a substance is by applying it to the eye or skin of an animal and recording its effects.) In 1980, Spira's coalition took out full-page newspaper ads to encourage the public to join in and boycott the cosmetics company Revlon; they wanted Revlon to immediately stop Draize testing. Those ads were paid for by the Millennium Guild, an organization headed up by a New York radio show host and animal rights activist, Pegeen Fitzgerald (Harriton, 1981). This was a coalition of convenience. Some of its members supported more humane methods of research; others wanted to end the use of animals in research altogether. However, in the immediate term, both goals were consistent with the action. Spira suggested that clear and broad-based expressions of disapproval could make companies end their inhumane practices through a process he referred to as reintegrative shaming (Munro, 2002). That is, to protect their reputation and their market share, organizations would stop using methods that many consumers found unacceptable. Shaming is reintegrative when it motivates a change in conduct to gain the approval of the group. However, shaming can also lead to rejection of the group's norms and a lack of desire for group membership, which becomes disintegrative shaming.

Keywords: Draize test, reintegrative shaming, Henry Spira

2.2 BRITISH ORIGINS

Most of the approaches to animal welfare described in this chapter originated in the United Kingdom, beginning in the late 1950s. Animal welfare became a subject of debate as public anxieties grew about more intensive and larger-scale animal use in farming and laboratory testing. Throughout the 20th century, Britain was the source of new ideas and movements to protect animals, with laws drafted to regulate and limit how animals could be used. At the other end of the legal spectrum, direct action groups such as the *Hunt Saboteurs Association* disrupted fox hunting, and the *Animal Liberation Front* (ALF) stole animals from laboratories, farms, and circuses. Some of these escapees (e.g., American mink) would go on to create havoc in the British countryside; other mink quietly liberated themselves! Later the ALF would become more extreme, both threatening and carrying out incendiary device attacks against stores and scientists targeted for their research (Figure 2.1).

Figure 2.1 This Coloradan fox might not be in the clear when it comes to encountering huntsmen resplendent in "hunting pinks". The Masters of Foxhounds Association of North America has a written code of practice that emphasizes "the chase". However, the sport is still called "hunting" because the hounds might not be of the same mindset. Disrupting lawful hunting activities is illegal in many US states, although there does seem to be an American Hunt Saboteurs Association. Photograph reproduced with the permission of Janet Patterson-Kane.

Over time, multiple theoretical animal welfare frameworks were developed to try and give structure to the discussions and regulations for animal-based activities. These frameworks varied, especially in how they integrated animal suffering with the role of science in animal welfare. Academics associated with English universities, especially Oxbridge (Oxford/Cambridge), developed many of the earlier theories. This chapter will focus predominantly on theories that sprang from two documents. The first is *The Principles of Humane Experimental Technique* (https://caat.jhsph.edu/principles/the-principles-of-humane-experimental-technique), which is usually shortened to *Russell and Burch.* This report led to the 3Rs animal welfare framework. The second document is the *Report of the Technical Committee to Enquire into the Welfare of Animals kept under Intensive Livestock Husbandry Systems* (https://archive.org/details/b3217276x), which is referred to as *The Brambell Report.* This investigation resulted in the development of the Five Freedoms animal welfare framework.

Vast quantities of text have been devoted to these reports and their meanings and implications, and this chapter will add a little more. But I would encourage you to consider reading at least part of the original text before being deluged with other people's interpretations. Both are available to read for free online. Once you become used to the writing style, you will see how the concepts drawn from these publications were originally expressed, reflecting the feelings and priorities of the authors. But before we delve into these documents, a quick word about models and theories.

Keywords: hunt saboteur, Animal Liberation Front (ALF)

"All models are wrong, but some are useful"

A model or framework is only valuable if it can be used. This nugget of wisdom is reflected in the aphorism "remember that all models are wrong; the practical question is how wrong do they have to be to not be useful". The quote is attributed to the statistician George Box (1987), who made a living by creating models and knew better than anyone that any statistical model will fall short in accurately describing the totality of the underlying data. After all, a model is still just a model.

One of the simplest types of statistical model is the linear relation. This asserts that for every unit increase in one variable, there will be a unit increase in another variable. For example, for every 15 grams of sugar, you can make one cone of cotton candy. If we predict a perfectly linear relationship, then grams of sugar (x-axis) and servings of cotton candy made (y-axis) should fall on a straight line. Often the relationship is imperfect, and some of the points fall away from the straight line. In these cases, the distance between the actual amount of sugar needed and the amount predictable by the line is considered the *error variance.* This difference may be due to random factors like mistakes in measurement, or other variables that are not accounted for. For example, you might use more sugar to fill a cone with cotton candy on humid days when it is less fluffy.

Just like a statistical model, a scientific theory tries to simplify and explain what is happening with a particular phenomenon. As Einstein commented (1936),

> The aim of science is, on the one hand, a comprehension, as *complete* as possible, of the connection between the sense experiences in their totality, and, on the other hand, the accomplishment of this aim by *the use of a minimum of primary concepts and relations.*

The goal of a theorist is to achieve an accurate representation of a natural phenomenon that includes not everything, but *everything that is truly important.* A successful theory helps those who use it to understand the underlying phenomenon and to accurately predict how to interact with it to achieve their goals. This requires 1) that what is included in the model is accurate and 2) that nothing significant is omitted.

Once derived, we have to consider how to use our theory. The way we treat animal welfare changes over time and place, in line with what we consider to be important and ethical (i.e., animal welfare science is situated). For example, the *Brambell Report* was commissioned to report to the British government as to how intensive farming might meet the ethical expectations of the public. Because it was designed for this purpose, it can be more usefully applied to pigs on British farms than the fate of the (now extinct) Chiriqui harlequin frog in the face of global warming.

That said, to create a shared understanding of what welfare is, and resist the biases and groupthink found in some settings, it is helpful that animal welfare theorists have an agreed-upon start point—a few foundational theories that we are familiar with. These can then be developed on an ad-hoc basis to create new theories that are a better fit for the current era and its most pressing needs (i.e., *are fit for purpose*).

Keyword: linear relation

2.3 *RUSSELL AND BURCH*

Russell and Burch is an excellent case study of how British experts were able to act decisively in making recommendations for animal welfare within the comparatively new and intensive research environments of the day. Russell and Burch co-authored their report with encouragement and guidance from the zoologist Sir Peter Medawar and the Universities Federation for Animal Welfare, which commissioned the document and selected the authors (Hubrecht and Carter, 2019).

Russell and Burch (1959) is a pithy and progressive manifesto that made no great impact when first released. However, its straight-talking style, sponsorship, and a revival of interest supported by HSUS (The Humane Society of the United States, now rebranded as Humane World for Animals) led to the report becoming influential. The approach taken became the template for regulating the use of animals for scientific purposes across much of the developed world. While this is the first animal welfare science framework being discussed in this chapter due to its publication date, it was not widely influential until the 1980s.

Although the language of this report may seem dated, the principles it established have been adopted, essentially unchanged, up to the present day. It's an interesting read, touching on topics relevant to the time, such as the decline in Guinea pig use for diagnosing tuberculosis (TB), as well as discussions that scientists shy away from nowadays, such as the impact of personality on a researcher's behavior toward animals. A key principle was that using animals for research is a privilege and not a right, and that this privilege depends on having the support of the wider community. *Russell and Burch* also asserts that procedures that cause suffering are justified only where risk or harm to the animal is minimized and balanced by the expected benefits (a *utilitarian* approach). Based on these assumptions, it was confidently predicted that the research community's goal was the 3Rs—to refine, reduce, and ultimately replace the use of animals, such that no animal suffering was caused in the pursuit of scientific advancement.

Keyword: utilitarianism

2.4 THE 3Rs

The 3Rs aim to:

- "Replace animals with other ways of gaining the knowledge required"
- "Reduce the number of animals used to achieve the results"
- "Refine the way animals are treated to minimize any impacts upon them"

(for more details see https://www.ufaw.org.uk/why-ufaws-work-is-important/welfare-of-animals-used-in-scientific-testing-and-research)

It is worth considering the background of the two people who were the inspiration and origin of the 3Rs. William "Bill" Russell (1925–2006) was a polymath with a long career as a sociologist. Rex Burch (1926–1996) was a microbiologist who pursued his career primarily outside academia and, while considered the

Figure 2.2 Burch might have won the confidence of the scientists he interviewed because of his expertise in laboratory work. Burch handled many Guinea pigs in his time, using them to test milk for the bacteria that causes tuberculosis. Imagine how many Guinea pigs had to be used for their name to become synonymous with being an experimental subject. This image is from Smith, A.C. (1915) *The Raising and Care of Guinea Pigs; a Complete Guide to the Breeding, Feeding, Housing, Exhibiting and Marketing of Cavies*. Kansas City, Mo., A. C. Smith. [Pdf] Retrieved from the Library of Congress https://www.loc.gov/item/15004969/ public domain collection.

junior partner, had more practical experience in laboratory science (Figure 2.2). It was noted that Burch "helped to channel the goodwill of scientists" (Balls, 2014). Russell belonged to an older academic tradition that roamed across discipline boundaries; he was an expert zoologist, a practicing psychoanalyst, well versed in the classics, and according to his *Guardian* newspaper obituary (https://www.theguardian.com/science/2006/aug/16/uk.obituaries), able to deliver his lectures in rhyming couplets! So, these two authors drew expertise from a broad base of sciences, social science, and beyond. Despite their collaborative success, Russell and Burch themselves had no lasting connection and would not meet again for over 30 years, when they were reunited at a 3Rs conference in 1995.

During the middle of the last century, academics in the United Kingdom had a quality, notable in Russell and Burch, of being able to craft novel and ambitious approaches to controversial issues. The duo consulted the relevant experts, used ethics to bridge the knowledge gaps, and created new foundational structures. They were forthright in their opinions of animal pain and suffering, our duty to animals, and the duty of science to society. There was also the belief that they were not wasting their time; they expected the government to adopt their recommendations and to act swiftly to address inhumane practices.

When this kind of principled position is used in applied settings and regulatory activities, there is a strong tendency to push back (this is especially true in the United States). However, the clarity and simplicity of the 3Rs has helped to resist this reaction. In the current iteration of their framework, a notable deviation from the original text is the omission of some of its more expressive terms, such as "inhumanity".

Keywords: The Three Rs, R "Rex" L, Burch, William Moy Stratton Russell

The three Ss

As people began to apply the 3Rs to different settings, they started to modify the concept to suit their purposes. One example is the 3Ss, presented by Carol Newton at a symposium in Washington, D.C. in 1975, and aimed at providing extra guidance for biomedical researchers. The 3Ss are Good Science, Good Sense, and Good Sensibilities. In 2016, Smith and Hawkins wrote a paper to flesh out the 3S concept (Smith & Hawkins, 2016).

Newton's considerations are meant to be used by researchers developing animal models and carrying out foundational research into living systems. They are a prompt to researchers to see the 3Rs as an opportunity to exercise some imagination and creativity as scientists. While the 3Ss concept is not widely known, it is included here to show that having an animal welfare goal is only one step toward achieving it. A conceptual tool like the 3Ss can help researchers to apply the 3Rs to their experiments. In the following chapters, there will be considerable coverage of how progress toward animal welfare goals can be achieved strategically through research and communication, laws and enforcement, and economic incentives, and as suggested by Newton, by the active and creative efforts of professionals.

More information about the 3Ss concept can be found on Norecopa's website (https://norecopa.no/3S). The 3Ss is an early example of a multi-dimensional model. This type of model seeks to provoke consideration of several qualitatively different aspects of a situation that might easily be neglected by people who may be busy, biased, or suffering from attentional blindness as they pursue the routine activities of their work.

Keywords: The Three Ss

2.5 CHANGING THE FRAME—THE 3Rs AND AGRICULTURE

One assumption of the 3Rs is that any animal suffering that occurs because of research is, put simply, a bad thing. This might seem like a self-evident statement, but it has some less obvious implications. Under the current 3Rs framework, animal suffering is tolerated because it is minimized and justified on a case-by-case basis by the potential benefits. And under this framework, the ultimate aim is to stop using animals for research. That said, while some use of animals in research has already been replaced, other areas are still being invented or are expanding. So, the period over which this might happen is often assumed to be very lengthy. Nevertheless, the existence of this assumption has the potential to be used strategically.

Take, for example, the suggestion to apply the 3Rs framework to animals used in agriculture. This suggestion has been made on a legal basis by Swiss lecturer and researcher Charlotte Blattner (Blattner, 2014) and is also promoted by Humane World for Animals in their brochure "Help Farm Animals: Follow the 3Rs" (published under their old HSUS logo). On the one hand, the suffering of animals needs to be understood uniformly across settings to avoid inaccuracies and biases in assessment. On the other hand, each context differs in its goals and community support, which will undeniably affect decision making. The research community embraced the idea that animal use in research is not a given but is currently tolerated based on the potential benefits for human society. On the other hand, farming is still widely accepted as wholesome and necessary. This feeling covers animal products, including those created by killing animals that are preferred for their taste and cultural meaning rather than necessity. In both cases, there are many nuances, with less support for intensive farming and expensive niche products such as veal or foie gras. However, except for certain robust minorities such as vegans or animal rights supporters, there is no widespread cultural consensus that the goal of society is to reduce or cease exploiting animals as agricultural commodities. And even within veganism, there is a range of observance, from not eating animal products to not using any product associated with animal exploitation.

Importing the 3Rs framework to agriculture has the appearance of seeking a careful and stringent way to protect animal welfare but carries with it the implicit purpose of reducing or eliminating the use of farm animals. This is not to suggest this goal is invalid, which is a matter of personal belief, but it is a less popular point of view across the mainstream culture of the United States, where Humane World for Animals operates. Instead, most welfare initiatives around farming aim to reduce suffering without challenging its worth or scale. Keep this in mind while considering the Five Freedoms theory, which will be discussed later and was oriented toward use in agriculture. But first, we'll take a break with a thought experiment.

Keyword: vegan

Thought experiment: Applying the *Belmont Report* to animals

In the United States, the ethical principles for the participation of *humans* in research came from the 1978 *Belmont Report* (available online at https://www.hhs.gov/ohrp/regulations-and-policy/belmont-report/index.html). These principles center on the concepts of respect for the participants, beneficence, and justice. Consider how the use of animals in research would differ if these principles applied to all species. For one expert perspective in favor of applying the *Belmont Report* to animals, see the following paper, which can be read online. Do you agree or disagree with this editorial, and why?

Ferdowsian et al. (2020). A Belmont Report for animals? *Cambridge Quarterly of Healthcare Ethics*, *29*(1), 19–37. Available online at: https://www.cambridge.org/core/journals/cambridge-quarterly-of-healthcare-ethics/article/belmont-report-for-animals/F4518E13F2FE89A7719C5082A7FB44F8

2.6 THE *BRAMBELL REPORT*

During the 1960s, the general public in the United Kingdom began to question their trust in animal farming. *Animal Machines*, published in 1964 by Ruth Harrison, had exposed people to the realities of new intensive farming practices that came to be known as *factory farming* (Harrison, 1964; Kirchhelle, 2021). The public disquiet about factory farming, framed as a moral or emotional criticism, was refuted by the agricultural industries as being unfounded or unfair.

In the following year, the British government sought to address these concerns through the work of a committee of experts, as was the typical practice

(Begemann et al., 2018). A more recent example of this is the 2010 *Bateson Report*, convened in response to the 2008 BBC documentary *Pedigree Dogs Exposed*, which highlighted the health and welfare problems experienced by some inbred pedigree dogs (Bateson, 2010). A technical report can take a concern seen as primarily emotional and perhaps not substantive and reframe it according to the scientific and ethical standards of the day. The *Brambell Report* represents an example of this transformation that was unusually effective and innovative for the times and, as with the 3Rs, could fast track new laws and regulations. The Brambell committee was assembled under the leadership of eminent zoologist Professor F.W. Rogers Brambell (1901–1970). Their *Report of the Technical Committee to Inquire into the Welfare of Animals Kept Under Intensive Livestock Husbandry Systems* is commonly referred to as the *Brambell Report*.

The report's findings became a touchstone for several concepts, foremost among these being that animal welfare was a vital issue in which science could and should play a significant role. The report also included an appendix by William Thorpe that presciently outlined the broad foundations on which this new science would be built, including existing information relating to behavior, motivation, preference, cognition, physiology, and veterinary medicine. It is significant that the *Brambell Report* explicitly states that scientific evidence that speaks to the subjective experiences of animals is indispensable. Specifically:

> welfare is a wide term that embraces both the physical and mental well-being of the animal. Any attempt to evaluate welfare, therefore, must take into account the scientific evidence available concerning the feelings of animals that can be derived from their structure and functions and also from their behaviour.

This directive shaped many later conceptual frameworks for understanding animal welfare and contrasted with the prevailing approaches to science that deemed subjective experience inaccessible to scientific examination. In the absence of a well-established scientific body of evidence, the committee further stated that, at the outset, it was ethically necessary to immediately consider animals' feelings when designing agricultural systems, rather than delay action until objective evidence was available (*the precautionary principle*). Moreover, they provided a framework for conceptualizing aspects of animal husbandry that might affect systems in potentially measurable ways. While the sentience of various animal species (defined as the ability to have feelings and sensations) has continued to be hotly debated, within the auspices of the *Brambell Report*, the matter was considered settled:

> animals show unmistakable signs of suffering from pain, exhaustion, fright, frustration ... and we accept that animals can experience emotions such as rage, fear, apprehension, frustration and pleasure

The UK government responded to the report with the *1968 Agriculture (Miscellaneous Provisions) Act*, which empowered them to create regulations to protect the welfare of farmed animals. This was a key moment in animal welfare science, and some consider that the *Brambell Report* led directly to animal welfare science emerging as a discipline (Carenzi & Verga, 2009).

Keywords: F.W. Rogers Brambell, Ruth Harrison, the Precautionary Principle

2.7 THE FIVE FREEDOMS

The release of the *Brambell Report* was followed by the creation of a government advisory group, that subsequently underwent several name changes until its final rebrand as the *Animal Welfare Committee* (AWC) in 2019. An early member of this group, John Webster, codified part of the *Brambell Report* into a free-standing animal welfare framework called the Five Freedoms, which was announced in a 1979 press release. The Five Freedoms framework strongly influenced animal welfare regulations across Europe. In addition, the Animal Welfare Committee structure was replicated across Europe and then unified in 2006 as the European Forum for Animal Welfare Councils. The Five Freedoms (AWC) are:

1. "Freedom from Hunger and Thirst"
2. "Freedom from Discomfort"
3. "Freedom from Pain, Injury or Disease"
4. "Freedom to Express Normal Behaviour"
5. "Freedom from Fear and Distress"

The Five Freedoms framework enjoyed an extremely high uptake, especially in the United Kingdom and much of Europe. Its success is attributed partly to its memorable and straightforward approach, which provided a clear basis for assessment and regulation. Furthermore, the proposed conceptual and regulatory frameworks gave leverage for the animal welfare sciences to satisfy the apparent need for empirical evidence. This helped establish the United Kingdom as the leader in animal welfare science at that time. The Five Freedoms also became a popular voluntary standard for agribusiness, including US-based producers, and were incorporated into voluntary certification programs (e.g., the United Egg Producers *UEP Certified Program*) and adopted by corporations such as the Subway sandwich chain and Tyson Foods. In these cases, industry was leveraging the Five Freedoms to reassure the public that their products come from animals that are humanely farmed.

Keyword: The Five Freedoms

The Five Freedoms: Criticisms and variations

Over time, the Five Freedoms framework has remained relatively unchanged, although further context and more specific documents about the corollaries and applications of this approach continue to be produced. While the AWC continues to present the Five Freedoms as a comprehensive approach, others have suggested expanding or changing the framework.

The main criticism of the Five Freedoms is that they focus on negative experiences at the expense of many potentially important factors (Mellor, 2016). To be fair, Webster also critiqued his own work after 25 years of reflection (Webster, 2016). A repurposing of the Five Freedoms has been suggested that restates them as the Five Provisions, with these indicating that although the total elimination of negative states is ordinarily impossible, we can, at the very least, act to minimize them. Theories that include a wider range of factors will be discussed in the next two chapters. The Five Provisions (adapted from Mellor, 2016) are:

1. Good nutrition: Provide ready access to fresh water and a diet to maintain full health and vigor

2. Good environment: Provide shade/shelter or suitable housing, good air quality, and comfortable resting areas
3. Good health: Prevent or rapidly diagnose and treat disease and injury, and foster good muscle tone, posture, and cardiorespiratory function
4. Appropriate behavior: Provide sufficient space, proper facilities, congenial company, and appropriately varied conditions
5. Positive mental experiences: Provide safe, congenial and species-appropriate opportunities to have pleasurable experiences

Keyword: The Five Provisions

We will mention one more model, the Five Domains, in Part III. While this model has a theoretical aspect, it is more commonly seen as an aid to assessment, and so is positioned as an *applied* methodology.

READINGS

Henry Spira and reintegrative shaming

Harriton, L. (1981). Conversation with Henry Spira: Draize test activist. *Lab Anim, 10*, 16–22.

Munro, L. (2002). The animal activism of Henry Spira (1927–1998). *Society & Animals, 10*(2), 173–191. https://doi.org/10.1163/156853002320292318

"All models are wrong, but some are useful"

Box, G. E. P., & Draper, N. R. (1987). *Empirical model-building and response surfaces* (Wiley Series in Probability and Statistics). Wiley.

Einstein, A.B. (1936). Physics and reality. *Daedalus, 132*(4), 22–25. https://doi.org/10.1162/001152603771338742

Russell and Burch

Hubrecht, R. C., & Carter, E. (2019). The 3Rs and humane experimental technique: Implementing change. *Animals (Basel), 9*(10), 754. https://doi.org/10.3390/ani9100754. PMID: 31575048; PMCID: PMC6826930.

Russell, W. M. S., & Burch, R. L. (1959) *The Principles of Humane Experimental Technique*. Methuen & Co Ltd.

The 3Rs

Balls, M. (2014). Rex Leonard Burch: Humane scientist and gentle man. *Alternatives to Laboratory Animals ATLA, 42*, 57–59. https://doi.org/10.1177/026119291404200510

The three Ss

Smith, A. J., & Hawkins, P. (2016). Good science, good sense and good sensibilities: The three Ss of Carol Newton. *Animals, 6*(11), 70.

Changing the frame—the 3Rs and agriculture

Blattner, C. (2014). 3R for farmed animals: A legal argument for consistency. *Global Journal of Animal Law, 1*, 1–26.

The Brambell Report

Bateson, P. (2010). *Independent inquiry into dog breeding.*

Begemann, S., Perkins, E., Van Hoyweghen, I., Christley, R., & Watkins, F. (2018). How political cultures produce different antibiotic policies in agriculture: A historical comparative case study between the United Kingdom and Sweden. *Sociologia Ruralis, 58*(4), 765785.

Brambell, F. W. (1965). *Report of the technical committee to enquire into the welfare of animals kept under intensive livestock husbandry systems.* HMSO.

Carenzi, C., & Verga, M. (2009). Animal welfare: Review of the scientific concept and definition. *Italian Journal of Animal Science, 8*, 21–30. https://doi.org/10.4081/ijas.2009.s1.21

Harrison R. (1964). *Animal machines: the new factory farming industry.* Vincent Stuart Ltd.

Kirchhelle, C. (2021). Slippery FACTs: The rise of a "mandated" animal welfare science. In S. Berger & H. Nehring (Eds.), *Bearing witness. Palgrave studies in the history of social movements* (pp. 175–202). Palgrave Macmillan. https://doi.org/10.1007/978-3-030-62792-8_10

The Five Freedoms

Farm Animal Welfare Council Press Statement. (1979, December 5). Retrieved from the Internet Archive https://archive.org/details/1979.-five-freedoms.-farm-animal-welfare-council.-brambell-comittee

Five freedoms, criticisms, and variations

Mellor, D. J. (2016). Moving beyond the "five freedoms" by updating the "five provisions" and introducing aligned "animal welfare aims". *Animals, 6*(10), 59.

Webster J. (2016). Animal welfare: Freedoms, dominions and "a life worth living". *Animals, 6*(6), 35. https://doi.org/10.3390/ani6060035. PMID: 27231943; PMCID: PMC4929415.

Chapter 3

Theories—Strain

3.1 COPING

Carpenter (1980) wrote:

> the welfare of managed animals relates to the degree to which they can adapt without suffering to the environments designated by man. So long as a species remains within the limits of the environmental range to which it can adapt, its wellbeing is assured.

The concept of *coping* (adapting without suffering) is central to another framework of animal welfare developed by Donald Broom. (Broom, a biologist, was the first holder of an academic Chair in animal welfare in 1986.) In its shortest form, this concept was: "The welfare of an individual is its state as regards its attempts to cope with its environment" (Broom, 1986). This simple idea had many implications and corollaries, the first of which being: "The indicators of poor welfare are of two general types, the one demonstrating that an individual has failed to cope with an environment, the other indicating the effort involved and the extent of an individual's attempts at coping" (Broom, 1986).

Broom's approach was distinct from those of many of his peers for two reasons. First, he indicates that welfare covers situations where the animal is not suffering, and that the experience of suffering cannot be measured scientifically (either directly or indirectly). Broom also explicitly asserted that animal welfare was not a matter of subjective states and could be poor without the animal experiencing suffering.

His list of how evidence can be used to assess welfare strongly emphasized physical states, albeit ones that are expressed and measured in diverse ways. These included: "reduced life expectancy, impaired growth, impaired reproduction, body damage, disease, immunosuppression, adrenal activity, behavior anomalies, and self-narcotization" (Broom, 1991). This concept chimes more with natural selection than with the wellbeing and comfort implied in the Five Freedoms. Centering the definition of welfare on physical expressions of coping rather than emotional states of suffering aligns good welfare more closely to

DOI: 10.1201/9781003375456-4

biology or even productivity (but in a more holistic sense). For example, where the goal is good biological function, the sacrifice of lifespan for immediate productivity would not be considered desirable.

Behavior is given less and later emphasis, and subjective states are not even mentioned. In this context, animal welfare becomes the responsibility of veterinary medicine and physiology, with less involvement from interdisciplinary and behavioral sciences such as *ethology* (the study of behavior from a biological point of view) and *psychopharmacology* (the effect of drugs on the mind and behavior).

Broom emphasized that coping was the dominant mechanism in understanding welfare, although suffering might still impact resilience. This coping approach reasserted the historical emphasis of science and objective evidence over the subjective, and was accompanied by advances in the understanding of *homeostasis* and its recovery (i.e., mechanisms that mobilize the body to deal with challenges and then return it to a balanced resting state or setpoint). While Broom's approach to welfare was the most widely adopted, it existed in a constellation of other definitions of welfare (some quite severe). These included: "a state of complete mental and physical health, where the animal is in harmony with its environment" (Hughes, 1976); "an animal is in a state of poor welfare only when physiological systems are disturbed to the point that survival or reproduction are impaired" (McGlone, 1993); and "good animal welfare is guaranteed when the regulatory range of allostatic mechanisms matches the environmental demands" (Korte et al., 2007).

Keywords: coping, Donald Broom

Homeostasis versus allostasis

In more recent research publications, the term *allostasis* has begun to appear in place of homeostasis. Allostasis refers to the processes employed to recover the desired physiological setpoint after stress. McEwen (2005) wrote that "allostasis also clarifies the inherent ambiguity in the term 'homeostasis' by distinguishing between the systems that are essential for life (homeostasis) and those that maintain these systems in balance (allostasis)."

Some researchers had felt that a better term was needed to focus attention on how the physiological processes that respond to stressors have both positive and negative effects. For example, chronic overactivity of the allostatic mechanism contributes to illnesses such as cardiovascular disease and stroke, and may also influence mood and anxiety disorders. These pathologies represent the allostatic load (i.e., the cumulative wear and tear of stress). It was also felt that it was important to demonstrate how the *resting* state of the body is not fixed but will vary over time, between situations, and between individuals.

Keywords: allostasis, homeostasis

3.2 STRESS

When Hans Selye (1907–1982) first studied the physiological stress response, it was known as the general adaptive response; it would be Selye who added "stress" to the medical lexicon. Selye's research with rats demonstrated that stress responses have an energetic cost that could not be met in states of *chronic* stress. The result was physical damage in the form of enlarged adrenal glands, atrophy of the thymus and pancreas, and ulceration of the stomach (Selye, 1938).

That a stereotypical stress response could be evoked by a broad suite of "demands" was a conceptual breakthrough. (Note that this is a broader concept than "fight-or-flight", which is the initial phase of the general adaptive response (see Part II). Physiological stress is now understood to be a broad domain of responses involving the nervous, hormonal, and immune systems. Stress responses also often have a psychological component, which popularized the idea of feeling *psychologically stressed*. This added a new term to our emotional vocabulary that is useful to describe some of the effects of fear (of an identifiable danger) or anxiety (without an immediate focus). However, it is very important to recognize that physiological stress can occur and contribute to welfare states in ways that the individual is not even consciously aware of.

Physiological stress responses are not always associated with negative experiences. They may also be associated with enjoyable activities like playing, exploring, and romance. We are familiar with activities that combine positive and negative sensations, such as riding a roller coaster or watching a horror movie. For animals, activities like play can have similar attributes. Boisterous roughhousing might lead to injuries, but in the context of an overall activity, it is still sought out and enjoyed.

Keywords: anxiety, physiological stress, Hans Selye

The best laid plans of rats and researchers

There are many standardized tests designed to assess levels of fear in animals such as rats and mice. These might be used to assess an anxiolytic drug or study the differences between rats with different genetic lineages or early life experiences. (The way that rodents are bred or raised can make them more susceptible to stress, and so they experience poorer welfare throughout life.)

One test of fearfulness is the open field, where the rodent is placed in a large box with no lid. A more fearful animal will stop moving (part of the *fight, flight, or freeze* behavioral responses associated with physiological stress). Other methods include a Light/Dark Box and the Elevated Plus Maze. They each use different methods to quantify how a more fearful rodent will prefer dark enclosed spaces and freeze or flee from bright and open ones. It is easy to assume that just because an apparatus is called a test of fearfulness, any score it gives you *is* fearfulness. This is not always the case. In the open field, a rat from a barren environment freezes at first, but after a while, it becomes even more active than a rat from an enriched environment. So, if you run the test for long enough, you may find that it is the fearful rats who are more active, while the bold ones get bored and take a nap.

There are standardized tests that are meant to measure other qualities, like intelligence, but which can be affected by fearfulness. The *Hebb–Williams Maze* is a set of barriers between an animal and a food source. The efficiency with which the animal finds the food is intended to be a test of cognitive abilities. Rats from barren environments, with a smaller and less complex brain, will perform poorly on this test. However, some researchers found that after the rats were motivated (using food deprivation or foot shock), their performance increased to the same level as that achieved with "enriched" rats. It is unlikely that being hungrier made the rats more intelligent. Rather, the food motivation overcame some other barrier to maze performance, such as fear of the maze itself (which, like the open field, is open at the top).

Another famous study showed that the expectations of the handler can also affect how well animals do in tests (Rosenthal & Fode, 1963). When students were issued rats, they were told that *their* rats were genetically bred to be "maze-bright" or "maze-dull". As expected, the maze-bright rats scored objectively better on the maze task. The problem was that the rats were all drawn from the same population. The experimenters' expectations have many ways to bias outcomes, some of which are unconscious and relate to how the animals are handled and treated.

In many settings, experimenter bias can be dealt with using "blinded" experimenters who are unaware which treatment group the animal is from. But in practice this can be quite difficult, and it is almost impossible in applied field-based research settings. For example, laboratory rats from barren and unenriched environments are readily identifiable based on muscle tone and how well groomed they are. The way researchers treat their subjects is also extremely important. This author (EPK) was asked to name her chicken participants in her MSc work. Chickens had to be given names that were neither derogatory nor had food connotations, the idea being that positive naming would lead to better treatment of our research subjects. So, thanks again to Gladys, Slim, Roadrunner, Scats, Struggle, and Spooky, aka Hens 31–36.

All of which is to say that an animal's response to stressors can be very important for assessing their welfare. But methods to assess these effects must be used cautiously, with an awareness that what a test is designed to do is not necessarily what it will deliver. This is true not only for tests of fear or intelligence, but for any assessment with a predetermined goal, including animal welfare assessments.

Keywords: open field, Hebb–Williams maze

3.3 UMWELT

In the 1980s, the Netherlands became a hub for animal welfare research, initially focusing on physiological stress and how an animal's wellbeing is determined by their surroundings. Ethologist Piet Wiepkema wrote: "animals can be conceived either as machines that produce and even reproduce independently from their environment or as living entities that have been adapted to a specific environment", which he referred to as their *umwelt* (Wiepkema, 1983).

Umwelt is a German concept that refers to the world as experienced subjectively by the individual. Wiepkema goes on to write: "animals fit into their environment like a key into a lock … the critical organ for all these interactions processes [sic] is the brain" (Wiepkema, 1983). Umwelt is the animal understanding its world through its personal sensory and perceptual lens. An echolocating bat will sense the world in a very different fashion than an alligator gently warming itself in the morning sun. The umwelt concept is idiosyncratic, influenced by cognition and memory, and was a departure from the British approach of *diagnosing* bad welfare from the outside (Figure 3.1). Now, animals might promote their own welfare (autonomy). The umwelt-informed perspective is holistic—a

Figure 3.1 Consider the umwelt of these animals. While human perception relies heavily on excellent daytime vision, olfaction is a major go-to sense for many animals. The flank of the horse is reputed to be more sensitive than the human finger-tip to touch, and you can see the curled lip on the white horse, indicating a good sniff is in progress. Consider all the spaces, sounds, temperatures, images, vibrations, and scents detected, however they are detected, by these animals. Then, add cognitive ability and overlay memories; a bush fire, an animal bite, a fight lost, or an offensive taste or smell that might influence behavior. For an excellent description of equine umwelt, you might read Carol Saslow's paper "Understanding the perceptual world of horses" (Saslow, 2002). Photographs reproduced with the permission of Steve Beardall.

synthesis of animals and their environments—and, like the *Brambell Report*, might be used to compare the stark environment of an intensive farm with a richer environment that better supports animal welfare.

The northern European approach was typified by its use of observational ethology and the influence of the German schools of philosophy. The discipline of ethology was pioneered by Niko Tinbergen in the Netherlands and Konrad Lorenz in Austria, and they shared the 1973 Nobel Prize (for physiology or medicine) with Karl von Frisch. A 50th anniversary article in 2023 gave an interesting perspective on ethology's fate since "the prize" (Font, 2023). A decade before winning the Nobel, Tinbergen published his *four questions paper* on the key questions that ethologists should busy themselves with (Tinbergen, 1963). These were causation, ontogeny, evolution, and function. Function, being much easier to measure than anything else, would become preeminent, with the other topics largely neglected (Colgan, 1990). Still, emerging schools of animal welfare science (in the 1980s to 2000s) in these parts of the world were more accepting of subjective data, which was still not the case in the United Kingdom or North America. Tinbergen's approach and ethology in general was to focus on observation and understanding rather than control. Ultimately, these welfare philosophies would become drivers in developing alternative agricultural paradigms such as organic production and regenerative agriculture.

Keywords: ethology, umwelt, Nikolass Tinbergen, Karl von Frisch

How technology influences strain models

Strain models were commonly explained using metaphors. For example, Descartes' proposition that non-human animals have no feelings was illustrated using the technology of the day, a moving statue using hydraulic pressure. Centuries later, in the 1950s, another hydraulic model was used by Konrad Lorenz to represent his notions of how instinctive behaviors occur (this is called the psycho-hydraulic model or sometimes, more flippantly, the flush toilet model).

In this model, an individual's potential to act (their drive) is represented by water filling a large tank. The tank has an outlet with a plug on a metal spring. This spring (release) is attached to a string suspending a tray (stimulus). The water is more likely to spurt out when the tank is full (drive is high) and the tray is more heavily weighted.

Hydraulic models treated behaviors as if driven by the *pressure* of motivation. Over time, research into cognitive processes eclipsed these models, although Lorenz was still, well into the 1960s, pushing the notion that the Olympic Games served as a "pressure release" for aggression. The change in focus to cognition was accompanied by an upgrade in metaphor, now based on electrical wiring. Initially, the brain was compared to an old-fashioned human-operated telephone exchange, with inputs connected to outputs by making choices. These choices might follow predictable rules but were also subject to changes and mistakes based on the foibles of the operator. The metaphor depended on the "telephone operator", which was criticized as being a homunculus (literally meaning "little man"). The behavior of this little person remained unexplained (Figure 3.2). Was there another little man in the little man's head, and so on, ad infinitum?

These days, computers and their software are used as homunculus-free analogies, although in this metaphor, the degree of input from the programmer or operator could still be seen as problematic. As we begin to imagine the existence of fully formed artificial intelligences, the role of emotion and any potential *welfare* of the technology may bring the metaphor full circle. That is, the metaphor of people being compared to very sophisticated machines may evolve into the reality of machines becoming sentient.

Keywords: homunculus, psycho-hydraulic model, infinite regression, Konrad Lorenz

Figure 3.2 The infinite regression homunculus problem—little people in our heads that make decisions for us, and little people in their heads, making decisions for them, *ad infinitum.* Image by Emily Patterson-Kane

3.4 MOTIVATION

We described this chapter as covering theories of *strain*. Strain is a motivator, and coping, stress, motivation, and preference all represent some form of tension that leads to action. The tension is often caused by having to make a choice that may alter welfare. Decision making may be complex, as the animal's perspective may not always be apparent. Other concepts can weave into this overall understanding of strain and response. For example, take the idea of needs. In order to cope, an animal must be able to satisfy their needs. So, you can discuss a situation either 1) from the perspective of how well the animal is coping or 2) from the perspective of what the animal needs. Freedoms refer to whether the animal is in a place that allows them to meet their own needs. Duties or provisions are how humans recognize that they must keep animals in spaces that permit these freedoms and needs to be met. The animal's impetus to seek a resource can be described as a drive or motivation. Endless other terms might apply, and more will be invented, but they all describe some aspect of this conceptualization of stimuli causing pressure, resulting in animals responding.

For example, a hungry animal will typically have a drive to seek out food. If they are free to move to a place where there is food, they do so to meet their nutritional needs. If they do not have access to a sufficient quantity or quality of food, then their caretaker has failed to meet their duty of care. The animal may be unable to access or benefit from food for a lot of reasons that can be expressed in a lot of ways, but the end result will be compromised welfare.

The specific way we describe this situation and our response to it reflect our thinking and biases. For example, a freedom focus might lead us to make food available 24/7 in such a way that the animal can healthily graze (e.g., bulked up with fiber or requiring effort to access), whereas a duty focus might lead to rules about how often food is made available, but perhaps still in a concentrated form that is eaten rapidly. Needs are most often considered to be organic needs, the basic requirements for biological functioning, such as food, water, temperature within a livable range, and breathable air. A broader array of needs (other than for survival) may be discussed in the context of reproduction, companionship, or other requirements for good welfare.

Keywords: needs (organic), motivation, drive

Whose drive is this anyway?

In the 1940s, Drive Theory was developed to explain the instincts animals follow to satisfy their needs. The term *drive* was defined by American psychologists Clark Hull and Kenneth Spence as a motivation that comes from within. Drive theory led to research into attachment in humans and later, to animals. It was also the basis for early research into the social facilitation of behaviors in humans and other animals.

But the term *drive* is used in other ways. For example, behavioral drivers are the unspoken rules of social norms, proposed by Taibi Kahler (Kahler, 1975). These are "Please Others, Be Strong, Hurry Up, Try Hard, and Be Perfect" (norms being informal rules and expectations in a society, such as people holding doors open for each other). The *drivers* approach to understanding and changing norms has been embraced by organizations such as UNICEF in their programs to increase vaccination uptake and eliminate forced labor (Petit, 2019).

And in yet another context, drive is a term that is used in psychoanalysis. Freud had a very distinctive approach, in which there are vital drives that support life, but people are also seen as having death drives, which cause behavior that harms themselves or others. As Freud famously said, "the aim of all life is death".

The point of this section is that many of the terms used in animal welfare science are not newly invented words. They are words that are defined in the dictionary, or defined in different ways in different discipline-specific dictionaries. In interdisciplinary fields like animal welfare, this can cause confusion. When a person says *drive*, they may be referencing innate motivations, social norms, or something quite metaphysical. It is hard to discuss a situation accurately when using different terms, but it is even more confusing if you are using the same words but fail to realize that you are using them to mean completely different things!

Keyword: norms

3.5 SUFFERING

Ethologist Marian Dawkins (doctoral student of Niko Tinbergen) challenged Broom's strain model and elevated suffering as the central and possibly only important part of animal welfare (where suffering is any kind of negative or unpleasant experience of a sentient being). Dawkins evolved her thinking on this through her career, and in fact, it may seem a little odd for this not to happen. Suffering was clearly front and center in her 1980 book *Animal Suffering* (Dawkins, 1980). In her 2021 volume *The Science of Animal Welfare*, the focus had switched to objectively measurable criteria, and specifically, measuring what animals want (Dawkins, 2021). These quotes from 1998 and 2017 exemplify the change: "a complete science of animal welfare would include a consideration of the conscious experiences of animals" (Dawkins, 1998) and "Animal welfare (well-being) can be defined objectively in terms of animal health and what animals want" (Dawkins, 2017).

The difference between these stances is more one of experience and perspective than of substance. In the first quote, assessing consciousness is a roadblock—a problem that needs to be solved. In the second, Dawkins takes a detour around it, and consciousness is no longer an obstacle to progress. Given that we're only really second-guessing what goes on in the animal mind, this appears to be a logical route to take. What remains unchanged is the idea that when we accurately measure what animals want, we can give them what is required for good welfare. Dawkins suggests that health and the animal's needs could act as a unifying definition or proxy for defining animal welfare.

A shared definition of this type could certainly aid communication and productivity in the field of animal welfare science. However, it is debatable whether animal welfare science can, or should, have this kind of central orthodoxy. The question for any model is: Does it exclude anything important? Is the model or definition relevant to every practitioner in the field? Being overly inclusive can cause a lack of coherence, and a field might devolve into confusion. Conversely, being too exclusive may provoke confident progress in the wrong direction due to an overly simplistic or biased understanding. Likewise, we should be aware of who is generating the definition. Does it come from a particular discipline, charismatic academic, or social tradition? Does it polarize the field or accommodate diverse opinions?

Keywords: suffering, Marian Dawkins

3.6 PREFERENCE

Measuring animal preference became a focus of researchers as they zeroed in on the animal's internal state (i.e., suffering or enjoyment) combined with their autonomy. Preference is an attractive area of study because it creates objective evidence, and we can intuitively connect it to animal welfare. That is, the animal is experiencing their welfare state, and they can communicate that state through their choices. If this is the case, it removes the need for our complex analyses of animal welfare, as the animal can do this for us.

A difficulty arises in that animals' methods of communication are often not the same as ours. So, we set about developing methods for animals to "show us" what they need to have good welfare. The simplest way to begin to tackle this was to give animals the opportunity to select the conditions they prefer, on the assumption that they prefer conditions that they enjoy, and enjoy conditions that are better for their welfare. There is considerable support for this position. For example, nutritional wisdom is often demonstrated when animals select their own food; they tend to select nutritionally suitable options (see Webster, 2009). In a broader sense, most species have evolved motivations that support their wellbeing through a mechanism of being biologically prepared to enjoy those options the most.

That said, there are limits to the ability of animals to make choices that benefit them, especially in artificial or unusual situations. For example, sugary and fatty foods would have been uncommon in nature and are generally overconsumed if provided freely. Some domestic breeds are also hardwired to be food obsessed. For example, some Labradors and flat-coated retrievers carry a gene mutation that creates insatiable hunger and a reduced metabolism (Raffan et al., 2016). These dogs are eating machines.

Preference testing, when provided in a manner that best activates welfare-preserving motivational systems, can be very informative. Preference can be assessed using many different techniques, for example:

- Dwelling time—when different environments are connected, where does the animal choose to spend all or most of their time?
- Choice—when offered different environments at a choice point, which one does the animal select?
- Economic demand—how hard will they "work" for an option by doing laborious or unpleasant tasks that can be increased? This was dramatically demonstrated in a study with mink that we will mention later in the book.
- Anticipation and frustration—higher-value options cause animals to look forward to them, and they express annoyance when thwarted from reaching them, expressing more variable and intense behaviors.
- Cognitive bias—animals from chronically stressful environments tend to interpret ambiguous situations as being negative or threatening.

In summary, Donald Broom's coping theory is discussed here as a highly effective approach that diverged from much of the Brambell Committee–inspired scholarship in the United Kingdom. However, the northern European approach emphasized animals' perceptions and desires. Marian Dawkins integrated the northern European ethos with her British academic culture to create a simplified animal-centered approach to welfare that struck a chord on the North American continent. Dawkins focused on animals' experiences and later, simply on what they wanted. These distinct approaches to animal welfare created new possibilities for the field and presaged new theoretical work, especially in Canada, the United States, and Australasia. These theories, emerging after 1980, moved away from strict definitions of what animal welfare is and edged toward a more descriptive and inclusive approach.

Keywords: *nutritional wisdom, cognitive bias*

READINGS

Coping

Broom, D. M. (1986). Indicators of poor welfare. *British veterinary journal, 142*(6), 524–526. https://doi.org/10.1016/0007-1935(86)90109-0. PMID: 3594185.

Broom, D. M. (1991). Animal welfare: concepts and measurement. *Journal of Animal Science, 69*(10), 4167–4175. https://doi.org/10.2527/1991.69104167x

Carpenter, E. (1980). *Animals and Ethics. A report of the working party convened by Edward Carpenter.* Watkins & Dulverton.

Hughes, B. O. (1976). Behaviour as an index of welfare. *In Proceedings of the Fifth European Poultry Conference, Malta* (Vol. 1005).

Korte, S. M., Olivier, B., & Koolhaas, J. M. (2007). A new animal welfare concept based on allostasis. *Physiology & Behavior, 92*(3), 422–428. https://doi.org/10.1016/j.physbeh.2006.10.018. Epub 2006 Dec 15. PMID: 17174361.

McGlone, J. J. (1993). What is animal welfare. *Journal of Agricultural and Environmental Ethics, 6*(2), 26–36.

Homeostasis versus allostasis

McEwen, B. S. (2005). Stressed or stressed out: what is the difference? *Journal of Psychiatry and Neuroscience, 30*(5), 315–318. PMID: 16151535; PMCID: PMC1197275.

Stress

Selye, H. (1938). Adaptation energy. *Nature, 141*(3577), 926–926. https://doi.org/10.1038/141926a0

The best laid pans of rats and researchers

Rosenthal, R., & Fode, K. L. (1963). The effect of experimenter bias on the performance of the albino rat. *Behavioral Science, 8*(3), 183–189. https://doi.org/10.1002/bs.3830080302

Umwelt

Colgan, P. W. (1990). Embattled ethology. *BioScience, 40*(2), 148–149. https://doi.org/10.2307/ 1311352

Font, E. (2023). 50 years of the Nobel Prize to Lorenz, Tinbergen, and von Frisch: integrating behavioral function into an ethology for the 21st century. *Frontiers in Ethology, 2,* 1270913. https://doi.org/10.3389/fetho.2023.1270913

Saslow, C. A. (2002). Understanding the perceptual world of horses. *Applied Animal Behaviour Science, 78,* 209–24. https://doi.org/10.1016/S0168-1591(02)00092-8

Tinbergen, N. (1963). On aims and methods of ethology. *Zeitschrift für tierpsychologie, 20*(4), 410–433.

Wiepkema, P. R. (1983). Umwelt and animal welfare. In S. H. Baxter, M. R. Baxter, & J. A. D. MacCormack (Eds.), *Farm animal housing and welfare* (pp. 45–50). Martinus Nijhoff Publishers.

Whose drive is this anyway?

Kahler, T. (1975). Drivers: The key to the process of scripts. *Transactional Analysis Journal, 5*(3), 280–284.

Petit, V. (2019). *The behavioural drivers model: A conceptual framework for social and behaviour change programming.* UNICEF Middle East and North Africa Regional Office, Amman: UNICEF.

Suffering

Dawkins, M. S. (1980). *Animal suffering: The science of animal welfare.* Chapman and Hall.

Dawkins, M. S. (1998). Evolution and animal welfare. *The Quarterly Review of Biology, 73*(3), 305–328. https://doi.org/10.1086/420307. PMID: 9737005.

Dawkins, M. S. (2017). Animal welfare with and without consciousness. *Journal of Zoology, 301,* 1–10. https://doi.org/10.1111/jzo.12434

Dawkins, M. S. (2021). *The science of animal welfare: Understanding what animals want.* Oxford University Press. https://doi.org/10.1093/oso/9780198848981.001.0001

Preference

Raffan, E., Dennis, R. J., O'Donovan, C. J., Becker, J. M., Scott, R. A., Smith, S. P., Withers, D. J., Wood, C. J., Conci, E., Clements, D. N., Summers, K. M., German, A. J., Mellersh, C. S., Arendt, M. L., Iyemere, V. P., Withers, E., Söder, J., Wernersson, S., Andersson, G., Lindblad-Toh, K., Yeo, G. S., & O'Rahilly, S. A. (2016). Deletion in the Canine POMC gene is associated with weight and appetite in obesity-prone labrador retriever dogs. *Cell Metabolism*, *23*(5), 893–900. https://doi.org/10.1016/j.cmet.2016.04.012. Epub 2016 May 3. PMID: 27157046; PMCID: PMC487361

Webster, J. (2009). Animal welfare and nutrition. In F. J. M. Smulders & B. Algers (Eds.), *Welfare of production animals: Assessment and management of risks* (pp. 113–132). Wageningen Academic Publishers. https://doi.org/10.3920/9789086866908_007

Chapter 4

Theories—Domains

4.1 THE THREE CIRCLES

Beginning around 1980, animal welfare sciences began to appear in Canada and the United States. In the United Kingdom, we already had a contrast between Broom's *coping* model and the *suffering* model favored by Dawkins. American theorists now exploded this approach by proposing that welfare could be defined using an amalgam of different factors, none of which was assumed to dominate. The first major multi-dimensional model for animal welfare was *The Three Circles*, formally proposed by Fraser et al. (1997). This theory proposed that three domains (as described in Figure 4.1) were generally sufficient to assess an animal's welfare, with no hard and fast rules as to what "counts" as animal welfare and what doesn't.

It can be argued that the Three Circles model does not function as a scientific theory because it is not truly falsifiable. Further, variables from each of the stated domains may, or may not, be important to the welfare of an animal in

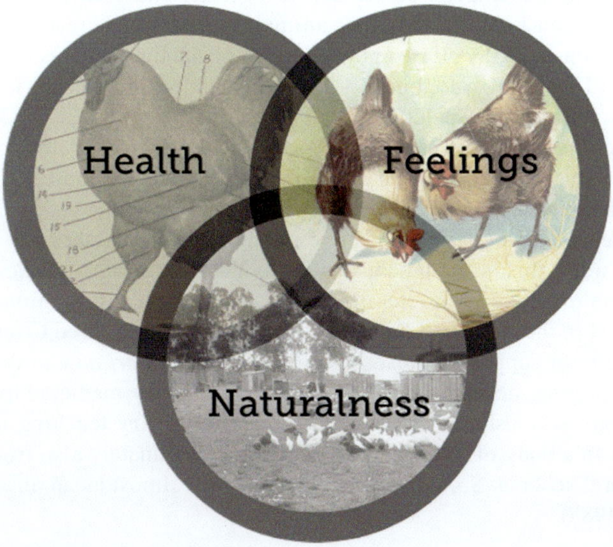

Figure 4.1 The three ethical circles of animal welfare included the following: "(1) that animals should lead natural lives through the development and use of their natural adaptations and capabilities, (2) that animals should feel well by being free from prolonged and intense fear, pain, and other negative states, and by experiencing normal pleasures, and (3) that animals should function well, in the sense of satisfactory health, growth and normal functioning of physiological and behavioural systems" (Fraser et al., 1997). Images in the public domain, including from the Library of Congress (https://lccn.loc.gov/17003753), and the Queensland State Archives (Digital image ID 2648).

DOI: 10.1201/9781003375456-5

any given situation. What the Three Circles model does achieve is to introduce a philosophical assertion that animal welfare is a multi-dimensional and inclusive concept, and that flexibility is more important than being precise and testable. The Three Circles and subsequent multi-domain models of animal welfare compromise some scientific norms as a trade-off in favor of increased public understanding. For some animal welfare scientists this is a step too far, and they remain wedded to the earlier theories.

Keyword: The Three Circles

Disciplines as domains

The domains outlined in the Three Circles align somewhat with major disciplines, which is no coincidence, as it was hoped that the theory might encourage interdisciplinary collaboration. For example, physical state and health tend to align with veterinary medicine, and behavior with psychology. Ethology can be seen as covering behavior and naturalness. The period when the Three Circles was most influential (roughly 1990–2010) also coincided with a time when these disciplines had different and rapidly changing approaches to the study of animal welfare. Traditionally, veterinary medicine placed a strong emphasis on physical health, and while veterinarians were often seen as animal welfare experts, the profession was not exactly embracing this role. Indeed, mention of animal welfare was only added to the veterinarian's oath in the United States in 2010.

A competing approach to animal welfare appeared most clearly from within ethology. Many ethologists were experienced with how behavior is shaped in complex ways (by environment, biology, and learning) and understood species-specific differences. Ethologists were less hesitant to embrace animal welfare as a multi-domain subject within the sciences and came to use "applied ethology" to indicate their shift in focus from wildlife to animals in captive or domesticated situations.

Consequently, the International Society for Applied Ethology (ISAE) became an important incubation space for animal welfare science, noting: "The study of applied animal behavior contributes to a greater understanding of the interactions between humans and other animals and helps to create a better balance between animal welfare and the requirements that humans have of other animals". The ISAE was founded at the instigation of veterinarian Andrew Fraser in 1966 as the Society for Veterinary Ethology (SVE); its 37 founder members were mostly Scottish. SVE was situated within the veterinary profession and focused on encouraging veterinarians to become more engaged with behavioral science. In 1970, its influential meetings had broadened in scope to include all ethologists, irrespective of any interest in veterinary medicine. In many ways, this development was symbolic of veterinary medicine failing to take the initiative in developing interdisciplinary ideas of animal welfare, as outlined in the *Brambell Report* and expressed in the Three Circles. While Andrew Fraser supported, unsuccessfully, keeping *veterinary* in the name of the organization, he lamented the failure of veterinary medicine to take animal welfare seriously, noting that "nor does veterinary teaching usually do justice to this body of knowledge; but this is unfortunately also true in many agricultural schools. So, our name has remained, almost as an anachronism" (Fraser, 1988).

ISAE's embrace of the subject of animal welfare provided an umbrella under which enormous amounts of animal welfare research and teaching occurred (Brown et al., 2016; Petherick and Duncan, 2016). This gave a behavioral emphasis to the study of animal welfare, and many of its leading advocates through the 1970s and 1980s would continue to come from the United Kingdom and have strong links with the ISAE. However, with time, and as ISAE became more international in scope, it drew in important researchers from North America and beyond.

Starting in the 1990s, it became apparent that other areas of science (as well as non-science disciplines) had a role to play in animal welfare, but these still tended to be sidelined relative to health and behavior. In recent years, veterinary medicine has seen an uptick of involvement and leadership in animal welfare science, which has rebalanced the relationship between the two and improved engagement with other sciences, notably those relating to social behavior, cognition, and neurology. Of note is the founding of the International Society for Anthrozoology (ISAZ) in 1991, with Dr Erika Friedmann serving as its inaugural President. ISAZ was founded in the United Kingdom, but as an international organization to study interactions between humans and other animals. ISAZ took over publishing the important multi-disciplinary journal *Anthrozoös* in 2001 and regularly holds meetings in the United States.

Keywords: anthrozoology

4.2 OTHER MULTI-DOMAIN MODELS

The models covered so far are all considered to be highly influential in animal welfare sciences and can be applied across a wide range of animals and settings. Other animal welfare models exist that are more specific in their purpose. A sample of these will be discussed below to show some differences in the subject areas that they include, how these are characterized, and their organization within each model. Each of these elements has powerful implications for what a model represents and how it should be used.

When a researcher begins to choose or create a model of animal welfare to use in a particular setting, they immediately start to categorize the information that is (or could be) available. They then assign this information different roles in a structure: for example, the three overlapping circles of the Three Circles model. When we look at models, their structures provide information about the assumptions being made and what they are designed to do. For example, in the Three Circles model, the aim is to derive an understanding of animal welfare that is more inclusive and represents, at a minimum, information relating to the animal's health, behavior, and naturalness. Any given piece of evidence may provide information about one, two, or all three of these domains, as represented by the overlaps in the Venn diagram. The overlap gives a visual representation of inclusion. Other multi-domain models also include and group together qualitatively different sources of evidence but make different choices about which domains to include and how they might interact with each other. They might be separate but balanced, as pillars, or tiered, introducing different levels of need. In other words, the model can be an abstraction of welfare that nonetheless conveys useful information by design.

Keyword: multi-domain model

4.3 DOMAIN INTERACTIONS

Four basic model structures and their functions will be described, with an illustrative example. The goal here is not to become intimately acquainted with each of these illustrative examples but to recognize how the structure visually represents something very important about the point of view being represented.

Venn: The Three Circles model shows a simple interaction between all of the domains as interlocked in a Venn diagram where they all partially overlap. Any data or information can be considered to fall into one, two, or all three circles.

Pillars: Another structure that may be used is the idea of pillars. Here, the areas are not seen as overlapping but instead contribute to an outcome—which is the "roof". There is a clear emphasis on all the pillar values being present to a similar degree to stabilize that outcome. In Kjosevski's pillar model (representing ethics, economics, and health), it was stated that "if the pillars are not balanced, then the implementation of AW [animal welfare] will be at the level of the weakest pillar" (Kjosevski et al., 2021).

Tiered: Other models have a tiered system where some domains indirectly influence welfare. This is often visualized using box and line diagrams, where each level affects the one above or to the right. One example of this is the pyramid, where good welfare is depicted as being at the top of the pyramid but is supported by the other layers. This design was used to describe a pyramidal hierarchy of dog needs (The Hierarchy of Dog Needs—Do No Harm Dog Training, https://www.donoharmdogtraining.com/hierarchy-of-dog-needs-presentation), based on Maslow's *hierarchy of human needs*. Organic needs such as food and water provide the most basic type of welfare, with other needs necessary to elevate welfare so that the animal can thrive. The welfare state runs through the center of the pyramid; it is merely sufficient for survival at the base and rises to a more ideal state at the top.

Cube: A cube can represent how different factors might interact in positive and negative ways using scales that interact with each other when plotted as the dimensions (height/width/depth) of a cube. Hickling (1994) used a "decision cube" to represent the welfare interests of two different populations of animals, how they might interact, and the tradeoffs involved. Specifically, the cube depicted the issues involved when protecting a population of animals valued by humans (e.g., livestock) from another considered to be a pest (e.g., possums that transmit tuberculosis to cattle).

Keywords: tiered model, cube model, pillar model

Exercise: Diagramming a model

Take a look at the *10 General Principles for the Welfare of Animals in Livestock Production Systems*—you can read it online:

Fraser et al. (2013). General Principles for the welfare of animals in production systems: the underlying science and its application. *Vet J*, *198*(1), 19–27.

Is this a domain model? If so, what would those domains be, and how might you visually represent them? What animals and settings do you think this approach was designed for, and what are its apparent assumptions and goals?

The published description covers inputs like genetics, physical environment, and handling, but is centered more on categories of output that are important for welfare. These outputs fall into three main areas (physical health, natural behavior, and suffering) and so would appear to follow a Three-Circle approach with more concrete guidance about what inputs to track and how to connect them to the circles.

4.4 REFLECTIONS ON WELFARE FRAMEWORKS IN A BICULTURAL SOCIETY

Animal welfare frameworks that are intended to be broad or universal in application also need to engage with the perspectives of different audiences. The Five-Eyes anglosphere of the United Kingdom, the United States, Canada, Australia, and New Zealand, will often describe themselves as multicultural, although the official recognition of this barely goes beyond language considerations. It is not widely known outside New Zealand that its foundation document (*Te Tiriti ō Waitangi*/the Treaty of Waitangi) is a partnership, and New Zealand is officially bicultural. In 2021, the Ardern government implemented a policy whereby science should incorporate *Mātauranga Māori* (Māori "ways of knowing") on an equal footing to Western science (this and other aspects of Indigenous knowledge systems will be touched on in various parts of this book).

So, how do we incorporate *Mātauranga Māori*? Based on the works of Sir Mason Durie (1984), a recognized Māori model of health and wellbeing also has multiple domains, arranged as the four walls of a *wharenui* (meeting house). Specifically, *Te Whare Tapa Whā* (health and wellbeing) is visualized as a house supported by the four pillars (walls) of *Taha Hinengaro* (mental wellbeing), *Taha Wairua* (spiritual wellbeing), *Taha Tinana* (physical wellbeing), and *Taha Whānau* (family and social connection). The New Zealand Ministry of Health website states:

> with its strong foundations and four equal sides, the symbol of the wharenui [a communal house of the Māori] illustrates the four dimensions of Māori well-being. Should one of the four dimensions be missing or in some way damaged, a person, or a collective may become "unbalanced" and subsequently unwell.

Woodhouse et al. (2021) investigated how Māori values might be incorporated into current models of animal welfare. One of the issues that they pointed out was the need to incorporate spirituality into a welfare model. For Western science, the immediate question becomes how to assess spirituality. Māori knowledge might provide some pointers. As suggested by Woodhouse et al., any weakening of an animal's life force (*mauri*) or soul (*wairua*) would likely impact natural behavior, affective states, and possibly health, with some of these concepts mapping to the Three Circles model. Improving the *mauri* of the environment (which has its own life force, and which would alter the animal's umwelt) would also benefit the *mauri* of animals, and unity with nature is a key component of Māori spirituality. Therefore, the environment and its richness could perhaps be used as a surrogate indicator for *mauri*. The animal's *mauri* also informs character and capabilities leading to natural behaviors. It quickly becomes clear that the deeply ingrained and situated aspects of spirituality in Māori culture make this domain more amenable to evaluation than one might have thought. Woodhouse and colleagues also point out the fact that on death, an animal may lose its life force, but an aspect of spirit is retained for several days. Therefore, the utmost respect should be paid to a newly deceased animal.

Keywords: bicultural, Mātauranga Māori

4.5 OTHER FACTORS

In the definitions and frameworks previously covered, there are many repeated themes between welfare and health, subjective experience, and natural behavior. There may also be a central or overarching aspect, such as coping or suffering, that exists on some kind of binary or scale. Other factors are often left as "implicit", seen as a source of error or a confounding variable, or are ruled "out of scope". One effect of this is to encourage a scientific understanding of animal welfare that is precise and veers away from "fuzzier" aspects that might focus public disquiet. These public concerns might relate to hygiene, wholesomeness, environmentalism, and ethics in general. A few important examples of potentially important or connected subjects are discussed below, including death, time, positive welfare, and populations.

4.6 DEATH

Many major animal welfare scientists, such as Donald Broom, consider that death is irrelevant to welfare. The reason is that dead animals don't have a welfare state. That is, if welfare is seen as involving a biological or emotional state, then death is mutually exclusive to having welfare—a dead animal is an object and not a being. Also, as animal welfare is generally seen as occurring at a specific point in time, the risk or certainty of future death is not seen as relevant unless it impacts the current lived experience. In summary: if welfare is only your current experience, something that will soon or inevitably cause your death but is not currently affecting you is irrelevant to your welfare, although it may be ethically significant in other ways.

Negating the influence of death is not a universally held opinion and is considered to be an inextricable part of animal welfare in other settings (Yeates, 2010; Jensen, 2017). For example, laws defining animal cruelty in the United States cover not only abuse and neglect but also killing an animal under conditions that could be characterized as cruel or unnecessary. When death is treated as part of welfare, it is generally for one of three reasons: 1) that death is counter to an animal's interests, 2) that death is associated with suffering, or 3) that death is associated with a failed duty of care.

Animal Interests: The idea that animals generally do not (or would not) want to die is associated with a rights perspective. Based on this understanding, animal welfare is impacted whenever an animal is killed for any reason other than to end their suffering (Bruijnis et al, 2013). It may also apply when an animal dies prematurely. For example, a farm with high lamb mortality would, all other things being equal, be seen as having poorer welfare than one where more lambs survive to weaning.

Association with Suffering: Death is also usually not instantaneous or without any negative experiences. So even if death as an event is not prioritized, it might indicate suffering. For example, lambs being taken by predators will cause suffering to the lamb and distress the ewe. Also, many techniques used for animal killing in contexts such as *slaughter* ("killing animals for the purposes of harvesting commodities such as meat or hides", AVMA guidelines for the euthanasia of animals, 2020 edition) or *depopulation* (protecting from a viral threat) fail to meet the veterinary standards for euthanasia. That is, while death itself might not relate to welfare, is it a bellwether for events that do cause suffering and might otherwise go unnoticed.

Human Care: Humans have a long history of moral judgment against those who kill animals for improper reasons (thrill killing, poaching, coercive control) or let them die through lack of competence (neglect, abandonment, hoarding, excessive culling, or other unplanned deaths). The implication is that the death of some animals might be associated with poorer care. For example, a farmer with high lamb mortality should investigate and remedy the problem. Could an infectious disease be to blame, is the farmer negligent, have the lambs been poisoned by some environmental hazard or plant, or is there some other competency issue?

The orientation to death that this author (EPK) has seen most often in recent years has two parts: 1) if the animal has the capacity and opportunity to continue to enjoy a good life, killing that animal is bad for their welfare; and 2) when killing an animal is deemed necessary, it must be done humanely. This generally popular approach in the United States is certainly not a universally held belief and is often not clearly articulated. However, it begins to help disentangle just how and why the time and manner of death could be included in an animal welfare assessment.

It may also help to understand how the risk of death may be weighed against other factors. For example, if it is bad to deprive an animal of years of good living, is it inherently worse for welfare to choose genetic lines of meat animals that reach market weight at a younger age? Does it matter if production animals are not biologically capable of living a long and natural life? Or, how would you balance the risk of euthanizing a pet earlier than necessary for "convenience" or because they are unwanted, versus later or not at all if this results in suffering?

Also, consider that there are many reasons why raising sheep in large, pastured areas is deemed good for their welfare (Figure 4.2). They get to experience natural flock dynamics, lots of natural behaviors, and more autonomy. However, it is also natural for lambs and even adult sheep to be killed and eaten by large predators and to die from weather extremes, falls, diseases, parasites, poisonous plants, and other misadventures. Therefore, it would not be unexpected for sheep on expansive pasture to be more likely to die from those causes than those kept indoors with climate control and monitoring (protection from these harms being one reason why intensive housing developed in the first place). So, when making an overall assessment of the welfare of sheep in different systems, how might rates, timing, and causes of death be captured and weighed? Is there a tradeoff for some animals having shorter but richer lives versus others having the longer but unenriched lives that a "safer" housing system could provide?

It remains important to appreciate that there is a diversity of positions about the connection between death and animal welfare that can easily lead to conflict between professionals as well as with animal owners, employers, or other agencies (see: no kill movement, faith-based opposition to euthanasia, outdoor cat colonies). As such, irrespective of whether you consider death itself to be an

Figure 4.2 Shetland lambs enjoying the outdoors, oblivious to their photographer. They're in their natural environment and are curious and playful, but could be attacked by a predator or otherwise hurt. Photograph reproduced with the permission of Janet Patterson-Kane.

aspect of welfare, it is certainly a closely related phenomenon. (It is worth noting that in human medicine, *euthanasia* refers to ending a life that is no longer worth living. In veterinary medicine, it *may* mean this, but more often, it means a method of killing that causes little to no suffering regardless of the reason for the killing.)

Keywords: euthanasia, slaughter

4.7 TIME

If we assume it is possible and reasonable to get a perfect measure of an animal's welfare, that welfare will vary from moment to moment. So, while an assessment at a single point in time may be accurate, it is of limited use on its own. Likewise, no animal lives in a perpetual positive state. Normal life has its ups and downs, disappointments and daring deeds. Negative states are normal when learning and during natural activities, as they are during enjoyable experiences (e.g., play, mating, foraging). The mix is what makes for a rich and full life experience.

Most animal welfare assessments attempt to make some judgment about not only how good or bad welfare is, but how this is affected by some event happening over time. The event might be, for example, tail docking or the adjustment to a lifelong captive habitat. This assessment is often performed somewhat intuitively, but there have been attempts to quantify impacts on welfare over time. One example is called *quality-adjusted life years* (QALY).

QALY was developed for human public health studies where a health event might impact two aspects of life; life expectancy and quality of life. The resulting score might rate each year of life as a value between 0 (not a life worth living) and 1 (perfect health). Some efforts have been made to extend this concept to the effect of animal disease but not to impacts on welfare more generally (Teng et al., 2018). Another important reason to consider the passage of time is that animals do not remain a blank slate after repeated events. The animal may weather challenges by learning and adapting, or might deteriorate, or both. And as noted by John Webster (2019), "sentient animals do not just live in the present".

One well-known example is the *critical period* for young animals. Sometimes called the sensitive period, this is a period in an individual's development when certain skills or characteristics are most easily acquired. Puppies are keen to learn about other individuals and their environment during the first 6 to 12 weeks of life. If they are kept in barren and isolated conditions during this time, it may not cause them any *obvious* suffering. (In fact, fear is an aspect of normal socialization and should be mild and transient when properly managed.) However, the under-socialized puppy is much more likely to develop a pronounced fearful or timid temperament that can persist post adoption. Ultimately, the puppy may not be able to cope with everyday experiences and will recurrently flee, hide, or cower in a corner. So, it is important to understand how experiences have one meaning to the animal as they occur, but other effects might be lifelong.

Temperament is not the only mechanism that contributes to chronic welfare states. For example, protective immunity that will prevent future suffering might arise after a transient but unpleasant illness or inoculation. Foods not eaten at an early age may be rejected by adults, as demonstrated by sheep that reject unfamiliar food even when faced with starvation. Stress or illness may suppress an animal's natural circadian rhythm and prevent them from performing natural behaviors such as crepuscular foraging or synchronized herd/flock behavior.

Welfare states in the past, whether ongoing or not, can leave a trace on or around the animal that remains measurable. Examples might include scarring on the body from past injuries, the expression of *biomarkers* of prior injury, psychological trauma (e.g., post-traumatic stress disorder [PTSD]), or the establishment of maladaptive behaviors such as stereotypic pacing or swaying (sometimes called zoochosis). (A biomarker is any substance that reliably indicates a current or previous biological state.) Stress can also leave heritable *epigenetic* effects that leave their chemical traces on DNA—the animal's genome and its associated proteins. In this way, a "stressed genome" can skew behaviors in ways that alter resilience, memory, learning, and anxiety, with these effects passing through several generations. How does this happen?

The genome is essentially the set of unique instructions for each cell that dictates its fate. The genetic code will determine what a cell does, where it is in the body, and whether it needs to replicate itself, die, or stay quiescent. DNA must be "read" to be used, and to read it, you must be able to access it. DNA is also covered with proteins, and if the structure of protein and DNA (collectively called chromatin) is compact, then it is difficult to access and read the DNA (compact chromatin is called heterochromatin). If the chromatin is more open and relaxed (euchromatin), then the code can be read. This author (TR) tends to think of chromatin like an accordion: expanded when open for transcription and collapsed when compact and invisible to transcription. Once DNA is accessible, it is the job of molecular machines to copy and interpret the genetic code (transcribe) in order to make messenger RNAs that are then translated to make proteins. There are countless instructions that are turned into RNA messages (mRNAs), and these become actionable when translated to make proteins. These messages may say "make insulin", or they might tell the body to dial down a stress response.

The link with stress is that stress can chemically modify the proteins that bind DNA or the DNA itself, and these changes can alter chromatin compactness.

This might mean that the "stop stress response" message never gets made. The upshot of this is that the DNA doesn't itself need to be mutated to alter how it is used—gene function can be indirectly controlled by its chemically modified jacket of proteins and by the chemicals that bind to DNA. An audit of these chemical modifications will give you the *epigenome*, with the Greek word *epi* literally meaning "above" the genome.

The classic animal/epigenomics study involved grooming in rats (Weaver et al., 2004). Well-nurtured rat pups grow to be much calmer than their neglected and more anxious counterparts. The altered epigenome (of genes in the stress response pathway) makes the poorly groomed rats more vigilant and hyper-stressed (advantageous in some settings). Other common epigenetic effects are metabolic changes in response to nutrition, or the lack of it. These effects can last a lifetime, in fact several lifetimes, which prompted this title from a group studying epigenetics in sheep—"Ewe Are What Your Grandparents Ate?" (Braz et al., 2022). These effects also mean that identical twins will become more dissimilar over time as they accumulate different life events that leave their chemical marks on their genomes.

The scientific instinct is to reduce animal welfare to something that is as close as possible to an objectively measurable state. This can be done by removing the dimension of time, as this factor introduces tremendous complexity. However, assessments almost always need to be generalizable to times other than when the assessment is made. So, this aspect is usually factored back into the assessment or is used to refine an interpretation in order to make more generalizable conclusions. Some organizations include the dimension of time as part of any meaningful definition of welfare. For example, the Denver Zoo posts its definition of wellbeing as "an animal's physical, mental and emotional health over a period of time".

Keywords: quality-adjusted life years (QALY), epigenome, critical period (sensitive period)

Competitive animal welfare assessment

Welfare assessments are often used to assess and compare the housing systems and management choices that we make for animals. One way to appreciate how this is done is by observing an *animal welfare assessment competition.* These competitions were developed in the United States based on livestock judging competitions, and animal welfare scientists can compete and demonstrate their skills.

There is also a long tradition of livestock judging in the United States that is associated with state fairs and other shows. This includes not only competitions to find the best animals but also organized training on how to raise, exhibit, and judge them. So people, especially children and youngsters, can compete at fairs and in 4-H clubs (youth clubs) to demonstrate how good they are at judging livestock. The *4-H Livestock Judging Manual* description of these contests describes how you must place the animals in order of quality and receive points for every pair put in the correct order. Additionally, you must explain your reasons for how you rated the animals, and: "Oral reasons allow you to justify your placings to an official judge. The official judge will score you on accuracy, completeness, length, presentation and delivery, and terminology".

This model was used to create competitions where animals are not only judged for their productivity-related factors but are given a full welfare assessment. Animal welfare assessment contests provide animals under different conditions with an expert appraisal of how well the competitor did at determining and justifying their opinion of an animal's welfare.

One prominent example of these contests is the annual AVMA Animal Welfare Assessment Contest (https://www.awjac.org/). This event allows teams and individual students to compete in their ability to assess the welfare of animals in different settings, including both live and audio-visually presented scenarios. The contest accepts entries in undergraduate, graduate, and veterinary divisions (Johnson et al., 2020), with judging performed by accomplished animal welfare scientists from many different backgrounds. Simplified online versions of some of these assessment scenarios can be taken online at https://www.avma.org/resources/animal-health-welfare/animal-welfare-assessments-test-your-knowledge.

Keyword: animal welfare assessment contests

4.8 THE UPSIDE OF BAD WELFARE

In many cases, the main concern of animal welfare research has been to detect and respond to situations that cause suffering. This reflects the history of animal welfare science in allaying the public's fears over animals used in intensive farming and research laboratories. Not surprisingly, then, many techniques were developed to be sensitive to negative welfare states, with the result that welfare might be considered good so long as suffering is absent.

Specific consideration of positive welfare states can have two general benefits (Mellor & Beausoleil, 2015). One is to ensure that our goal is not a flattened neutral state, but that animals can lead a full, vigorous, and enjoyable life. But secondly, it is important to consider that animals have good and bad experiences at the same time and in different times and situations. So, while welfare is seen as a single unitary state, it might incorporate good and bad aspects simultaneously: for example, an animal that feels unwell but is enjoying huddling comfortably with others. Even when sources of suffering are not fully addressed, animals are likely to be able to benefit from positive experiences and opportunities. (The idea of being satisfied even in the face of inevitable suffering is better recognized in philosophies such as Taoism.)

There is evidence that experiences that are unpleasant can provoke higher states of welfare. For example, after seeing a predator nearby, a group of primates might be frightened, but after acting together to drive it away, they will show more positive behaviors and interactions, suggesting improved welfare. One important "mixed" experience is social play. Many animals, especially juveniles, find play one of their most enjoyable activities (Figure 4.3).

However, there is a good reason to predict that an unscheduled outcome of play will be falls, grapples, social defeats, and even minor injuries. This could be called "training for the unexpected" and allows young animals to rehearse life strategies that might actually help them to survive real life-threatening scenarios (Spinka et al., 2001). Learning to negotiate, behave fairly, and cooperate are all skills associated with play (Allen & Bekoff, 2005). As a result, a playing

Figure 4.3 Rough and tumble. These coyotes are enjoying playing and sparring, but you can see how things could get out of hand. Photographs reproduced with the permission of janetkessler/coyoteyipps.com.

animal is likely to experience unpleasant events but is also initiated into a social group of peers. Small challenges lead to animals becoming fitter, more agile and skilled, and perhaps achieving a sense of accomplishment, which, collectively, improves their chances of surviving and thriving later in life.

That said, some individuals will literally die as part of play activity when they attract the attention of a predator or are seriously injured. In short, while play is considered an indicator of good welfare and is a highly enjoyable activity, even this is not a purely positive experience and is not beneficial for everyone. The inherently mixed nature of experience is important to keep in mind, as many conceptions of animal welfare are overly simplistic (for example, in suggesting that unpleasant or aversive experiences can or should be completely eliminated).

Keyword: mixed experience

4.9 POPULATIONS

Just as we need to assess how stable or variable an animal's welfare is over time, we also need to understand how it might vary across individuals in a group. The individual's status as a member of a group, and relative to other group members, can also directly affect whether welfare is considered important to different human stakeholder groups. For example, an African elephant has great conservation value but might destroy farmers' crops and be seen (more or less) as a pest. Physical characteristics can also be used to discriminate between individuals. An elephant's tusks are so valuable that their size and presence are rapidly decreasing in many elephant populations due to selective poaching that eradicates the tusked individuals.

Many disciplines study animals primarily as populations rather than individuals. Examples of these disciplines include agriculture, conservation (preservation of species and their habitats), zoology, anthrozoology, and sociology. Some of these population-based approaches will tend to emphasize the health, productivity, or survival of the population or species over the welfare of the individual. Even where individualized, and with welfare carefully considered, it can be difficult to balance the interests of animals within populations that have competing interests.

For example, in many cases, humans have introduced animals into new environments where they are more likely to become pests that negatively impact livestock or wildlife. Even feral peacocks have become a problem in urban America. Unscheduled population changes can also be the result of geopolitical upheaval. For example, the collapse of the USSR provoked a switch to a free market economy. An enfeebled government was exploited by the Russian Mafia, with many Russians suffering severe financial hardship. In rural areas, people abandoned their farms and turned to hunting and poaching to survive. A notable beneficiary of relaxed oversight were the wolves, which started to move west and to multiply (Bragina et al., 2015).

Keyword: conservation

A pragmatic view of invasive species

Indigenous land stewards have proven to be immensely successful in protecting the biodiversity of their territories. For many of these groups, use of the term "native" and "alien" has been a painful reminder of how they have been treated by Anglo European colonizers and expelled from their own ancestral territories to be resettled on foreign land. Indigenous stewards can favor relational approaches to the management of invasive species. While eradication may be necessary in some circumstances, that might not be the best course of action when taking into account cultural needs and the needs of other plants and animals. There might be the recognition that the invasive species used to belong somewhere, has a value somewhere, and may positively contribute to an ecosystem (Wehi et al., 2023). A more pluralistic vision of species management would examine all of the interrelationships that may be forged or broken. For example, Tasmanian blue gum (*Eucalyptus globulus*), which arrived in Southern California in 1865, has become the favored roost for monarch butterflies (Figure 4.4 top panel) over native Monterey pines and cypresses; great horned owls also nest in them. What has been the broader impact of the eucalyptus on native species, and what happens when the eucalypts die? Change management can also be used to mitigate against the shock of the new, with the gradual replacement of native species that fail to thrive but occupy ecological niches needed to support others.

Sometimes, an animal is deliberately introduced that proves to be problematic for a native species. Mountain goats were first introduced in Colorado in 1947 by the Parks and Wildlife service to encourage hunting and are now displacing the native Rocky Mountain bighorn sheep (Figures 4.4 bottom panel and 4.5). We might also see arbitrary timestamps introduced for flora and fauna that are

Figure 4.4 Top panel: Monarch butterflies roosting on Monterey pine in California. There is now debate about whether the monarchs expanded into America together with milkweed plants and tree-felling colonists. This adds another layer of complexity to the arguments of nativists. Bottom panel: Mountain goats (*Oreamnos americanus*) on the Front Range of the Rocky Mountains in Colorado, USA. These mountain goats are displacing bighorn sheep (*Ovis canadensis*) and, above the treeline, outcompete them for access to mineral licks exposed by melted ice. Both images reproduced with the permission of Janet Patterson-Kane.

considered native, with attempts to "go back in time" and recreate those ecosystems in places where they are now functionally defunct. What is the correct timestamp, and what happens to an animal or species on the wrong side of it? If an invasive species successfully adopts the ecological niche of a now extinct species (as a functional proxy), should it be eradicated?

Our perception of a species' value can shift rapidly based on how it is framed. For example, a 2023 video showed an orangutan at the Perth Zoo throwing a possum high into the air and out of their enclosure. A spokesperson for the zoo first treated the matter somewhat lightly, referring to the possum as a "pest" that had died and saying, "sometimes the orang-utans evict the odd possum themselves". The zoo faced a backlash on two counts. One is that the possum is wildlife native to the area and so cannot really be characterized as a pest species. (While a pest species is generally defined as one that is damaging to human interests and activities, the protection of endemic wildlife is presumed to be part of a zoo's intent.) The other is that while many who viewed the viral

Figure 4.5 A line drawing of a Rocky Mountain (RM) bighorn sheep—with features fading as the influence of this animal declines on the slopes. This iconic animal is the mammalian symbol for Colorado Parks and Wildlife, with Colorado home to more RM bighorns than any other state. Reproduced with the permission of Janet Patterson-Kane.

video found the "yeeting" amusing, it seemed incongruous for a zoo spokesperson to be so indifferent to the welfare of the possum. The spokesperson then released a partial retraction and claimed, somewhat conveniently, that possums are "robust" and that the hurled possum had survived (https://www.perthnow.com.au/news/wa/perth-zoo-retracts-claims-possums-are-pests-after-video-shows-orang-utan-throwing-one-from-enclosure-c-11892561).

Even in settings where a given animal's life or welfare has been consciously deprioritized, like a pest being poisoned, it is important to continue to acknowledge that these animals do have welfare states that are ethically significant. They still enjoy their lives when well, and suffer when harmed. Otherwise, to avoid cognitive dissonance, people will be biased toward assuming that any harm they have decided to do to an animal is minimal, causing an imbalance in the ethical decision-making process. (*Cognitive dissonance* is the strain caused by holding contradictory thoughts: for example, that it is wrong to cause animal suffering, but that it is correct to do something that will cause animal suffering.)

Continuing to acknowledge animal suffering even when it has been deemed necessary (e.g., to protect the welfare of other animals) provides motivation to carefully consider alternatives and refinements, or to invest funds necessary to research and invest in other control options. That said, a half-hearted eradication technique can be the worst of both worlds, causing suffering to many animals and yet continuing to allow the unwanted population to breed and relocate back to the previously cleared areas. On occasion, though, truly successful and humane alternatives to killing one species of animal to protect another have been found.

Figure 4.6 Clockwise from top left: Sisters Eudy and Tula (named after the Latin name for the little penguin—*Eudyptula minor*) were the first two penguin guardian dogs, and they served their penguin charges for ten and eight years, respectively. You can see one of the little penguins hiding top right. Bottom right: Tula and Eudy trained up the next generations of guardian dogs, so their legacy carries on. The last picture shows a boardwalk patrol. Reproduced with the permission of Warrnambool City Council, Victoria, Australia.

The predation of the little penguin (*Eudyptula minor*) by foxes on Middle Island off the coast of Victoria, Australia, has been stopped in its tracks by using guardian dog species (Figure 4.6). Maremma dogs are an Italian breed of livestock guardian dog. They guard the penguins when they come ashore to breed, and when not on duty, they keep an eye on free range chickens on the mainland. A local farmer called Swampy Marsh had used a Maremma dog called Oddball to guard his chickens, and so he suggested that guardian dogs might be good at protecting the penguins. Dogs are banned from the island, but an exception was made for the Maremmas, who became fiercely protective of their turf and "their" penguins. It's win-win, with the penguins doing well, dogs with jobs, foxes not being culled, and a 2015 film that won a few awards.

Keywords: cognitive dissonance, pest species

4.10 DISCUSSION—ENOUGH ABOUT THEORY

If you need to evaluate the welfare of an animal, you will probably use some theory and pick some specific things to measure: for example, indicators of health, behavior, naturalness, and happiness of an animal. The previous sections have provided you with some additional options for devising that assessment. And, of course, there's always the non-specialist observer who can point out the one crucial thing of importance that you've missed (Weary & Robbins, 2019). In Part III, we will cover how to create and use an animal welfare assessment in a range of different settings. But first, we need to go into a bit more depth about physical state, naturalness, and subjective experiences after a short chapter on definitions.

READINGS

Three circles

Fraser, D., Weary, D. M., Pajor, E. A., & Milligan, B. N. (1997). A scientific conception of animal welfare that reflects ethical concerns. *Animal Welfare*, *6*, 187–205.

Disciplines as domains

Brown, J. A., Seddon, Y. M., & Appleby, M. C. (Eds.). (2016) *Animals and us: 50 years and more of applied ethology*. Wageningen Academic Publishers. https://doi.org/10.3920/978-90-8686-828-5_6

Fraser, A. (1988, October). Letter published in the SVE Newsletter 39.

Petherick, J. C., & Duncan, I. (2016). The International Society for Applied Ethology: Going Strong 50 Years On. In J. A. Brown, Y. M. Seddon, & M.C. Appleby (Eds.), *Animals and Us: 50 years and more of applied ethology*. Wageningen Academic Publishers. https://doi.org/10.3920/978-90-8686-828-5_1

Domain interactions

Hickling, G. J. (1994). Animal welfare and vertebrate pest management: Compromise or conflict? In *Animal welfare in the twenty-first century: Ethical, educational and scientific challenges — Proceedings of the ANZCCART Conference, Christchurch, New Zealand* (R. M. Baker, D. J. Mellor, & A. M. Nicol, pp. 119–124). Glen Osmond, Australia: ANZCCART.

Kjosevski, M., Nikolovski, M., Ilievska, K., Pendovski, L., & Ilieski, V. (2021). The three pillars of applied farm animal welfare. *Veterinarski glasnik*, *75*(2), 99–111. https://doi.org/10.2298/VETGL210629014K

Reflections on welfare frameworks in a bicultural society

Woodhouse, J., Carr, A., Liebergreen, N., Anderson, L., Beausoleil, N. J., Zobel, G., & King, M. (2021). Conceptualizing Indigenous human-animal relationships in Aotearoa New Zealand: An ethical perspective. *Animals*, *11*(10), 2899. https://doi.org/10.3390/ani11102899. PMID: 34679920; PMCID: PMC8532608.

Durie (1984). https://mentalhealth.org.nz/te-whare-tapa-wha

Death

Bruijnis, M. R. N., Meijboom, F. L. B., & Stassen, E. N. (2013). Longevity as an animal welfare issue applied to the case of foot disorders in dairy cattle. *Journal of Agricultural and Environmental Ethics*, *26*(1), 191–205.

Jensen, K. K. (2017). How should death be taken into account in welfare assessments? *Journal of Agricultural and Environmental Ethics*, *30*(5), 615–623.

Yeates, J. W. (2010). Death is a welfare issue. *Journal of Agricultural and Environmental Ethics*, *23*(3), 229–241.

Time

Braz, C. U., Taylor, T., Namous, H., Townsend, J., Crenshaw, T., & Khatib, H. (2022). Paternal diet induces transgenerational epigenetic inheritance of DNA methylation signatures and phenotypes in sheep model. *PNAS Nexus*, *1*(2), pgac040.

Teng, K. T., Devleesschauwer, B., Maertens De Noordhout, C., Bennett, P., McGreevy, P. D., Chiu, P. Y., Toribio, J. L. M. L., & Dhand, N. K. (2018). Welfare-Adjusted Life Years (WALY): A novel metric of animal welfare that combines the impacts of impaired welfare and abbreviated lifespan. *PLoS One*, *13*(9), e0202580. https://doi.org/10.1371/journal.pone.0202580

Weaver, I. C., Cervoni, N., Champagne, F. A., D'Alessio, A. C., Sharma, S., Seckl, J. R., Dymov, S., Szyf, M., & Meaney, M. J. (2004). Epigenetic programming by maternal behavior. *Nature Neuroscience*, *7*(8), 847–854. https://doi.org/10.1038/nn1276

Webster, J. (2019). Sentient animals do not just live in the present. *Animal Sentience*, *25*(10). https://doi.org/10.51291/2377-7478.1443

Competitive animal welfare assessment

Johnson, C. L., McKinney, L. J., & Patterson-Kane, E. G. (2020). Effects of participating in the annual Animal Welfare Assessment Contest on veterinary students' self-perceived knowledge of and attitudes toward animal welfare science and their career choices. *Journal of the American Veterinary Medical Association*, *256*(2), 239–244.

The upside of bad welfare

Allen, C., & Bekoff, M. (2005). Animal play and the evolution of morality: An ethological approach. *Topoi*, *24*(2), 125–135.

Mellor, D. J., & Beausoleil, N. J. (2015). Extending the 'five domains' model for animal welfare assessment to incorporate positive welfare states. *Animal Welfare, 24*(3), 241.

Spinka, M., Newberry, R. C., & Bekoff, M. (2001). Mammalian play: Training for the unexpected. *The Quarterly Review of Biology*, *76*(2), 141–168.

Populations

Bragina, E. V., Ives, A. R., Pidgeon, A. M., Kuemmerle, T., Baskin, L. M., Gubar, Y. P., Piquer-Rodríguez, M., Keuler, N. S., Petrosyan, V. G., & Radeloff, V. C. (2015). Rapid declines of large mammal populations after the collapse of the Soviet Union. *Conservation Biology*, *29*, 844–853.

Pragmatic views of invasive species

Wehi, P. M., Kamelamela, K. L., Whyte, K., Watene, K., & Reo, N. (2023). Contribution of Indigenous peoples' understandings and relational frameworks to invasive alien species management. *British Ecological Society, 5*(5), 1403–1414.

Discussion—enough about Theory

Weary, D. M., & Robbins, J. A. (2019). Understanding the multiple conceptions of animal welfare. *Animal Welfare, 28*, 33–40.

Chapter 5
Definitions

5.1 THE PURPOSE OF DEFINITIONS

A definition of a word, such as from a dictionary, is generally simpler than a model and does not need to be scientific in nature. Due to their scope and purpose, these definitions tend to be broad and imprecise. However, definitions of animal welfare are often simplified versions of models like the ones discussed in the previous chapters, or are at least influenced by them (Hewson, 2003). In applied settings, our focus is not necessarily on which of these definitions are correct, but how they affect our ability to work together. The best place to start is to understand how animal welfare is defined and operationalized by influential people and groups. Animal welfare definitions are often the very first thing to look at when working with a new collaborator or group. In doing so, you can get a feel of that entity's philosophy, focus, and goals. Is the scientific method represented as part of their mission or tactics, whether directly or not; is it *evidence based*, and does it name or exclude specific disciplines? You can also see which theorists are referenced and which groups they align their definition to, whether it is aspirational or pragmatic, specific or broad. Always check explicit definitions against other statements, budgets, reviews, and concrete activities. Do they do what they say they do? You may find that their definition is not always observed in their own activities or is out of date, but it is typically a very good place to start.

Some specific, representative examples of animal welfare definitions are discussed below. Many of these are highly influential organizations in the field of animal welfare science. Others are examples of definitions with a more specific scope and influence. What is more important is not necessarily these specific examples but to set up a habit of understanding how definitions of animal welfare not only influence your own work, or the work of specific science programs, but show coherence across the field of animal welfare in general. When there are specific definitions to consult, it is helpful to be familiar with them and to understand their implications. On other occasions, these definitions may be implicit or unclear but can be deduced using a similar approach.

DOI: 10.1201/9781003375456-6

The usefulness of knowing or deducing a group's definition of animal welfare may include some of the following:

- Maintaining a working knowledge of important animal welfare entities
- Meeting an organization's requirements (e.g., for job or grant applications)
- Understanding which entities are suitable for engagement and collaboration
- Understanding whether they have a coherent scope and goals, and what these are

5.2 INTERNATIONAL EXAMPLE—WOAH

The intergovernmental organization (IGO) World Organization for Animal Health (WOAH) was founded in 1924 under the name Office International des Epizooties (OIE). WOAH's goal is to improve animal health globally, focusing on animal diseases, and the organization was founded to combat disease outbreaks such as "cattle plague" (*rinderpest*), which could wipe out entire herds (Figure 5.1). These dramatic events were disastrous for cattle-dependent cultures, who might also have supernatural beliefs about the causes of the plague. An international eradication effort was begun by WOAH in 1924, and the last confirmed outbreak was in Kenya in 2001. WOAH understands that one of the reasons why animal disease is internationally significant is the impact that it has on animal welfare, which is defined as follows: "Animal welfare means the physical and mental state of an animal in relation to the conditions in which it lives and dies".

This definition (WOAH terrestrial code, accessed online October 22, 2024 and first adopted in 2019) makes clear that in addition to being a physical state of good health and functioning, welfare also includes mental states and the ability to express behaviors that support good mental and physical states. The definition is then further qualified to include the absence of negative emotional states or sources of frustration, with animals that feel safe and comfortable. The WOAH's 2024 "vision paper" concerning animal welfare explicitly describes the Five Domains model (we will meet this in Part III), and emphasizes the need to place animal welfare in the context of human society, which has an

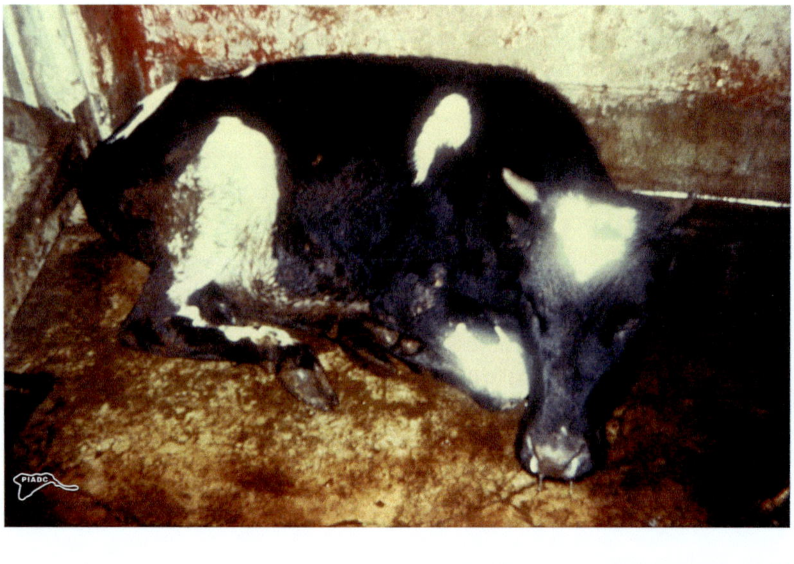

Figure 5.1 This unfortunate animal has rinderpest disease and is in the classical milk fever position that is assumed shortly before death. This photograph (1982) is from the "*Illustrated Manual for the Recognition and Diagnosis of Certain Animal Diseases*" (produced by the Mexico-United States Commission for the Prevention of Foot and Mouth Disease). The image is provided by the US Centers for Disease Control (CDC), Dr Jerry J. Callis from the specialist Plum Island Animal Disease Center in New York (PIADC), and Dr Brian W.J. Mahy of the CDC. (Source: CDC Public Health Image Library https://phil.cdc.gov/default.aspx. This public domain image is available free of charge at the CDC link; its use does not constitute any endorsement by the CDC.

obligation to the animals that it commodifies and otherwise affects. As with many definitions of welfare, it is immediately clear that the WOAH definition is multi-domain. Various terms used in this definition reflect how the animal's state responds to how they are cared for (duty of care). For example, provision of veterinary care contributes to the physical state of being healthy, and provision of shelter contributes to the mental state of being comfortable. Other dimensions or dichotomies that are present include stimulating versus safe, life versus death, and prevention versus treatment. All of these suggest a rather complex constellation of considerations that contribute to the central issue of whether an animal's physical and mental state is poor or good.

The WOAH uses a council structure to provide detailed codes of practice that national governments can sign up to or use as standards. They focus on practices that protect animal health primarily by preventing or reducing the transmission of disease. Over time, their mission and scope have broadened, but their most common focus remains agriculture.

It is significant that an organization with a long history of dealing with animal disease makes prominent use of the concept of animal *welfare* rather than just health. Animal disease is undoubtedly important because of how it affects economic systems and kills animals. However, the WOAH is acknowledging that the role of animal disease in causing suffering is inextricably part of what makes disease an important issue to address through a well-resourced and an internationally coordinated response. This is one of many indications that welfare is a culturally pervasive aspect of any activity that involves animals even when not explicitly part of the specific topic being addressed.

It is also significant that the WOAH chooses to define animal welfare in a way that pivots on the state of the animal as influenced by multiple complex factors. This description of animal welfare would make intuitive sense to many people when they read it and might even seem self-evident. Indeed, many organizations that are tasked with affecting the care of animals choose to make their goals simple and therefore easier to achieve.

Overall, this definition is phrased in a relatively informal manner but shows evidence of an underlying structure that could be characterized as trying to align health and welfare, centering the experience of individual animals, and encouraging consideration of a great many contributing factors and relevant situations.

Consistent with this approach, the WOAH places this definition within and at the beginning of its *Terrestrial Code*, which is an extensive and detailed set of standards relevant to veterinary public health worldwide, inclusive of diagnoses, transportation, hygiene, and species-specific management in different disease settings. So, while broadly applied, the overall definition seeks to align how each situation can be handled with a view to optimizing animal welfare.

Keyword: rinderpest

5.3 NATIONAL EXAMPLE—AZA

Up until 2023, the *Association of Zoos and Aquariums* (AZA), an American 501(c)3 non-profit, used a definition of animal welfare that, interestingly, included the dimension of time. Specifically, the definition stated: "Animal

Welfare: [is] an animal's collective physical, mental, and emotional states over a period of time, and is measured on a continuum from good to poor". The accompanying explanation of the definition mentions the need to express species-typical relationships. You would expect animal conservation and exhibit spaces to have the greatest motivation and potential for providing a naturalistic environment. These spaces not only serve the umwelt of the animal but also contribute to potential reintroduction to the wild. Describing this as *species-typical* suggests a preference for ethological-scientific approaches over more holistic but ambiguous terms like *naturalness*. (In ethology, *species-specific* is the more commonly used term.)

However, even as this book was being written, their definition was revised as part of their Strategic Framework for the Wellbeing of Animals (2023; https://www.aza.org/about-animal-wellbeing, accessed Feburary 2025). At this time, they moved to the term "Wellbeing" rather than welfare, as part of a more relatable approach to animal care that bridges public and scientific understandings. Their definition for wellbeing folds back in the element of time as "Wellbeing – a state of being comfortable, healthy, or happy; achieved by AZA facilities giving animals lifelong opportunities to thrive".

Keyword: species-specific

5.4 SMALL ORGANIZATION—UBC SPROUTS

Many smaller or more local organizations adopt or endorse a definition of animal welfare. These are often taken from authorities or more overarching organizations such as the WOAH, the American Veterinary Medical Association (AVMA), or government agencies such as the United States Department of Agriculture (USDA). These patterns of copying can illustrate how a particular definition suits the different philosophies and missions of particular organizations, and by inference, which more influential organizations they take their lead from.

However, others will develop their own definitions based on having a specific or unique perspective, focus, or work. Take, for example, the *UBC Sprouts food co-op* (https://www.ubcsprouts.ca/). This organization does not provide meat but defines "animal welfare" in relation to the distribution of animal by-products from outside providers. The UBC Sprouts definition of animal welfare is "an animal's quality of life; animals in healthy, pain-free, and comfortable farms".

This essentially swaps *quality of life* for *welfare*, which might be considered a bit of a conceptual homunculus. Quality of life (QOL) is a concept taken from human caretaking and is heavily weighted toward how an individual experiences their own life. The use of QOL in a top-level definition of animal welfare is unusual and suggests an organization that has carefully considered the issue.

Many animal welfare scientists (e.g., Donald Broom, 2007) would consider the terms "animal welfare" and "quality of life" to have essentially the same objective meaning. However, there are softer differences in perspective that can be very important in reflecting an organization's authentic mission. In the case of UBC Sprouts, the goal is to provide accessible, ethically sourced and made food in a university campus area. This definition suggests an organization providing a wide range of vegan food, without veganism being an explicit goal, with a free meal and produce initiative for those in need. Their practice is more of a natural solution to providing food that is culturally sensitive and considers the welfare of animals in a way that shows equivalence to human welfare.

5.5 FOREIGN NATIONAL EXAMPLE—ANSES

Some version of the Three Circles, Five Freedoms, or Coping is the normal basis for a working definition of animal welfare. The use of quality of life, which is a distinctly different theory, would count as a notable variation. Another example would be definitions where welfare is seen as only describing positive states, as indicated by the French Agency for Food, Environmental and Occupational Health & Safety (ANSES; anses.fr; posted online April 2018 and accessed October 22, 2024, https://www.anses.fr/en/content/anses-proposes-definition-animal-welfare-and-sets-foundation-its-research-and-expert). Their definition states that: "The welfare of an animal is its positive mental and physical state as related to the fulfilment of its physiological and behavioural needs in addition to its expectations. This state can vary depending on the animal's perception of a given situation." This position is logical, based on the general use and etymology of the term *welfare* from faring or doing well. So, the underlying meaning of the word is to indicate a positive state of being healthy or happy. In a field that is interdisciplinary and engaged with by the public, there is an advantage to staying as close as possible to the intuitive meaning of a word.

When working according to this positive-only definition of welfare, it is not reasonable or accurate to say that an animal has poor welfare, as this would be saying that you have a "bad good thing". Users of this positive-only definition would say welfare can be good, or it can be "compromised". One benefit of assuming that welfare refers to positive values is that it implicitly reinforces the idea that a positive welfare state is the default state.

Throughout this book, we take the more common approach that welfare is a state that ranges from very bad to very good. Apparent contradictions in terms are not that uncommon, and they often make sense within a given context, for example, a "working vacation" or a "jumbo shrimp". Or, more directly, it is quite natural in the English language to say a person is in "poor health" despite the default meaning of "health" being the positive condition. And there is an advantage to using direct terms to communicate the welfare scale in terms of both valence (bad versus good) and intensity (somewhat versus very). Either approach can be used without much danger of being misunderstood.

When you enter a new space, it is often possible to quickly *read* how animal welfare science is being approached. For example, when I (EPK) last opened the Wikipedia page on animal welfare science, I saw that the historical events mentioned are mainly British, the assessment methods were limited to two domains ("Animal welfare science uses a variety of behavioral or physiological measures or indicators"), and Donald Broom is quoted. While no explicit definition of animal welfare is provided, this curation shows me that this page has been created and edited mainly by people enmeshed in the British tradition and theoretical approach of Professor Broom. There is nothing incorrect about that, but it represents only one aspect of the field.

The Same Soul

In Lewis Carroll's *Through the Looking Glass* (1871), Humpty-Dumpty says: "When I use a word, it means just what I choose it to mean — neither more nor less". With this in mind, never assume that words will hold the same meaning between people or across time. Some of our oldest texts have the most disputed meanings. For example, the weighty words of ancient Greek philosophers are often used to lend the gravitas of history to a message.

In 2012, the organization *PeTA* (People for the Ethical Treatment of Animals) ran a billboard that said: "[T]he soul is the same in all living creatures, although the body of each is different", attributed to Hippocrates (aka Hippocrates of Kos, 460–377 BCE, considered by many to be the father of modern medicine). It makes a nice point rather pithily, but …. You shouldn't put too much faith in people using words according to their dictionary definitions, even at the best of times. When the words in question come from another country and another era, and are translated from another language—even less so.

Even trusting the 1931 translation from which this line is probably drawn, consider the context (Hippocrates, 1931). This is from Hippocrates Volume IV: "Male and female have the power to fuse into one solid, both because both are nourished in both and also because soul is the same thing in all living creatures, although the body of each is different" (p. 267). But if you back up a few sections, you will also find this: "For the suitable joins the suitable, while the unsuitable wars and fights and separates itself. For this reason a man's soul grows in a man, and in no other creature. It is the same with the other large animals" (p. 241).

If you review the whole section, you will see there is quite a lot of nuance in just how much the stuff of the soul is seen as universal versus the souls of individuals being distinct partly due to species. And in truth, animals other than humans were barely touched upon as a subject. (Animals do tend to appear mainly as minor footnotes in the works of many great philosophers.)

Hippocrates' views on medicine reached down through multiple major world cultures and color modern attitudes in powerful ways. However, a proper understanding of how this impacts the issue of animal welfare science would require much more space than can be found on a billboard. And to the best of my knowledge, no scholar has yet attempted it.

5.6 DISCUSSION

There are several terms used as close synonyms for animal welfare, one of which is wellbeing (also spelled well-being). Some theorists attempt to create shades of difference between welfare and well-being (Appleby and Sandøe, 2002). Well-being is a term more closely associated with humans, and it has been suggested by Gonyou (1993) that well-being be used to describe the current state of the animal, while welfare is a more general concept that includes the past, present, and future. AZA chooses the term as more intuitive for the public, with a simpler and more relatable meaning. However, currently, welfare and well-being continue to be used interchangeably (or at least inconsistently) in most settings.

Most people will not refer to a scientific definition for either word but will simply use their intuitive understanding. Most terms used in animal welfare science continue to have normal dictionary and lay usage. This is another reason why having and sharing definitions of key terms can be helpful, as is keeping your definition as close as possible to common lay usage. Where a word has no definition, or the definition is not actively used, be alert to the potential for ambiguity or conflict. (For example, neither *animal welfare* nor *welfare* is included in the list of definitions for the United States Animal Welfare Act and some other legislation.)

In the previous chapter, we mentioned the International Society for Applied Ethology (ISAE). Their mission statement clarifies that the focus on animal welfare is a result of a larger goal rather than a fixed part of their identity. So, while it is likely that the ISAE will always include the direct study of animal welfare, this may no longer be the key area where research is required to overcome obstacles and create progress. The ISAE states that its mission is: "Advancing animal-focused ethical research and education in academia and public to eventually achieve the inclusion of non-human animals in everyday moral theory and practice".

Exercise: personal definitions

This author's (EPK) own definition of animal welfare in the mid-1990s was to provide clarity for my PhD thesis and a philosophical North Star among a sea of other definitions that existed at the time. (Note: this definition, shown below, is also a product of its times in describing animals as "it" when I would now use "they".) This definition was my own, was distinct from more famous definitions at that time, but was also not completely different from them. It was not, and is not, my intent to suggest that this definition be used by other people. But writing your own definition may help you, as it helped me, to focus my thoughts on my philosophy of animal welfare and what I wanted to do in my research career.

Animal welfare is compromised:

> When the animal is significantly impaired in its ability to adapt to the contingencies of reinforcement and punishment operating in its chronic environment.
>
> When the contingencies operating in the environment will not support a varied behavioral repertoire without abnormal behaviors, regardless of the animal's efforts to adapt to it.
>
> (Patterson-Kane, 1999).

Like anyone else, my approach to things has changed over the years. It has been interesting to return to this exercise to see how my understanding of animal welfare has evolved. The definition I use today is simpler, and broader, and not as connected to my foundational discipline of behavioral psychology. "Positive animal welfare is a state of being that occurs when an animal has rich engagement with a supportive environment" (Patterson-Kane, 2023).

I can see how developments in the field have affected me in that I no longer need definitions to be as specific and prescriptive. Also, I note the importance of positive psychology without defining "welfare" as an innately positive state. I would suggest that you take some time to review and consider the definitions of welfare provided by the frameworks covered in the previous chapter. Does one of these definitions fully align with what you think and feel animal welfare is? Or does some combination or modification of them? I would encourage you to write your own definition of animal welfare and consider how this might affect your personal and professional goals.

READINGS

The purpose of definitions

Hewson, C. J. (2003). What is animal welfare? Common definitions and their practical consequences. *The Canadian Veterinary Journal*, *44*(6), 496–499. PMID: 12839246; PMCID: PMC340178.

International example—WOAH

World Organisation for Animal Health. (2019). *Terrestrial Animal Health code* (28th ed.). World Organisation for Animal Health.

Small organization—UBC Sprouts

Broom, D. M. (2007). Quality of life means welfare: How is it related to other concepts and assessed? *Animal Welfare*, *16*(S1), 45–53. https://doi.org/10.1017/S0962728600031729

The Same Soul

Hippocrates, H. (1931). Hippocrates, Volume IV: Nature of Man (Loeb Classical Library, No. 150).

Discussion

Appleby, M. C., & Sandøe, P. (2002). Philosophical debate on the nature of well-being: Implications for animal welfare. *Animal Welfare*, *11*(3), 283–294. https://doi.org/10.1017/S0962728600024866

Gonyou, H. W. (1993). Animal welfare: Definitions and assessment. *Journal of Agricultural and Environmental Ethics*, *6*, 37–43.

Exercise—personal Definitions

Patterson-Kane, E. G. (1999). *Assessing and enriching the cage environment of laboratory rats.* Open Access Te Herenga Waka-Victoria University of Wellington. Retrieved May 4, 2025, from https://openaccess.wgtn.ac.nz/articles/thesis/Assessing_and_Enriching_the_Cage_Environment_of_Laboratory_Rats/16959307/1

Part II

Assessment—Body and Mind

“What Is It Like to Be a Bat?” is the title of a famous article about animal consciousness written by Thomas Nagel. This paper questioned the notion that subjective experience, in all its complexity, can be reduced to physical processes or properties alone. Concepts of mind and consciousness, the physical state, and a natural life are tackled in Part II. Vintage pen and ink bat illustration (#2379559671) by Hakan Kutlu/Shutterstock.com.

DOI: 10.1201/9781003375456-7

Chapter 6

Physical State

6.1 INTRODUCING ASSESSMENT

What defines an assessment is that some attribute, in this case welfare, is estimated so that reliable and potentially actionable data are retrieved. Each welfare assessment must be tailored to suit the attributes being measured, how they are measured, and your overall goal. The organization for these sections matches the domains that we discussed earlier in the Three Circles—physical state, naturalness, and subjective experiences. While the use of this model is somewhat arbitrary, these categorizations should act as an aide-memoire, starting with physical state.

Keyword: assessment

6.2 PHYSICAL HEALTH AND ITS MEASURABILITY

Physical state, including physical health, is clearly important to animal welfare and is also, usefully, amenable to immediate and objective measurement. The ability to measure things clearly and reproducibly is the *lifeblood* of science, as it opens the door to statistical analysis, which helps us to make sense of our work and present it to others. In considering physical health, we must remind ourselves that its measurability ranges from the obvious, such as a bruise, lameness, or "red tear staining" (iron-containing porphyrins that stain the corner of the eye of an animal such as a dog or a rat), to more covert changes, like suppressed immunity or an aberrant brain structure. Animals can also mask injury or illness, making it more challenging to detect. Increasingly subtle physical indicators have been investigated for associations with animal welfare, such as small degrees of *asymmetry* (Salgirli Demirbas et al., 2023; Knierim et al., 2007); for example, paired features like horns being a bit different or having a paw preference, like "handedness" in humans (aka motor bias). The assumption is that these imbalances are caused by stressors.

DOI: 10.1201/9781003375456-8

What follows are selected examples of how the state of an animal's physical body is important to their welfare. It focuses on work that is widely influential in the field. There will be biases toward approaches and methods that are well established rather than cutting-edge, as areas of study such as physiology, neurology, and biomarkers are currently undergoing rapid change. Animal welfare scientists competent in these developing disciplines will likely be in demand in the future.

Keyword: asymmetry

6.3 HISTORICAL IMPROVEMENTS IN ANIMAL HEALTHCARE

The first thing most people think of in the domain of physical health is injury, disease, and disorder, and how those might compromise welfare. Traditionally, the veterinarian or veterinary nurse/technician has been the point of contact when assessing physical welfare. However, the nexus between science and animal health occupations is increasingly blurred and might now include animal agriculture, animal breeding, animal feed development and production, and others. It is also worth pausing to consider how various disciplines relating to animal health and husbandry have evolved and changed the lives of animals.

Veterinary medicine arose largely around the care and treatment of horses and livestock at a time when people depended heavily on them for their livelihoods. (The word *veterinary* is traced to the Latin *veterinum*, which means "beast of burden".) This developing field tackled the immense problems of animal plagues like rinderpest (described at the end of Part I) and black leg (*Clostridium chauvoei*). Due to improvements in animal diets, care, and vaccinations, these diseases, as well as zoonotic diseases that used to pass routinely from animals to humans (e.g., rabies), are now uncommon.

In the 1960s, the focus of the veterinary profession began to shift to companion animal medicine, and now 70% of veterinarians (as of December 2023) work in companion animal practice. In three generations of veterinarians, the client base has flipped, so that veterinarians must now settle in urban areas to pay off their student debt (approx. $180K in 2022 for students with loans [data from the American Veterinary Medicine Association]). As a result, rural Americans can struggle to find large animal vets (Phillips, 2024). An excellent article published by a trade magazine made some good points about the issue, including the following: female veterinarians going broke because clients won't use them, rural students with a passion for animals but not the grade point average to go to veterinary school, a shortage of veterinary school places for locals because of lucrative out-of-staters, and farmers and ranchers who can ill afford corporate prices for vet care, so just do it themselves.

Solutions are needed, and building more vet schools isn't the answer if our vets keep quitting. Veterinary medicine is more specialized than ever, and perhaps there is space for *vet practitioners* who could be highly trained for specific jobs. This is a complex question, beyond the scope of this book, but there are solutions. As is, veterinary medicine is a profession on the edge.

Example: Hydatid disease

Bland scientific descriptions of hydatid disease (cystic echinococcosis) fail to reveal its horror. But suffice it to say, countries dealing with this parasite have paid it quite some attention. This zoonotic infection in humans is caused by a parasitic tapeworm, which causes cysts to form that can be fatal, depending on where they lodge in the body. The infographic (Figure 6.1) shows the life cycle of the tapeworm (*Echinococcus granulosus*), shuttling between its usual sheep and dog hosts. Not surprisingly, New Zealand, with its long history of sheep farming, had a major issue with the disease. The caseload "high" was reached in the mid-50s, then declined following 45 years of control, until, in 2002, the disease was declared "provisionally" eradicated. This is quite a feat, and one achieved by other geographically remote island states like Tasmania and Iceland. The New Zealand government used inspectors to dose dogs with tapeworm meds if needed, allowing an impromptu welfare assessment at the same time. In 2015, the WHO (the World Health Organization) estimated that echinococcosis kills approximately 19,000 people annually (worldwide), and where it is endemic, people are warned to avoid dogs.

Thankfully, hydatid disease is no longer a serious risk to humans in many developed countries. The tapeworm life cycle is readily broken by the routine use of animal dewormers, using prepared dog food (not raw offal), and in agricultural areas, preventing dogs from scavenging sheep carcasses. Unfortunately, disease resurgence is also rapid when control practices slide, as happened in Kazakhstan after the collapse of the Soviet Union (USSR) in 1991 (Abdybekova et al., 2015). You might wonder why we bother to discuss this at all if *Echinococcus* parasites only usually cause asymptomatic disease in dogs. The reason is that the way we control this disease has greatly influenced *how safe* we feel around dogs.

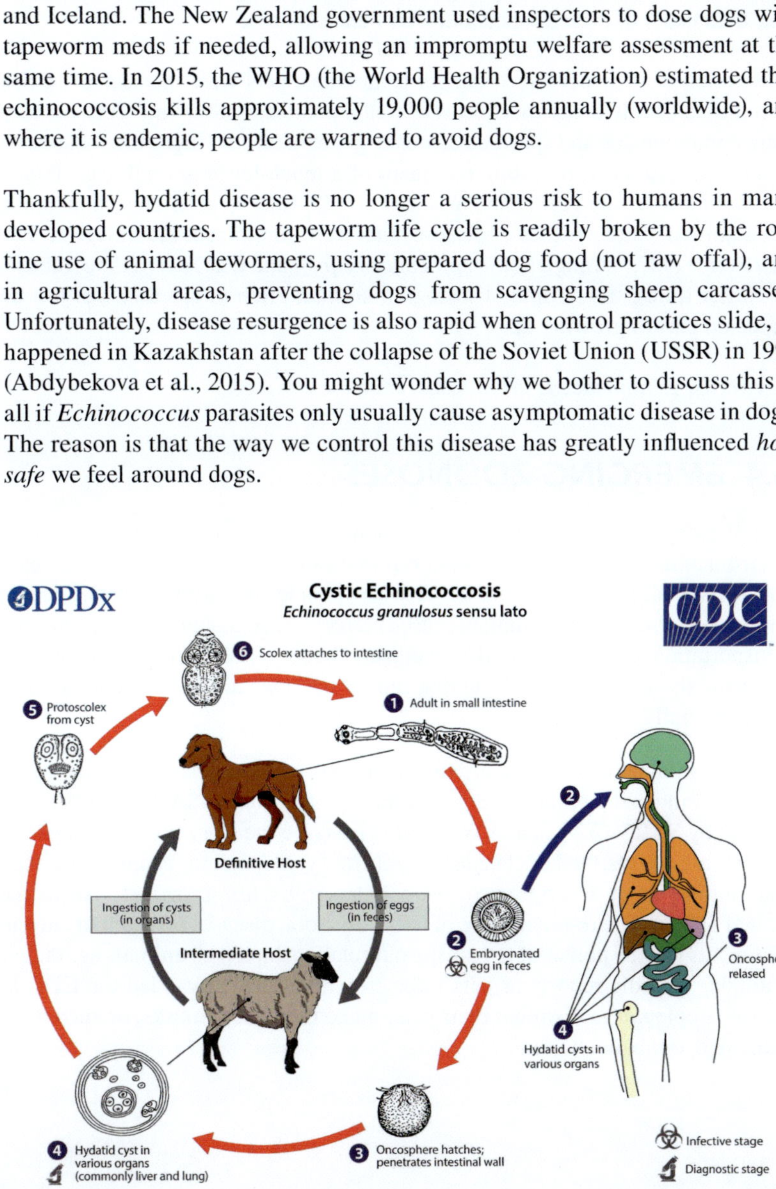

Figure 6.1 The life cycle of the parasite that causes cystic echinococcosis. The full details are available online at https://www.cdc.gov/dpdx/echinococcosis/index.html, along with those for alveolar echinococcosis. You can see that humans are an accidental host—a diversion of the cycle. The hydatid cysts that form at stage 4 of the canonical life cycle can lodge in multiple different organs in the human body. Image source: CDC-DPDx, a website maintained by the CDC's Division of Parasitic Diseases and Malaria. This public domain image is available free of charge from the CDC; our use of it implies no endorsement by the CDC.

Human behavior is the key element here. People, and especially parents, feared dogs, given that hydatid disease could be fatal after one brief accidental contact with dog feces. Without proper care, dogs still carry ticks, fleas, worms, ringworms, and many other unpleasant and zoonotic conditions, including some that remain lethal to people and animals, such as rabies. Even today, people who are repelled by the thought of dogs living in the home or cuddling up to their owners on the couch probably carry a memory and tradition from times when such contact was genuinely risky. This can lead to animals being made to live outside the household, which is (with some exceptions) associated with lower levels of attachment and care. However, an understanding of what was normal in the past can provide simple explanations for behaviors that persist to this day. Likewise, immigrants in the United States who recently migrated from countries where free-roaming dogs are still a hazard might demonstrate more fear than usual of even leashed and well-behaved dogs.

In addition to their perceived danger to humans, pets were themselves at risk of multiple deadly diseases. Stories of tragic pet death in the mid-20th century were commonplace and spawned an entire genre of animal literature focused on childhood trauma surrounding the death of a much-loved animal, e.g., *Where the Red Fern Grows* (1961) and *Sounder* (1969). In several of these books, the main animal character dies of diseases that were common at the time, e.g., *The Red Pony* (1937, "strangles" [*Streptococcus equi*]) and *Old Yeller* (1956, rabies). In earlier times, the trauma of losing an animal, often to disease, was just an expected part of growing up.

Keywords: Echinococcus granulosus, endemic

6.4 EMERGING ZOONOSES

One of today's greatest threats is from zoonoses (diseases that can be transmitted from animals to humans). Conditions that cause zoonotic disease to develop and spread often involve contact between people and wild or domesticated animals. *Anthropogenic* (human-caused) effects, such as long-distance animal transportation, crowded animal housing, and human incursion into pristine natural habitats, are a potent trifecta that can set the stage for the breakout of novel and potentially deadly disease.

Current examples of zoonotic diseases include brucellosis, rabies, plague, Lyme disease, influenza, severe acute respiratory syndrome (SARS), COVID-19 (caused by SARS-CoV-2), and West Nile virus (see the story box and watch the TEDx talk [https://tedx.ucla.edu/project/tracey_mcnamara_canaries_in_the_coal_mine/] for a description of West Nile virus's first recorded visit to the United States). The re-emergence of lethal zoonotic diseases resistant to current treatments has the potential to reshape our relationships with animals again. For example, sporadic reports of pets infected with SARS-CoV-2 led the CDC to advise people *not* to disinfect their pets, make them wear masks, or surrender, euthanize, or abandon them.

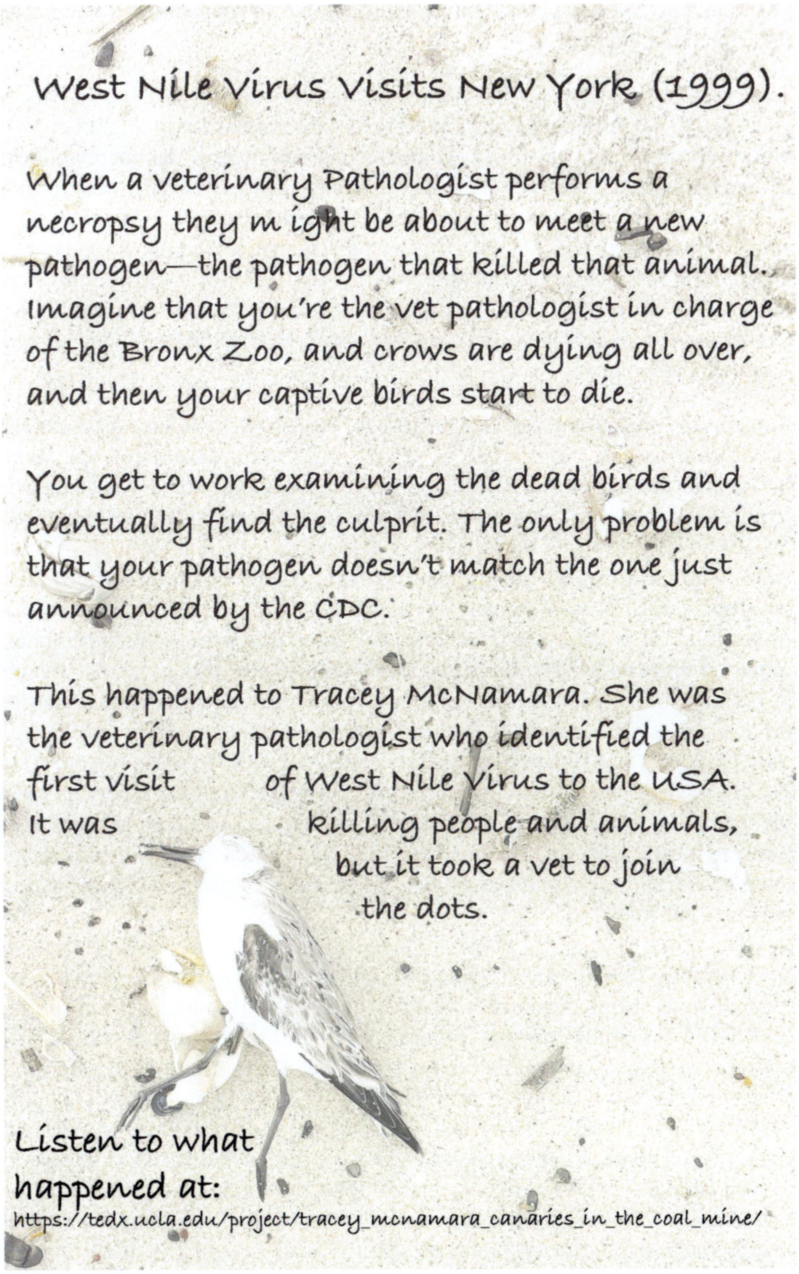

The development of antibiotic-resistant bacteria due to the abuse of antibiotics in farming is another stubborn problem. The *Alliance to Save Our Antibiotics* report (2024) revealed the extent of the abuse of antibiotics by UK farmers as promises to keep up with tougher EU directives post Brexit did not materialize. Loopholes remain in the most recent regulations, and the fear is that antibiotic-resistant infections will surge, and our backstop "last resort" antibiotics will fail. This crisis dovetails with a second, self-inflicted one, which is a post-Brexit shortage of veterinarians, who are in greater demand than ever because of the increased bureaucracy surrounding animal movements into and out of the United Kingdom.

Another looming threat is the highly pathogenic subtype of avian influenza virus A (H5N1), also called *bird flu*. This disease is common in migratory wild birds and causes relatively mild symptoms in them. In large commercial flocks of domestic poultry, the outlook is very different, and high mortality rates ensue. This has led to frequent poultry farm depopulations (eradicating the entire flock at a given location), which from a welfare perspective, is the premature death of tens of millions of chickens and other poultry. Extreme culling procedures were meant to be a last ditch effort to stop disease spread, but concentrated animal feed operations (CAFO, aka *factory farms* with animals in confinement) now routinely depopulate.

Euthanasia is warranted because the disease has such a high mortality rate and causes considerable suffering. However, the sheer scale of these cull operations means that the euthanasia methods are likely to fail the standards ordinarily required for veterinary euthanasia. At the time of writing, H5N1 is also spreading across US dairy cow herds, which is a worrying new development (Abbasi, 2024), and farm cats that drink unpasteurized colostrum and milk have died (Burrough et al., 2024). The entry of the virus (spillover) into such a key animal in our food chain increases the chance of human exposure, with an economic impact if animals need to be culled. Although disease was generally thought to be mild in dairy herds, there is a new twist, with mammary tissue getting infected (Ríos Carrasco et al., 2024), which also begs the question of how big milk producers like Holsteins might fare compared with Jerseys. The concern is that if the virus becomes endemic in cattle, then novel mutations may increase the threat of contagion to, and then among, the human population. This is not a threat to be taken lightly, as human mortality rates often exceed 50%, although that may be an overestimate, as asymptomatic people or those with mild illness may not seek help.

In November 2020, the Danes culled 14 million farmed mink, their entire commercial stock. Mink fur farm workers with COVID-19 may have transmitted SARS-CoV-2 to mink (reverse-zoonosis). The virus then mutated in mink with spillover back to humans (Jahid et al., 2024). The Danes, anxious that the *mink* variant might lessen the efficacy of any COVID-19 vaccines, decided to eradicate their farmed mink. The agriculture minister would later resign, as the rush to cull was unlawful. Jahid and colleagues concluded that "mink are the only known animal species found to spread SARS-CoV-2 to humans and other contact animals". This begs the question of whether mink pelt production on such a scale represents an unacceptable biosecurity issue irrespective of any welfare objections.

There are other opinions as to the ethics of culling approaches, especially when they appear to be based on flimsy evidence (Figure 6.2). A different position follows a *One Health* tradition, which considers human and non-human health as interdependent (Degeling et al., 2016). The One Health approach operates under the oversight of the One Health Commission, and you can read about their various initiatives at https://www.onehealthcommission.org/.

Keywords: anthropogenic, depopulation, spillover (infection), One Health, zoonosis

Figure 6.2 A happy Holstein, but at what cost? For years, the British government has pursued an ineffectual badger culling policy to protect cattle from tuberculosis while lacking a robust scientific evidence base for its approach. It took until August 2024 for the government to announce a reappraisal; a widely viewed documentary was aired the same month by a concerned celebrity-citizen scientist, who thinks he's cracked the problem. You can watch the documentary for yourself at *The Badgers, the Farmers and Me* (https://www.youtube.com/watch?v=AxsKzS_NwGQ). Image by Emily Patterson-Kane.

Culling Animals as a Geopolitical Strategy

Culling is a crude method to stamp out disease and can also serve as a "societal cudgel". Animals and plants of social and cosmological importance are often destroyed to control and assimilate a group. This scenario has played out repeatedly for the Dine'é (Navajo) of North America. (Navajo people use *Dine'é* to refer to their Nation, and *Diné* to refer to themselves ["The People"].) When the Spanish arrived in the New World in the 16th century, they came for gold and left behind their sheep, goats, and horses. Diné shepherds adopted the sheep, and the Navajo Churro, as they came to be known, thrived in the high deserts and scattered woodlands of the Four Corners region of the Southwest. They ate sagebrush and whatever else they could forage, and their fiber was suited to spinning and felting. In 1863, Kit Carson was sent to evict the Diné from their lands, kill their livestock, and imprison them. After four years of internment, the Diné were released to a reservation carved out of a portion of their ancestral lands. By the 1930s and the Great Depression, the land was now a dust bowl, and windblown dirt was threatening to silt up the Hoover dam project on the border between Arizona and Nevada.

Overgrazing was blamed, and so a federal Livestock Reduction Program was imposed that sanctioned the slaughter of half a million sheep. The Churro sheep were a particular target, as they were considered scruffy—an inferior breed. The actions of the federal government were designed to sever the Diné from their traditions, and many have been trapped in intergenerational poverty ever since. The Bureau of Indian Affairs was described as working hard to turn Diné shepherds into housewives, part of a shameful episode of social engineering (Weisiger, 2007). (Traditionally, women raised the sheep. They would weave Churro fiber into rugs and textiles, which provided them with some financial security.) Remember this section when we discuss green criminology in Part IV and social norms in Part V. Ultimately, the Churro was brought back from the brink of extinction by Lyle McNeal, a professor of animal science at Utah State University. Starting in the late 1970s, scattered Churro survivors were rounded up from mesas and arroyos and brought under the protection of the Navajo Sheep Project (https://www.navajosheepproject.org).

Keywords: culling, Diné, Navajo Churro, Lyle McNeal

6.5 ACCESS TO VETERINARY CARE

This is an enormous topic, which is too large to do justice to here and that mirrors many of the issues that people have in receiving adequate healthcare themselves. It seems to be an enduring problem that the ability to benefit from veterinary breakthroughs over the last few decades can be prohibitively expensive. Even people who are relatively well off now treat their trips to the veterinary clinic with trepidation. Efforts to improve access to care involve charitable organizations, volunteer veterinarians, and vet techs. Lending funds to pet owners in need also makes financial sense, as research shows that repayment rates are quite high, even from clients barred from traditional sources of credit (Cammisa & Hill, 2022). Other potential obstacles to accessing veterinary care include a lack of transport, shared language, or an understanding of the need for care in the first place. Less affluent areas also suffer the phenomenon of becoming "veterinary clinic deserts". A recent (January 2024) CBS news story described volunteers helping the Diné to look after their animals, where, for 27,000 square miles of territory, there were only three veterinarians and an estimated half a million cats and dogs.

Animals that are not owned, such as feral and wild animals, or those barred from certain types of care (e.g., livestock in organic agriculture), often fall outside the usual veterinary care demographic. Debate over when and how to provide care and what level of care is appropriate is discussed in the context of *spectrum of care*, which expands the usual treatment options to a range of acceptable outcomes, all above the ethical minimum. These options allow providers to better accommodate what may be preferred, is suitable, or is affordable. It should not be assumed that novel options represent poorer outcomes. For example, the rapid spay/neuter methods developed to address situations like large breeding feral cat populations routinely produce superior outcomes due to (practiced) surgical skill and reduced time for the animal under anesthesia.

Conventional veterinary care may come into conflict with other belief systems if the goals of each cannot be harmonized. Cultural competency is becoming an increasingly important part of veterinary practice and can help practitioners navigate the requirement to accommodate diversity and maintain the standard of care. (*Standard of care* is a legal term for a professional achieving a minimum acceptable result.) A few examples of *spectrum of care* are given below:

Treatment of livestock in organic systems either without the use of chemicals or with the use of appropriate withdrawal periods

Treatment of animals in accord with the client's spiritual beliefs. For example, end of life care without euthanasia, or negotiating whether companion animals are subject to a prohibition against blood transfusion

Conflicts relating to the performance of non-therapeutic modifications—i.e., "cosmetic" or convenience surgeries—when these may be necessary to keep the animal in their home

Keywords: standard of care, spectrum of care

6.6 PHYSIOLOGICAL STRESS

Historically, in our understanding of physical health and its effects on welfare, we were limited to what we could see. As instruments were developed, we began to accurately record physiology (e.g., using thermometers), with each new development widening our knowledge of animal anatomy, physiology, and pathology. One good example relates to changes in physiology due to stress. (Physiology is the branch of biology that studies the operations of living tissue at both the physical and the chemical level.) The concept of chronic stress was introduced in Part I with Hans Selye (Selye, 1936). But Selye was aware of other work showing how the animal had a coordinated and acute *fight or flight* response, which ended with a return to the resting state of homeostasis (Cannon, 1939).

A key pathway of the physiological stress response (that underpins the flight or fight response) is the HPA axis (Mormède et al., 2007). The acronym HPA denotes the hypothalamus (H; close to the brainstem), which is physically connected to the pituitary (P) and chemically connected to the adrenal glands (A). Collectively, these hormone-secreting glands are triggered by stress to initiate and propagate a cascade of reactions that ultimately result in the secretion of the steroid hormone cortisol by the adrenal glands. Different parts of the body have cortisol receptors, enabling a stress response to induce a fight or flight–compatible physiology (alertness, raised blood sugar and pressure levels, and other processes). High blood cortisol levels ultimately feed back to dampen the stress response, so that cortisol can switch stress responses both on and off.

The HPA axis is very widely employed in understanding an animal's welfare state. For example, high levels of cortisol in body fluids (from a blood draw, saliva swab, urine, or milk from lactating animals) would indicate a recent stress event. Cortisol laid down in other tissues, such as hair, can be used to assess stress over longer periods of time. Cortisol's usefulness as an indicator of active stress has led to its description as "the stress hormone".

This hormonal stress measure was very useful in assessing stress levels during procedures such as castrating piglets. Unlike stoic sheep, the piglet will respond and vocalize vigorously when disturbed. (The disturbed piglet has been described as sounding like a fire alarm going off.) A cortisol response helps distinguish between this typical response and the higher levels of distress that are caused by the experience of pain. Cortisol-based evidence was key in convincing scientists and the agricultural industry to consider providing analgesia (substances that prevent pain) or to replace surgical castration with immuno-castration or slaughter before sexual maturity (Prunier et al., 2006).

An acute cortisol response will occur within minutes of exposure to the stressor, with cortisol levels peaking and then declining as the animal re-establishes homeostasis. This pattern is easiest to appreciate when making repeated measures after presenting a single short stressor. These repeat measurements will reveal how hormone levels fluctuate, with daily and seasonal cycles. There are also marked differences in cortisol release between species, breeds, and individuals, as well as in the same individual over time. A repeated stressor may cause a greater response, but eventually, the response will become blunted as the animal becomes less able or willing to respond, especially when the stressor is inescapable. Variation in blood cortisol levels that track the circadian rhythm (levels generally increase when an animal wakes up and remain elevated

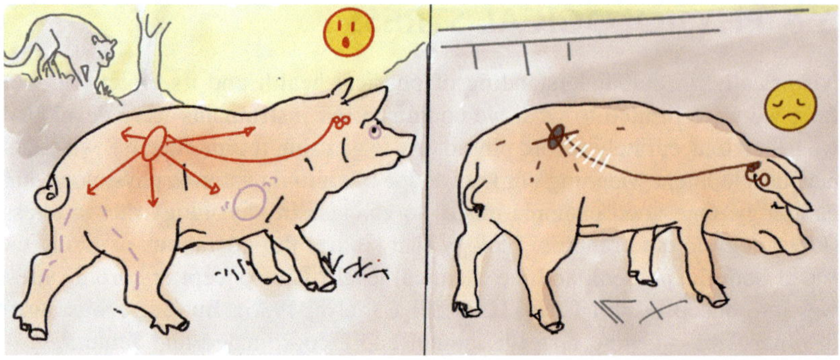

Figure 6.3 (Left) When the pig senses that she is in danger (noticing a mountain lion), small parts of the brain swing into action to alert the adrenal glands to release cortisol (indicated by the red lines). This steroid hormone prepares the body for minutes or hours of strenuous action. This includes dilating the pupils of the eyes, increasing the heart rate, and releasing glucose to help muscles work better. This acute stress response is normal, but it can become deregulated by chronic stress. (Right) When pigs experience stress over a long period of time, the increased allostatic load can become damaging. This animal is withdrawn and disengaged from the environment. A disrupted HPA axis can have knock-on effects for any acute stress response that might be needed (e.g., vigilance). Image by Emily Patterson-Kane.

during their active period) may be less apparent or even absent in a chronically stressed animal. Therefore, counter-intuitively, highly stressed animals, such as depressed pigs that become disengaged from their environment and are no longer coping, may show depressed levels of cortisol or fail to show the expected patterns of fluctuation (Figure 6.3). For these reasons, cortisol levels can be challenging to interpret when stress is chronic, and eventually, prolonged stress will leave its mark on the body, such as shrinkage of the thymus (an organ that supports immune function) and changes to DNA chemistry.

Keywords: analgesia, cortisol

6.7 NEUROSCIENCE

Every part and process of the physical body is relevant to understanding animal welfare. The brain, as the main integrator of sensory information into lived experience, is a particularly important organ. There are disciplines of study specific to the role of the brain, including neuropsychology and neurophysiology. Neuroscience is a general term for the scientific study of brain structure and function and the wider nervous system.

Historically, our inability to study the brain at the same level of detail as we can study directly observed behavior has caused considerable difficulties. For example, when B.F. Skinner led the science of behavior, our understanding of neural and physiological systems was remedial. We only began to image neurons in the 1950s with the invention of electron microscopy (Palade & Palay, 1954). In terms of physiological systems and processes, much of the theorizing was occurring in a vacuum of robust experimental evidence, and third-person measurements were largely impossible. Skinner's approach was to dispense with speculation, which was, at that time, rampant. Other unhelpful, if not flat out destructive ideas, were circulating in the human realm, including the notion of the "refrigerator mother". These were mothers who were blamed for their child's autism because they were supposedly cold and distant. While human psychologists might wield their fringe theories with abandon, Skinner was developing a granular understanding of the structure of behavior while treating the physical body as a simple black box. (A *black box model* is one where you study input or outputs but not the structure between them. For example, in behaviorism, the characteristics of the environment are seen as causing behavior irrespective of what is inside the person or animal.)

Behaviorist approaches (conditioning) were critical for the development of psychology as a science in the mid-20th century but began to wane in the face of the Cognitive Revolution and a focus on what might be happening *inside* the black box of behavior. Critics like Noam Chomsky tore into behaviorist approaches (Virués-Ortega, 2006). Even Skinner's own former grad students would publish their classic 1961 paper (Breland & Breland, 1961), reporting that the learning curve induced by reward was more of a detour from instinct, and that the animal would inevitably "revert to type" (see Part V). The equivalent expansion of knowledge in neuroscience is arguably happening now with the current challenge of blending cognitive sciences and neurosciences, and applying this knowledge to the study of animal welfare. At its most conceptually difficult is the "hard problem" coined by Chalmers in 1994 (and published in Chalmers, 1995), which asks how we can convert the brain's structure and function into *mind*. As we will see, science *might* have made a hard problem impossible, and we'll see why later.

The early starting point for neuroscience was a quantification of the conditions in which brains fail to thrive, a key example being barren or impoverished environments with very little stimulation in which opportunities for goal-directed behavior are rare. Very early research showed that the brains of rats raised in small barren cages were visibly less well developed. Understanding how this happened, which was ultimately linked to the function of neurotransmitters and neurotrophic factors, likely subjected many more rodents to poor standards of welfare.

There is also evidence that the brain will bias its development to structures that sense what little stimulation is available. Temple Grandin (1989) found that piglets in barren housing seek stimulation through repetitive and functionless (stereotypic) behaviors, where they rub their sensitive snout on their enclosures or on other pigs ("belly nosing"). This establishes the notion of the brain as a "plastic" receiver of information and not just a producer of it. Over time, more of the pig's brain becomes connected to snout stimulation instead of other sensory inputs. This explains why stereotypic behaviors, especially those that develop in juveniles, persist in animals even when moved to more enriched environments.

The alignment of cognition and neuroscience is especially critical, and the same dilemma exists today in terms of picking the right model and underlying concepts as it did for Skinner. In 2004, Berridge wrote: "yet, if our motivational concepts are seriously wrong, our quest for closer approximation to brain-behavior truths will be obstructed as much as if we had no concepts at all" (Berridge, 2004). Neuroscience may ultimately provide more objective knowledge about the molecular basis of sentience or consciousness. Extensive work has been done, for example, in tracking how brain damage (deliberate tissue damage is called ablation) can alter the perception and response to pain, and how brains that appear to be very different are able to carry out many of the same functions. More on this later.

Keyword: neuroscience

Exercise: Ethics and foundational research

It is important to note that psychology and neuroscience share a problematic history of using animals in research. There are many studies that are famous not only for the discoveries made but also for the suffering caused to animals. This research often involved the deliberate injury of animals to observe their debilitation and then recovery. Studies of this type might involve blinding, paralysis, or selective removal of part of the brain. There is considerable debate about whether this research is consistent with current ethical standards, and where it is not, whether it should continue to be used in educational curricula. When it is used, it is important to discuss the studies' ethical implications as well as the scientific advancements they contributed to.

Two prominent examples are listed below. You might want to read one of these experiments as originally published and then Google them using the short title provided and review the associated animal welfare controversies. Consider how you would assess whether this research was justified, could be justified if carried out differently, and what the effect might be if this research had never happened at all (i.e., the *counterfactual* scenario). This includes how the research has directly informed animal welfare sciences in conveying the sensitivity, resilience, and needs of animals, and especially primates.

1) **Short title:** Psychology: "*the Harlow monkey experiments*" and maternal attachment/social needs (Harlow et al., 1965; Harlow & Zimmermann, 1958).

2) **Short title:** Neuroscience: "The Silver Springs monkeys" and neuroplasticity (Taub et al., 1973; Taub, 1976).

Now that you understand a little more about these ethically problematic papers, you might also want to research "Neuralink and animal welfare" or pull this paper: "Livingstone, M.S. (2022). Triggers for mother love. *Proc Natl Acad Sci U S A, 119*(39)". You should begin to get an idea of how entrenched and pervasive these issues are. You might also reflect on how "lab lore" can impact animal welfare, as we will discuss in Part V.

Keywords: Edward Taub, Harry Harlow

6.8 AFFECTIVE NEUROSCIENCE

There are many areas of neuroscience that are significant for understanding animal welfare. One of these is affective neuroscience—that is, developing an understanding of the neural basis of felt emotion. (*Affective* means relating to feelings of emotion, while *affect* is usually meant to describe how an individual is outwardly expressing their emotions. For example, a person with a "flat affect" is one who appears dull or stoic.)

One of the simplest ways to explore the connection between the brain and emotions was with electricity. Electrical stimulation of certain areas of the brain would directly elicit the expression of powerful emotions in humans and other animals. The ability to stimulate emotional states also allows us to measure those states, even in situations where other indications such as behavior or physiological stress might be inaccessible or ambiguous due to their valence. *Valence* and *hedonic state* are values associated with emotions, with pleasurable emotions rated as positive (i.e., a positive valence/hedonia) and repulsive/unpleasant emotions as negative. Jaak Panksepp was a groundbreaking researcher in this area. This Estonian-born neuroscientist first coined the term *affective neuroscience* and demonstrated that positive emotions elicited by electrical brain stimulation (EBS) could be used to reward behavior.

In contrast to behaviorist psychology, Panksepp suggested that learning occurred primarily due to internal emotional states rather than the environment. But in contrast to later cognitive psychology, he considered this process to be the same in humans and other mammalian species (Panksepp, 2011; Panksepp & Biven, 2012). Panksepp proposed that the primary emotional motivators are seeking/expectancy, fear/anxiety, rage/anger, lust/sexual excitement, care/nurturance, panic-grief/sadness, and play/social joy. (Any curiosity-driven scientist will understand the allure of *seeking* via discovery. For the animal, "foraging" might approximate "seeking".) Panksepp clearly found his a lonely field of research, and as he commented at the beginning of a TEDx Talk (https://www.youtube.com/watch?v=65e2qScV_K8), "science does not answer 'why' questions; science answers 'how' questions". The question "why do we have feelings?" is, according to Panksepp, because they support our survival. If you listen to that TEDx Talk, you'll hear a description of how EBS of the "seeking" system in seriously depressed human patients restored their enthusiasm for life (Schlaepfer et al., 2013).

This neuroscience-based approach to emotion asserts that the mammalian non-human animal feels emotions in the same way humans do and for the same reasons. Panksepp also emphasized the benefits of positive experiences such as play. The benefits of studying human and animal affective neuroscience as the same subject were reflected by Panksepp's award of the Baily Endowed Chair in Animal Well-Being Science at Washington State University's Veterinary College. Neuroscience continues to be very important in understanding how we move beyond (just) avoiding suffering to supporting lives that include pleasure and joy. Berridge and Kringelbach (2008) note that "a particularly important topic for affective neuroscience is to understand how brains generate pleasure and other psychological components of reward because reward is important in daily life. Pleasure is essential to a normal sense of well-being".

Just as the study of physiology educated the public about new ideas like stress, neuroscience is becoming more widely understood. For example, many people are now aware of dopamine as a neurotransmitter associated with positive feelings. Neuroscience is also helping us to appreciate how individuals may combine multiple positive and negative feelings to reach an overall evaluation of valence. While certain parts of the brain seem to use different modes to determine overall valence (Berridge & Kringelbach, 2015), the wider neurological system spreads across and outside the brain. These structures are influenced by chemical messengers, with feedback coordinated by molecules such as neurotransmitters, neuropeptides, and hormones. This has led some to propose that novel neurophysiological biomarkers could be developed to interrogate animal welfare (Tilbrook & Ralph, 2017). Neuroscience and the social movement of *neurodiversity* have also now combined to show how brain systems are influenced by environmental cues, experience, and a myriad of other factors to yield human neuro-individualization (Goldberg, 2023).

Keywords: affective, valence, Jaak Panksepp

The mind–body problem

Many cultures hold that sentient beings are split into the physical body and an intangible aspect, which is variously named mind, spirit, soul, or "self". Historically, this spiritual aspect was expected to have some kind of literal substance, which, according to a 1907 study by a Massachusetts physician (Duncan

MacDougall), weighed in at 21 grams—the exact weight lost postmortem. By the Roaring Twenties, a time of dramatic social change in America, most people believed that the spirit had no physical form.

The notion of a non-physical mind is called *Cartesian dualism*, named after the contributions of René Descartes (whom we met in Part I). But how does an intangible mind control the physical body? This central issue is the *mind–body problem*. Descartes theorized that the pineal gland was the nexus between mind and body because of its central location in the brain and ready access to the "spiritual substance" that was assumed to flow around the body. The pineal gland as "the seat of the soul" is not considered one of his better ideas, because no matter how special it looks, the pineal gland is still just flesh like any other part of the body. In earlier times, it was the heart, the only organ that very obviously looks alive, that was thought to be the seat of all emotions (Brandt & Huppert, 2021).

These days, people have a range of philosophies about what generates our conscious experience. One perspective is that the entirety of the unconscious and conscious qualities of the mind derives from brain and body functions, and that the mind/soul as some kind of physical object or resident ghost simply does not exist. However we conceptualize the mind–body problem, the central question remains, as noted by Colin McGinn (1999), "how did evolution convert the water of biological tissue into the wine of consciousness?"

Keywords: mind–body problem, Cartesian dualism

6.9 ROUTINE SURGERIES

For the moment, let's focus back on the physical body and a more obvious challenge to welfare, cutting or burning something off: specifically, the surgical removal of enervated tissue in a way that will cause pain if analgesia is not provided. These surgeries are not controversial when performed therapeutically but stray into ethically gray areas when carried out for aesthetic reasons (for looks), for convenience (for the human handler/caretaker), or preventatively (to prevent a worse outcome from occurring).

Routine surgical modifications of animal bodies are a hotly debated welfare issue. Examples would include ear cropping (dogs), tail docking (dogs, horses, cows, pigs, pet monkeys), horn or horn bud removal (goats, cows, deer), castration (many species), declawing (cats, large exotics), and beak tip or tooth removal (chickens, piglets, large exotics). These surgeries are commonly referred to in the United Kingdom as *mutilations*, a term that is generally avoided in the United States. Where the goal of the surgery is purely aesthetic, support for these practices is generally lower or in decline. Modifications such as ear cropping or tail docking of companion animals (dogs) are largely seen as cosmetic, and owners defend these surgeries on the basis of looks, tradition, and "a right" to subject their pets to them with effective pain management. They might even argue that the modifications are in the animal's best interests.

Where does the evidence take us when considering these claims? This author's analysis (EPK) suggests that tail docking could simply not be justified in terms of preventing tail injury (Patterson-Kane, 2017). Then and now, many highly qualified people will disagree and defend tail docking. They might take issue with how I chose to answer the question in my study, or how I interpreted my data. Others would say my position was not strong enough, and that I underestimated the pain and other direct harms caused by the surgery. Regardless of the supporters or critics of this work, the mere existence of such technical reports helps us to connect the discussion to the evidence base and hopefully move us toward greater consensus over time.

It is important to understand good faith opinions and investigate why people opt for surgery for their pets and what might be needed to render those opinions obsolete. For example, body parts that are not removed might be injured in more serious and uncontrolled ways (hunting dog tails in rough terrain, pig tails bitten by other pigs, horse tails caught in carriage harnesses). Also, animals with body parts that could be used as weapons may use them to harm others (being gored by a bull horn, piglets biting others to compete for access to a nipple while nursing).

Where there is some solid evidence that leaving the body part on might compromise welfare, the exercise becomes a risk assessment. This might include trimming hair from the tail or ensuring enclosures cannot trap and injure body parts. One widely agreed example of a surgery with better options is the removal of pet monkey tails to make it easier to fit them with diapers. Many feel that this amputation caters far too much to the convenience of the owner over the bodily integrity of the animal.

Sometimes, the reason for ethical concern is how the surgery is performed, i.e., how radical it is. For example, some farmers crop sheep tails to a few inches in length to help prevent flystrike (maggot infestation of soiled wool and skin). In these settings, total tail removal is avoided, as it is associated with higher rates of rectal prolapse. However, in the United States, sheep that are exhibited often have the tail totally removed so no part of it projects from the skin. This is most likely because it gives the animals a more "muscular" profile that is thought to give them a competitive edge. While some disagree, I would suggest that the shortened tail is a convenience surgery that helps sheep avoid a serious health problem while living on the range in a highly natural way, while the short dock for closely monitored show animals is more for aesthetic reasons.

Tradeoffs can also be complicated in other examples such as the removal of horns from cattle. Cattle can and do kill farm personnel every year and are more dangerous when horned. Therefore, in the United States, most cattle have their horn buds removed as calves, before the bud can attach to the bone of the skull. Horn bud removal is called "disbudding" and is much less involved and stressful than dehorning, although still unpleasant. An alternative to disbudding is to introduce polled (hornless) genetics, although these cattle have traditionally had lower productivity traits than horned sires. The ability to use precise gene editing to introduce the polled gene without eroding genetic merit will be a game changer for the industry: in other words, engineering a hornless animal without lower productivity.

Even when pain-free methods can be used to omit an animal's body parts, animal welfare objections can be raised. This is because body parts, like a dog's tail, have functions in balance and communication, and many are convinced that animals should be appreciated in something approaching their natural state. Others believe that dog bodies have already diverged so much from their ancestral type that it hardly matters if an ear is cropped or a tail docked. One exception to the trend away from non-therapeutic surgeries is the reproductive sterilization of cats and dogs (also called neutering, from the Latin word *neuter*, meaning neither male nor female). In the United States, "spay/neuter" is used to describe female/male sterilization and is seen as broadly beneficial for managing pet populations and behaviors, and for disease prevention. In Norway, it is illegal to sterilize a dog without a medical reason, and sex-related behaviors are seen as normal, with breeding controlled by responsible ownership—preventing access to a bitch in season. Neutering is viewed as mutilation, and behavioral issues (for males) can be addressed by chemical castration.

In the United States, spay/neuter for dogs, colloquially known as *fixing an animal*, is now a more nuanced conversation, and less radical options, such as an ovary-sparing spay, are sometimes preferred. Spay/neuter might be convenient for municipalities (especially for cat control), but the benefits of the six-month spay/neuter for some larger dog breeds are less obvious (Hart & Hart, 2021), and a less standardized approach may be appropriate. These procedures can also increase rates of obesity, with multiple other issues reported, although the data are controversial (Kutzler, 2020). In some cities, there is less of a choice about whether to fix an animal. The city of Denver currently has some of the strictest spay/neuter rules in the country, and all cats and dogs over six months of age need to be sterilized.

Rather than resorting to cosmetic surgery, a desired physical trait can be introduced genetically, but this is extraordinarily difficult in breeds with a closed genetic pool (e.g., pedigree dogs). However, Dr Bruce Cattanach, a geneticist and dog breeder, succeeded in importing the bobtail gene (very short tail) into boxer dogs by out-crossing to corgis. Good homes were arranged for all the "in between" dogs that did not look like boxers, which he called "borgis". By back-crossing, he was able to produce naturally tailless boxers of show quality, and eight generations later, the descendants from this out-cross were only 0.5% corgi.

The UK Kennel Club allowed some of these bob-tail boxers to be registered as pedigree boxers in 1998, something that would not usually be possible (i.e., they opened up the stud book). At the time, the Kennel Club knew that poor health in overly inbred or exaggerated pedigree dogs was becoming a hot button issue for pet owners. Other organizations have used crossbreeding to improve health (e.g., for the Norwegian Lundehund) or have prohibited breeding from unhealthy bloodlines, such as the purebred Cavalier King Charles Spaniel (in Norway).

Keywords: polled, spay/neuter, Bruce Cattanach

Example: Tennessee Walking Horse and the "big lick"

Sometimes, a breeder will not use genetics to introduce the look that they are after and might resort to devices and instruments that may be banned. One very contested example of an exaggerated show animal relates not to their physical form but to their behavior. The Tennessee Walking Horse has three natural gaits: flat walk, running walk, and canter. When any of these gaits involves a very high and exaggerated step forward, this is called the "big lick". This is rewarded in the show ring and so became a goal for breeders.

Breeding and training will encourage the big lick; however, more direct interventions on the horse's body can also be used. These include "stacks", which are large, sometimes weighted shoes placed on the front hooves. And finally, the practice of *soring* developed. Soring refers to the practice of putting caustic chemicals on the horse's legs and is used alongside other procedures such as *pressure shoeing*, where the horseshoe is attached in a way that it is deliberately painful, or the horse is left for a long period of time with heavy chains around their ankles. These and other strategies cause sensitivity and pain, provoking a more abrupt high step—the high step that wins competitions. Horse soring is illegal, but enforcement is difficult. The long-awaited Prevent All Soring Tactics (PAST) Act, effective February 2025, will hopefully improve enforcement and end the soring saga.

Keywords: big lick, soring (gaited horse)

6.10 ARTIFICIAL SELECTION

An individual's genetics strongly affect their needs and capabilities. In the next section, the role of breed or type will be discussed as a factor in determining how an animal naturally lives, its ideal state, and what causes an individual to suffer or thrive. Evolutionary theory provides a foundational understanding that animals with a shared evolutionary history, or a convergent evolutionary pressure, will tend to have more comparable internal subjective experiences. Charles Darwin expressed this sentiment in his 1871 book, the *Descent of Man*: "The lower animals, like man, manifestly feel pleasure and pain, happiness and misery" (Darwin, 1871).

Artificial selection through domestication creates new species with distinctly different needs from the source stock, typically including a dependence on human care for good welfare. Domestication is a subtype of selection where a sub-species of a wild type adjusts to being taken care of by another species, for example by adapting to different food types and being less reactive when handled. It is most often thought of in relation to human caretakers, but non-human examples exist as well, such as leafcutter ants that farm fungus gardens.

Given that the genome of human-domesticated animals is largely determined by us, we are obligated to consider and factor in our care for these animals when we design our breeding strategies (see the ancient contract in Part IV). When it comes to intentional breeding, there are at least two approaches. One is to create a *landrace*, where certain cultures and communities keep animals for a local purpose with no strict requirement for heritage or non-task-related traits (e.g., the color of a racehorse). The other requires the animal's recorded heritage to go back many generations and be exclusively vetted and confirmed as purebred (e.g., an American Kennel Club [AKC] show dog). In the latter case, inbreeding becomes an inevitability, although it can be slowed by eschewing practices such as *line breeding*, where close relatives are bred to propagate their reliably shared traits.

The genetics of domesticated animals are frequently affected by the human tendency to go to extremes when seeking profit or excellence in competition and when pursuing beauty. Take, for example, the art of classical dance. In ballet, performance is thought of as unchanging in its adherence to the choreography of the work. In practice, the dancers' leg positions have become more extreme over time to demonstrate ability and as the historically restrictive attitude to female behavior relaxed (Daprati et al., 2009). (It used to be unusual for the leg to be raised above the hip, or the female dancer to appear noticeably muscular, as this was considered improper or unfeminine.) Along with these more extreme positions, the dancers are more likely to exceed the body's ability to perform without pain and irrevocable injury (Russell et al., 2011).

Just as our culture encourages extremes in ballet positions, it also tends to reward the exaggeration of desired traits beyond what is optimal for animal welfare. If you look at photographs or taxidermy of pedigree breeds of dog or cat, it is unmistakable that almost all breeds with a long history started out looking more similar to each other than they do now. Differences often emerge between lines raised to be shown versus those bred for performance. Breeders seeking a more moderate and functionally robust type must resist the temptation to seek out extremes. The "evolution" of the extreme can be seen if you ever have the chance to visit the Natural History Museum at Tring (UK), formerly known as the Rothschild Zoological Museum. This museum houses a remarkable taxidermy collection that, by comparison with today's standards, demonstrates vividly how some dog breeds have been changed by 19th- and 20th-century breeders.

Over time, short noses got shorter, eyes bulged, skin flaps were more pronounced—the list goes on. Extreme traits for many species increase their risks of health problems to the point where they become routine. Breeds start to be unsustainable and increasingly need artificial insemination (e.g., production turkeys), elective cesarean section delivery (e.g., pygmy goats), or corrective surgery. At this point, aesthetic ideals clash with biological function. Inbreeding itself, a requirement of many pedigrees, eventually results in a loss of robustness (or health) called *inbreeding depression*.

The goal of a domesticated animal is to fit into its role in human society while benefiting from good welfare. A wolf would not do well as an American lap dog or a sheep herder. A theoretical optimal level of modification from the ancestral wild type may exist that allows a dog to slot into its role as a companion animal. Or, as others would suggest, keeping dogs as pets should never be considered acceptable because an agreeable compromise can never be achieved.

There are some immediate reasons to prefer extremes of function, such as the fastest horse or the sow who has the most piglets. These extremes become more problematic when assessed using only immediate or short-term metrics. For example, pig genetics skewed toward increasing the number of piglets born, but not the number that survive to be weaned, will result in sickly extra piglets who die, and so the goal of selling more meat is missed. Likewise, the dairy cow that produces the most milk in her first year of production is putting too much energy into milk at the expense of her own body—she is likely to have a poorer body condition and a shorter lifespan. Smaller groups of people may gradually come to find extreme body forms highly attractive, even when outside observers find them strange. For example, if you go on YouTube and look for American German shepherd dogs being shown in the ring, you might find their hind legs and gait (the way they walk) rather unusual. Enthusiasts of the breed will be quick to explain how this look, rather different from what you would have seen 10 or 20 years ago, is healthy and ideal. On the other hand, European German shepherd dogs have a different breed standard that deliberately avoids the sloped back and shortened hind legs.

Over time, many traits associated with inbreeding have come to be a desirable feature of the breed, e.g., white markings or being all white in color. If you go online and look at images of Pembroke corgis pictured with Queen Elizabeth II as a young girl, or in old movies like Disney's *Little Dog Lost* (1963), you will notice their all-brown heads and little to no white on their chest. It is now very rare to find a Pembroke corgi without a white muzzle and chest, and dogs missing these white markings are likely to be mistaken for mixed breeds.

Keywords: artificial selection, breed, inbreeding depression

6.11 ECOTECHNOLOGY AND CRISPR

So far in this chapter, we've discussed some very low-tech methods of altering animal genetics that can have profound effects over a long time scale. You can speed up change by line breeding, superovulation, sperm preservation and shipping, artificial insemination, and the use of a sire sometimes described in the meat industry as a terminal sire (a male whose sperm is widely used to breed animals going to slaughter). But if you want to turbocharge change, you can just edit the genome; perhaps by adding, inactivating, or repairing a defective gene.

Most people have seen fluorescent fish (Glofish®) in their local pet stores. These are a lucrative spinoff from a project to use fluorescent fish as live pollution detectors. The genomes of these fish had been changed to include a gene that encodes a fluorescent protein. Gene editing is now incredibly precise, much more so than it used to be, and all because of an editing tool called CRISPR-Cas9 (clustered regularly interspaced short palindromic repeats and CRISPR-associated protein 9). To edit genomic DNA, the double helix must be cut and unwound a little so that molecular editing machines can get to work. Cutting the DNA in the right place is key, and the awkward sounding acronym for CRISPR-Cas9 describes the guide mechanism (the guide RNA) that turns a "dumb" molecular scissor (Cas9) into a guided snip. This discovery won the 2020 Nobel Prize in Chemistry for Emmanuelle Charpentier and Jennifer Doudna and is *the* breakthrough behind precise gene editing (reviewed by Li et al., 2023).

Releasing an animal with a modified genome into the wild can result in its engineered gene/s spreading quickly by natural breeding (Esvelt et al., 2014) in a process called a gene drive. As you might imagine, there are endless useful applications for this kind of modification. Using gene editing to modify wild white-footed mice (*Peromyscus leucopus*) so that they are heritably resistant to the Lyme disease that they carry (which is dangerous for people and pet dogs) has already been pitched to the locals of Nantucket and Martha's Vineyard (Buchthal, 2019). These island communities make obvious targets for the controlled spread of a modified mouse. CRISPR modified mice would reduce suffering from the disease and the need for lethal eradication methods to control it. However, this intervention might alter host availability for *Borrelia* bacteria (the pathogen that causes Lyme disease) in unexpected ways. Theoretical outcomes might include changes in white-footed mouse fitness and its interactions with other species, other species becoming the primary reservoir for Lyme disease, and changes in Lyme disease infection rates and symptomatology.

Before this technology is deployed, a great deal of study and consultation would be needed, for example with geneticists, medical professionals, ecologists, and community members. A more immediate candidate for change is to modify beef cattle so that they are more heat tolerant; this gene edit was recently approved by the Food and Drug Administration (FDA) (March 2022) with a low risk rating. This means that these animals can skip the premarket approval process normally required for a genetically modified animal. The direction of travel seems to be to fast-track gene edits that match naturally occurring mutations that could otherwise have been inherited using classical breeding strategies. This technology is worth a fortune, but the scientific achievement is overshadowed by long-running legal disputes over the patent rights.

Keywords: CRISPR-Cas9, gene drive

The law of unintended consequences and beta-agonists

Sociologist Robert K Merton (1910–2003) coined the term "unintended consequences", and this alerts people to the fact that any intervention that has an intended impact will also have other outcomes. These might fall into the areas of unintended benefits, unintended drawbacks, or "perverse results", which makes the problem even worse. This same principle appears in the curricula of different disciplines. In pharmacology, for example, one of the first things you are taught is that all drugs have multiple effects, including dose-dependent effects and side effects. So, any therapeutic substance advertised as having no conceivable side effect should be treated with great suspicion. The human tendency to take anything we like and want more of it can collide with the law of unintended consequences. Take, for example, a feed supplement called ractopamine hydrochloride (RAC), which is a β-adrenergic receptor agonist (a β-agonist). When it is given to pigs raised for meat, they will develop bulkier, less fatty muscles (Mills, 2002), making them more valuable on the market. This was good news for the US pork industry, as they had a new and effective growth promoter. RAC was approved for swine in 1999, cattle in 2003, and turkeys in 2008.

RAC was banned in Europe, China, and Russia for mixed reasons. Regulators worried about chemical residues in food products, and some countries used this as an excuse to erect trade barriers. Around the same time, it became clear that RAC had serious side effects. Pigs were more easily fatigued and prone to metabolic acidosis and collapse, and RAC was found to cause anxiety in animals and people. Even more alarmingly, there were reports that cattle fed with the β-agonist zilpaterol hydrochloride could suffer "hoof sloughing"—the hoof wall would spontaneously fall off, requiring euthanasia of the animal (Thomson et al., 2015). Most of these effects became apparent only after the animal had been shipped for slaughter and could have been minimized by simply not exceeding the approved dosage.

It soon became apparent that the use of RAC was out of control; it was being used illegally for species without approval, in countries where it was banned, and at ridiculously high doses. Even human athletes were abusing beta-agonists and getting caught in anti-doping tests. Over-dosing for animals, while hard to prove, was widely suspected, based on producers wanting to maximize their profits. The human tendency to get carried away with a good thing was on full display and interacted disastrously with the law of unintended consequences, as higher doses revealed alarming side effects. The use of RAC is not well understood by the public and remains controversial. The difference between its effects when used as recommended, and how it ends up being used, is an obvious cause for concern in meat production. In a recent review on this subject (Abbas,et al., 2022), this comment was striking: "currently, there is no evidence in the scientific community to prove that RAC is safe". While this is not a very helpful statement, it is clear that RAC remains a controversial and widely used food additive in the United States—a market where the onus is on the consumer to beware and be aware.

Keyword: ractopamine

READINGS

Physical health and its measurability

Knierim, U., Van Dongen, S., Forkman, B., Tuyttens, F. A., Spinka, M., Campo, J. L., & Weissengruber, G. E. (2007). Fluctuating asymmetry as an animal welfare indicator – a review of methodology and validity. *Physiology & Behavior, 92*(3), 398–421. https://doi.org/10.1016/j.physbeh.2007.02.014

Salgirli Demirbas, Y., Isparta, S., Saral, B., Keskin Yılmaz, N., Adıay, D., Matsui, H., Töre-Yargın, G., Musa, S. A., Atilgan, D., Öztürk, H., Kul, B. C., Şafak, C. E., Ocklenburg, S., & Güntürkün, O. (2023). Acute and chronic stress alter behavioral laterality in dogs. *Scientific reports, 13*(1), 4092. https://doi.org/10.1038/s41598-023-31213-7

Historical improvements in animal health care

Phillips, C. J. C. (2024). Farm animal welfare-from the farmers' perspective. *Animals: An Open Access Journal from MDPI, 14*(5), 671. https://doi.org/10.3390/ani14050671

Hydatid disease

Abdybekova, A., Sultanov, A., Karatayev, B., Zhumabayeva, A., Shapiyeva, Z., Yeshmuratov, T., Toksanbayev, D., Shalkeev, R., & Torgerson, P. R. (2015). Epidemiology of echinococcosis in Kazakhstan: An update. *Journal of Helminthology, 89*(6), 647–650. https://doi.org/10.1017/S0022149X15000425. Epub 2015 Jul 10. PMID: 26160276.

Emerging zoonoses

Abbasi, J. (2024). Bird flu outbreak in dairy cows is widespread, raising public health concerns. *JAMA, 331*(21), 1789–1791. https://doi.org/10.1001/jama.2024.8886. PMID: 38718040.

Alliance to Save Our Antibiotics. (2024). *How to end the misuse of antibiotics in farming*. https://saveourantibiotics.org/media/2166/how-to-end-the-misuse-of-antibiotics-in-farming-full.pdf

Burrough, E. R., Magstadt, D. R., Petersen, B., Timmermans, S. J., Gauger, P. C., Zhang, J., Siepker, C., Mainenti, M., Li, G., Thompson, A.C., Gorden, P. J., Plummer, P. J., & Main, R. (2024). Highly Pathogenic Avian Influenza A(H5N1) Clade 2.3.4.4b virus infection in domestic dairy cattle and cats, United States, 2024. *Emerging Infectious Diseases, 30*(7), 1335–1343. https://doi.org/10.3201/eid3007.240508

Degeling, C., Lederman, Z., & Rock, M. (2016). Culling and the common good: Re-evaluating harms and benefits under the one health paradigm. *Public Health Ethics, 9*(3), 244–254. https://doi.org/10.1093/phe/phw019. Epub 2016 May 3. PMID: 27790290; PMCID: PMC5081039.

Jahid, M. J., Bowman, A. S., & Nolting, J. M. (2024). SARS-CoV-2 outbreaks on mink farms-a review of current knowledge on virus infection, spread, spillover, and containment. *Viruses, 16*(1), 81. https://doi.org/10.3390/v16010081. PMID: 38257781; PMCID: PMC10819236.

Ríos Carrasco, M., Gröne, A., van den Brand, J. M. A., & de Vries, R. P. (2024). The mammary glands of cows abundantly display receptors for circulating avian H5 viruses. *Journal of Virology*, e0105224. Advance online publication. https://doi.org/10.1128/jvi.01052-24

Culling animals as a geopolitical strategy

Weisiger, M. (2007). Gendered injustice: Navajo livestock reduction in the new deal era. T*he Western Historical Quarterly, 38*, 437. https://doi.org/10.2307/25443605

Access to veterinary Care

Cammisa, H. J., & Hill, S. (2022). Payment options: An analysis of 6 years of payment plan data and potential implications for for-profit clinics, non-profit veterinary providers, and funders to access to care initiatives. *Frontiers in Veterinary Science, 9*, 895532. https://doi.org/10.3389/fvets.2022.895532. PMID: 35978709; PMCID: PMC9376616.

Physiological stress

Cannon, W. B. (1939). *The wisdom of the body*. W.W. Norton and Company.

Mormède, P., Andanson, S., Aupérin, B., Beerda, B., Guémené, D., Malmkvist, J., Manteca, X., Manteuffel, G., Prunet, P., van Reenen, C. G., & Richard, S. (2007). Exploration of the hypothalamic–pituitary–adrenal function as a tool to evaluate animal welfare. *Physiology & Behavior, 92*(3), 317–339. https://doi.org/10.1016/j.physbeh.2006.12.003. Epub 2007 Jan 16. PMID: 17234221.

Prunier, A., Bonneau, M., von Borell, E. H., Cinotti, S., Gunn, M., Fredriksen, B., … Velarde, A. (2006). A review of the welfare consequences of surgical castration in piglets and the evaluation of non-surgical methods. *Animal Welfare, 15*(3), 277–289. https://doi.org/10.1017/S0962728600030487

Selye, H. (1936). A syndrome produced by diverse nocuous agents. *Nature, 138,* 32.

Neuroscience

Berridge, K. C. (2004). Motivation concepts in behavioral neuroscience. *Physiology & Behavior, 81*(2), 179–209.

Breland, K., & Breland, M. (1961). The Misbehavior of Organisms. *American Psychologist, 16*, 681–684. https://doi.org/10.1037/h0040090

Chalmers, D. J. (1995). Facing up to the problem of consciousness. *Journal of Consciousness Studies, 2*, 200–219.

Grandin, T. (1989). *Effect of rearing environment and environmental enrichment on behavior and neural development in young pigs* [Unpublished doctoral dissertation]. University of Illinois at Urbana-Champaign.

Palay, S. L., & Palade, G. E. (1955). The fine structure of neurons. *The Journal of Biophysical and Biochemical Cytology, 1*(1), 69–88. https://doi.org/10.1083/jcb.1.1.69. PMID: 14381429; PMCID: PMC2223597.

Virués-Ortega, J. (2006). The case against B. F. Skinner 45 years later: An encounter with N. Chomsky. *The Behavior Analyst, 29*(2), 243–251. https://doi.org/10.1007/BF03392133

Exercise: Ethics and foundational research

Harlow, H. F., Dodsworth, R. O., & Harlow, M. K. (1965). Total social isolation in monkeys. *Proceedings of the National Academy of Sciences of the United States of America, 54*(1), 90–97.

Harlow, H. F., & Zimmermann, R. R. (1958). The development of affectional responses in infant monkeys. *Proceedings of the American Philosophical Society, 102*, 501–509.

Taub, E. (1976). Movement in nonhuman primates deprived of somatosensory feedback. *Exercise and Sport Sciences Reviews, 4*(1), 335–374.

Taub, E., Perrella, P., & Barro, G. (1973). Behavioral development after forelimb deafferentation on day of birth in monkeys with and without blinding. *Science, 181*(4103), 959–960.

Affective neuroscience

Berridge, K. C., & Kringelbach, M. L. (2008). Affective neuroscience of pleasure: Reward in humans and animals. *Psychopharmacology (Berl), 199*(3), 457–480.

Berridge, K. C., & Kringelbach, M. L. (2015). Pleasure systems in the brain. *Neuron, 86*(3), 646–664. https://doi.org/10.1016/j.neuron.2015.02.018

Goldberg, H. (2023). Unraveling neurodiversity: Insights from neuroscientific perspectives. *Encyclopedia, 3*(3), 972–980. https://doi.org/10.3390/encyclopedia3030070

Panksepp, J. (2011). Cross-species affective neuroscience decoding of the primal affective experiences of humans and related animals. *PLoS, One, 6*(9), e21236.

Panksepp, J., & Biven, L. (2012). *The archeology of mind. Neuroevolutionary origins of humans emotions.* W. W. Norton & Company.

Schlaepfer, T. E., Bewernick, B. H., Kayser, S., Mädler, B., & Coenen, V. A. (2013). Rapid effects of deep brain stimulation for treatment-resistant major depression. *Biological Psychiatry, 73*(12), 1204–1212. https://doi.org/10.1016/j.biopsych.2013.01.034

Tilbrook, A. J., & Ralph, C. R. (2017). Neurophysiological assessment of animal welfare. *Animal Production Science, 57*(12), 2370–2375.

The mind–body problem

Brandt, T., & Huppert, D. (2021). Brain beats heart: A cross-cultural reflection. *Brain, 144*(6), 1617–1620.

McGinn, C. (1999). *The mysterious flame: Conscious minds in a material world.* Basic Books.

Routine surgeries

Hart, L. A., & Hart, B. L. (2021). An ancient practice but a new paradigm: Personal choice for the age to spay or neuter a dog. *Frontiers in Veterinary Science, 8*, 603257.

Kutzler, M. A. (2020). Possible relationship between long-term adverse health effects of gonad-removing surgical sterilization and luteinizing hormone in dogs. *Animals, 10*, 599. https://doi.org/10.3390/ani10040599

Patterson-Kane, E. G. (2017). *Canine Tail Docking. Independent report prepared for the ministry for primary industries.* ISBN No: 978-1-77665-633-632. https://doi.org/10.13140/RG.2.2.22912.33284

Artificial selection

Daprati, E., Iosa, M., & Haggard, P. (2009). A dance to the music of time: Aesthetically-relevant changes in body posture in performing art. *PLoS One, 4*, e5023. https://doi.org/10.1371/journal.pone.0005023

Darwin, C. (1871). *The descent of man.* D. Appleton.

Russell, J. A., Shave, R. M., Kruse, D. W., Koutedakis, Y., & Wyon, M. A. (2011). Ankle and foot contributions to extreme plantar-and dorsiflexion in female ballet dancers. *Foot & Ankle International, 32*(2), 183–188.

Ecotechnology and CRISPR driven methods

Buchthal, J., Evans, S. W., Lunshof, J., Telford, S. R., & Esvelt, K. M. (2019). Mice against ticks: An experimental community-guided effort to prevent tick-borne disease by altering the shared environment. *Philosophical Transactions of the Royal Society B: Biological Sciences*, *374*(1772). https://doi.org/10.1098/rstb.2018.0105

Esvelt, K. M., Smidler, A. L., Catteruccia, F., & Church, G. M. (2014). Concerning RNA-guided gene drives for the alteration of wild populations. *Elife*, *3*, e03401.

Li, T., Yang, Y., Qi, H., Cui, W., Zhang, L., Fu, X., He, X., Liu, M., Li, P. F., & Yu, T. (2023). CRISPR/Cas9 therapeutics: Progress and prospects. *Signal Transduction and Targeted Therapy*, *8*(1), 36. https://doi.org/10.1038/s41392-023-01309-7. PMID: 36646687; PMCID: PMC9841506.

The law of unintended consequences and beta-agonists

Mills, S. E. (2002). Biological basis of the ractopamine response. *Journal of Animal Science*, *80*(E-suppl_2), E28–E32.

Abbas, K., Raza, A., Vasquez, R. D., Roldan, M. J. M., Malhotra, N., Huang, J. C., Buenafe, O. E. M., Chen, K. H., Liang, S. S., & Hsiao, C. D. (2022). Ractopamine at the center of decades-long scientific and legal disputes: A lesson on benefits, safety issues, and conflicts. *Biomolecules*, *12*(10), 1342–1357. https://doi.org/10.3390/biom12101342. PMID: 36291550; PMCID: PMC9599871.

Thomson, D. U., Loneragan, G. H., Henningson, J. N., Ensley, S., & Bawa, B. (2015). Description of a novel fatigue syndrome of finished feedlot cattle following transportation. *Journal of the American Veterinary Medical Association*, *247*(1), 66–72. https://doi.org/10.2460/javma.247.1.66

Chapter 7
A Natural Life

7.1 NATURAL LIFE

What if biotechnology could create a pig that was physically sound and euphorically happy as a legless and immobile tube of meat? I imagine that most of you would be horrified by the idea and seriously doubt that the pig had good welfare. Surely, a living tube of meat is no longer a pig, or any kind of animal! In this chapter, we will cover topics relating to how the animal's natural state intersects with wellbeing. The natural conditions of animals were first scientifically studied through the observation of wild animals or free-roaming domestic animals. Specific disciplines then began to understand how the animal interacts with the environment to meet all its needs, engaging in important behaviors that are often innately enjoyable, such as hunting and foraging, play and exploration, and social interactions.

Animal welfare is improved when animals perform adaptive behaviors rather than simply having resources given to them. The phenomenon of contra-freeloading (meaning: choosing to work for something when you could get the same reward for free) is an example. When an animal has the choice between eating freely available food and natural foraging behavior, they will often choose the latter (Lindqvist, 2009). This shows that performing the foraging behavior has its own intrinsic value, whereas providing everything for an animal "on tap" deprives them of the enjoyment and benefits they would get from doing some of those things.

There are also many cases where the natural method is shown to contribute to better physical health. For example, it is natural for elephants to walk very long distances to get the various resources that they need, including regular water immersion. In a setting like a zoo, where this is not possible, both the foot pad and nails need to be actively trimmed, as they are not naturally worn down (Roocroft & Oosterhuis, 2001). Many intrinsically motivated natural behaviors are part of how the animal looks after itself and reduces the need for artificial husbandry to stay in good health.

Keyword: contrafreeloading

DOI: 10.1201/9781003375456-9

7.2 NATURAL HISTORY

Narrative descriptions about plants and animals are called "natural history", and as the sciences developed (between 1860 and the early 1900s), these practices and their materials coalesced into separate fields, including ecology and ethology. Natural history persisted for some time as a foundational course for a degree in biology but has now largely faded away. However, popular books for a general audience, such as *The Flamingo's Smile* (Gould, 1985) and *When Elephants Weep* (Masson, 1995), can still be described as natural history. A new wave of natural history writing and documentary is currently underway, with works similar to scientific/investigative journalism, such as *The Light Eaters: How the Unseen World of Plant Intelligence Offers a New Understanding of Life on Earth* (Schlanger, 2024), or experiences of nature with a personal meaning, such as *Bird Cottage* (Meijer, 2018).

Geoffroy Delorme's *Deer Man: Seven Years of Living in the Wild* (2022) is striking in the bond that developed between man and deer, as evidenced by the grainy black and white photographs of deer displaying a notable lack of fear. There is no pretension to any scholarly endeavor, but rather, an intimate portrait of Delorme's extended "kinship group" and how deer taught him how to survive in the woods. Language such as a deer "teaching" Delorme would never be permissible in the academic record, as that language is fiercely policed for any perceived anthropomorphism—which has itself become a rather toxic word.

While somewhat unstructured, early natural histories are a trove of observational data, especially for rare events and long-gone species and settings. Often poorly researched, these studies still offer a rich base of anecdotal reports, and like any anecdotal report, they range from insightful to the fanciful or delusional. For example, you will find reports seriously suggesting that American snakes will hold their tail in their mouth and roll over the ground like hoops. It was also widely believed that rats would steal eggs by having one lie on its back holding the egg and the others dragging them away by the tail. Both rat wagons and hoop snakes appeared to be tall stories/folklore while being presented as fact in first-person accounts (Figure 7.1).

Figure 7.1 The notion of "rat-egg wagon transport" is a long one, according to E.W. Gudger (1935), with origins in 1291 AD Persia. This article can be read online at archive.org and includes, in its last paragraph, a promise from the author to keep an open mind. This public domain image is from an illustrated American newspaper and art journal printed in 1887.

However, it is equally important to appreciate that absence of evidence is not evidence of absence, and we should not shrink away from studying phenomena that are reproducible, even where the underlying mechanism is obscure. For many years, we had no idea of how homing pigeons managed to find their way home, but they did, and so we thought they might be psychic or have some other supernormal power. Likewise, anecdotal reports to the editor of the medical journal, the *Lancet*, described dogs sniffing lesions that turned out to be cancerous (Church & Williams, 2001). Now, the unimaginable is coming to be known as quantum biology. Pigeons have a quantum compass in their heads (magnetoreception; Xu et al., 2021), and according to Turin (2002), the sense of smell might involve vibrating scent molecules (quantum tunneling; Rinaldi, 2007). Perhaps natural histories are not so bizarre after all, and my dog, with her olfactory superpowers, experiences life as a symphony of scent.

Keyword: natural history

7.3 BODILY INTEGRITY

In many settings, welfare has been reduced to health and happiness. Returning to the subject of legless pigs, would it be acceptable to deliberately produce such an animal? The idea is not entirely outlandish, as many conditions associated with being immobile do lead to more efficient production of meat. For example, pigs housed under warm tropical conditions and low light conditions will gain weight more rapidly because they spend more time innactive. Natural mutations that produce legless pigs have been known since the 18th century, were scientifically documented in 1938, and were rather glibly described as "streamlined pigs" in at least one paper. While these pigs died of presumed starvation, it was now at least theoretically possible to produce legless pigs by artificial selection or genetic modification. Most consumers show some degree of disgust or anxiety at the prospect of such a highly modified animal. Examples that have been widely discussed also include the featherless chicken and livestock without heads (Yeates, 2018; Fraser 2001). The objection that a legless animal would suffer could potentially be addressed through appropriate supportive housing and might be completely absent for a headless animal. The upshot of these thought experiments was to confirm the notion that people intuitively assign value to bodily integrity that acts to limit the commodification of animals.

With humans, bodily integrity is generally interpreted to mean the individual's right to determine what happens to their own body. Humans may choose to modify their own bodies (so they do not function as well), and autonomy or self-determination is typically the more important consideration, although this becomes more contested as these choices become more active versus passive, and more extreme in terms of biological functioning (e.g., amputation or self-euthanasia). Overall, human bodily integrity equates with the idea of bodily autonomy or sovereignty, which animals are typically not accorded, as these rights are ceded to their human owners.

As already discussed, neutering is a particularly contested and culture-specific modification, even to people holding strong animal rights positions, due to the *perceived* benefits to the animal and other animal populations. In relation to animals, there is a spectrum of opinions about whether animals have rights of this type, with Steven Wise having proposed that we extend common law rights to animals that include liberty and bodily integrity (discussed in Davison-Vecchione & Pambos, 2017).

Overall, the idea of bodily integrity is often used differently with regard to non-human animals in prioritizing their biological function over autonomy. It may refer to how well the animal is able to interact with its environment, for example by losing faculties like locomotion, temperature control, or one or more of the senses. A critical loss of this kind of bodily integrity may mean a perceived loss of meaningfully "being an animal". A discussion about whether an animal retains bodily integrity is sometimes used when deciding whether to euthanize an animal. Solving or mitigating a welfare challenge by significantly reducing an animal's bodily integrity is generally seen as unethical. This is sometimes referred to as the *blind hens' challenge* (discussed in Sandøe et al., 2014) because hens that cannot see do less damage to each other through pecking (a common problem in poultry raising).

Keywords: blind hens challenge, bodily integrity (of animals), Steven M. Wise

7.4 ETHOLOGY

Ethology continues to be a haven for the close observational study of animals (without experimental intervention). Ethology is a discipline at the heart of developing an understanding of how an animal's natural manner of living is the context for how it suffers or enjoys its life, struggles, or thrives. Over time, ethologists have increasingly embraced the study of animals in captive settings, even designing interventions or full experiments.

One of the tools ethologists use and develop is the *ethogram*. This is a thorough description of all the behaviors observed by members of a species within the context in which these behaviors occur. A complete ethogram will show behaviors only rarely encountered, for example the female rat's estrous ear wiggle. An ethogram can be used as a basis for comparing an animal in captivity with their counterpart in the wild and potentially identify deficits in the built environment. Reference ethograms are harder to apply to domesticated species; however, in many cases, the behavior of a wild ancestor is still relevant. Where, even under ideal conditions, differences emerge between the wild and domestic phenotype (behavior, morphology, and physiology), then other indications of welfare will be needed to improve conditions. For example, the domesticated chicken (*Gallus gallus*) descends from the Burmese jungle fowl, which dwells in forest environments. Despite their differences in appearance, many behaviors remain as important for the domesticated bird as for the wild one. For example, dustbathing (Figure 7.2) is thought to help in removing parasites and grooming plumage (Olsson et al., 2005). Providing dustbathing substrate to chickens allows them to perform a functional behavior that is likely to be satisfying insofar as they can maintain their own plumage, which is then more effective as a thermal barrier against cold stress.

Studying behavior often necessitates some degree of identification. Methods for marking animals that look similar can also affect their behavior, sometimes in unexpected ways. In one study, male birds with leg bands that matched their plumage colors began to advertise their "services" to females other than their mates (Johnsen et al., 2000). In another study, Australian Magpies assisted others in detaching tracking harnesses (Crampton et al., 2022). Methods for

Figure 7.2 Dustbathing chicken at the Hillside Animal Sanctuary, UK. Image by Dave Crosby (online at https://www.flickr.com/photos/87316606@N00/2823277506) and provided under the CC BY-SA 2.0 license (https://creativecommons.org/licenses/by-sa/2.0/deed.en).

remote tracking that were initially presumed to be harmless have proven to be otherwise. For example, removing toes from wolves to identify them by their tracks, and attaching bulky tracking devices to animals, can impact quality of life and survival. Even monarch butterflies are tracked with stickers. How do "wing stickers" impact balance and resilience when migrating long distances? Remuneration for returning tags to participating organizations is another ethical issue, as this can incentivize capture. The ethical review of wildlife research now (supposedly) weighs how the methods of observation will impact the animal versus the benefits that might be gained, with a particular reference to the presumed benefits for the animal (Figure 7.3).

Keyword: ethogram

7.5 REWILDING

Influential geneticist and ethologist D.G.M. Wood-Gush (1922–1992) carried out extensive observations of domesticated animals in naturalistic and natural environments. Wood-Gush is notable for having been admitted as an associate member to the Society for Veterinary Ethology in 1970, before they accepted non-veterinarians. Wood-Gush also created one of the first graduate degree courses in animal welfare (the University of Edinburgh Master's Degree in Applied Animal Behavior and Animal Welfare, launched in 1990). But Wood-Gush is better known for his iconic series of studies on the behavior of domestic animals after they are released into expansive free-ranging settings.

In 1978, a group of pigs were released into a naturalistic enclosure near Edinburgh. In this relatively enriched and unconstrained social setting, the behavioral potential of these pigs was more fully revealed (Stolba &

Figure 7.3 Overloading wild animals with tracking devices impacts their welfare. The upper panel shows a coyote carrying ear tags and a collar. The collars should eventually fall off, but sometimes they don't. The bird in the lower panel is called a chough, and the inset shows the four bands on its legs. The coyote images were reproduced with the permission of janetkessler/coyoteyipps.com. Janet Kessler uses a "hands off" tracking approach in her observations of coyotes in San Francisco. The chough image is reproduced with the permission of Steve Beardall.

Wood-Gush, 1989). This behavior allowed researchers to develop a much better understanding of the complexity of pig-to-pig interactions. Specifically, they exhibited sociological complexity and structure, mediating which pigs came together as groups and how offspring were raised. Wood-Gush then actively worked to reconcile the needs of animals with their commercial farmed environment, for example using the "Pig Park" observations to propose a group pen housing system for farming swine (Wood-Gush, 1985).

Work of this type, studying the ethology of domestic species in relatively extensive settings, revealed that there was much more to their behavior than is apparent in smaller, simpler settings. But the limitations of this adaptability were seen in a different series of studies involving domestic chickens released into an uninhabited island. The first group of chicken "migrants" was almost entirely wiped out over the winter by predation from American mink (*Mustela vison*), an accidentally rewilded species not native to the area (Duncan et al., 1978). The chicken population was supplemented, and the chickens began to show signs of adapting to the threats by hiding their nests more effectively. This demonstrated that natural behavior is not always immediately obvious and instinctual but may take the form of biological preparedness to learn survival strategies as needed. Ultimately, enough of the population survived for the study of their feeding and reproductive behaviors (Savory et al., 1978).

Currently, there are many spontaneously produced populations of feral domestic chickens and swine that provide interesting opportunities for ethological study. However, they are also starting to interact with wild populations in a way that may have problematic consequences. For example, in many areas, the four species of wild jungle fowl coexist with domestic fowl and interbreed with them (sometimes called "genetic pollution" on the assumption that a *pure* wild type is superior; Gering et al., 2015). Genetic pollution of the Scottish wildcat (*Felis silvestris)* by domesticated and semi-feral cats is a similarly vexing problem for those attempting to preserve this elusive species. In the wild, this cat has now been declared genomically extinct. What was a "drip" of domestic cat DNA entering into the wildcat genome (a process called introgression) turned into a torrent, and the modern wildcat population has become a hybrid (Howard-McCombe et al., 2023). Does this mean that two branches in the tree of life just fused—hybridization driving speciation (Rosser et al., 2024)? Recently, some of the last captive animals of purer wildcat lineage were released into a national park in the Scottish Highlands in the hope that they will re-establish wildcat primacy. We'll have to wait and see what happens.

Keywords: rewilding (re: domesticated animals), D.G.M. Wood-Gush

7.6 ENVIRONMENTAL ENRICHMENT

There are many approaches to trying to bring "the wild" or the important aspects of it into the captive environment. One of these is referred to as *environmental enrichment* and is the most popular approach with captive wild animals, especially in zoos and for laboratory animals. There have been almost as many arguments about how to define environmental enrichment as there have been about *animal welfare*. It has been suggested that because the baseline captive environment for many animals is clearly inadequate (e.g., being too small or barren), correcting these deficits is hardly enriching. Perhaps, environmental mitigation or amelioration might be more appropriate terms, although they have yet to catch on (if you do use these terms, it is important to also use *environment enrichment* as a keyword so people can find your work).

Environmental enrichment is a research theme that goes back to the 1920s. Early exponents were Robert Yerkes and Heini Hediger, and later, Hal Markowitz would become influential in the field. Psychology research showed that more complex environments produce animals that are less fearful and do better in cognitive tests. When applied in settings like zoos, enrichment could promote species-specific behaviors and support good welfare. These different goals, and the difficulties in achieving them, are reflected in the following descriptions of environmental enrichment: "The provision of stimuli which promote the expression of species-appropriate behavioural and mental activities in an understimulating environment" (Reinhardt & Reinhardt, 1998) and, "we propose here that there may be no single definition of enrichment; instead, we suggest that enrichment can be re-defined for each species and perhaps for each individual" (Mellen & MacPhee, 2001).

Across disciplines, a good overall definition would be that environmental enrichment is an increase in the complexity of a habitat with a goal of improving the welfare of the occupant. This complexity can be provided by adding features, social interactions, sources of novelty and change, training, the ability to control aspects of the habitat, or foraging or hunting opportunities, or in many other ways. Almost anything can be considered environmental enrichment, although in practice, there tends to be a focus on objects and experiences that can be provided within the enclosure, or within relatively modest increases in enclosure size.

Other options require greater resources and may be controversial. For example, in some countries, animals like elephants and chimpanzees may be taken from enclosures to interact with trees and other features in parks, but in the United States, this would typically be seen as too risky for any human bystanders. However, it is not uncommon for smaller animals like penguins to be allowed excursions outside their enclosures.

Keyword: environmental enrichment

Example: The dystopian mouse "utopias"

John Calhoun created what he described as a utopia for mice, where they had an abundance of food, a favorable climate, and no disease or predation. This was the opposite of rewilding—all organic needs were met, but the environment was relatively barren or unenriched. In these utopias, the population would increase, getting increasingly crowded. Calhoun replicated this environment over and over, reproducing broadly the same outcome. The population never rose to the highest level that could be supported by the food and water available, and, initially at least, this was referred to as "A method of self-control of population growth among mammals living in the wild" (Calhoun, 1949).

His focus then switched to how overcrowding led to abnormal behaviors, including aggression, neglect, and population collapse (Figure 7.4) . Calhoun came to name this collapse *the behavioral sink*. These findings were reported in an influential paper published in 1970 titled "Population Density and Social Pathology" (Calhoun, 1970). The behavioral sink was seen as a portent for things to come when any species is overcrowded in the presence of abundant resources. Calhoun's writing style was a magnet for social commentators. Who could resist reading this single-author paper, published in 1973, "Death squared: the explosive growth and demise of a mouse population" (Calhoun, 1973). Calhoun's research resonated with social anxieties about urban overcrowding and societal collapse and was seen as a prophecy that humanity could not thrive as an urban species (Ramsden & Adams, 2009). Calhoun himself began to show how well-structured environments resist these catastrophic declines and

Figure 7.4 681 days of "Universe 25". You can see the withdrawn mice collecting in the middle of the floor-space—demonstrating the "behavioral sink". Photo by Yoichi R Okamoto, White House Photographer. Public domain via Wikimedia Commons.

flagged the benefits of accessible sources of information for people at a time well before the existence of the internet.

What is noticeably missing from this utopia is any notion of naturalness. Open space is for rodents the worst kind of space, and not a setting for wellbeing or a good background against which to assess crowding. The mouse utopias offer fertile ground for considering when animals can be accurately or productively used as a model for people. Calhoun's work is now rarely mentioned, but it inspired the design of high-density living spaces in ways that are compassionate and even recognize the needs of humans for naturalness in our living spaces. The mouse utopias were quite extraordinary (in an unnerving way) and revealed the desperate need that animals have for social space when physical space is limited (Calhoun, 1971). With no ability to manage reputations, run away, hide, migrate, or avoid pregnancy, these utopias were really horror shows.

Keyword: behavioral sink

7.7 LOSS OF BEHAVIORAL INTEGRITY

Holes in behavioral functioning can be seen in some animals that have been optimized for productivity. Take, for example, swine raised for meat. Multiple syndromes have been noticed at slaughter that suggest a failure to cope with stress, leading to metabolic and behavioral collapse. That is, the pigs literally fall to the ground. In the United States, collapsed animals that cannot stand and walk are referred to as *a downer* or *non-ambulatory disabled* (NAD), and their slaughter is banned. This is because they might be sick and so are barred from the food chain. Even if they do not collapse, highly stressed pigs may result in PSE pork (pale, soft, and exudative). This is a lower-quality pork product that was first noted in the 1960s but became more common when producers selected pigs that are fast growing and very lean (more muscle, less fat). These pigs also tend to be more excitable and easily stressed (Grandin, 1994). Rates of PSE have been declining in recent years due to further genetic selection against this trait.

Currently, there is an exception to the downer requirement for pigs that gives them a recovery time to see if they can return to an ambulatory state. Some animal advocates, like Farm Sanctuary, feel that this accommodation should be withdrawn to incentivize transporters and slaughterhouses to use lower-stress methods that prevent collapse from occurring in the first place. (Farm Sanctuary has also been a leader in fighting laws that would criminalize under-cover investigations that report animal cruelty—the Ag-Gag laws.)

A loss of behavioral competency does not need to be based on compromised health. For example, some modern sheep breeds are prone to mismothering and abandoning their lambs, and some breeds of egg-laying hen are more prone to smothering each other. In a supportive environment, these breeds continue to be highly productive, but placed in a natural setting, they might struggle to survive at all. It is possible that some modern breeds are in danger of becoming behavioral orphans, no longer able to thrive in the wild environment of their species, thus severing the main way in which their "naturalness" needs can be effectively understood.

Keyword: PSE pork

7.8 SPECIES-SPECIFIC BEHAVIORS

One concept coming from ethology has become particularly prominent in animal welfare sciences. This is the *species-specific behavior*—behaviors that have a relatively uniform appearance and are performed by most individuals belonging to a species. Some may be unique to the species (Figure 7.5), while others are shown by some species but not others. Prominent examples include the hammering of a woodpecker foraging for food, or the use of complex grammatical language by humans. It is not that other animals do not eat or communicate, but that they do so using noticeably different patterns of behavior.

The study of species-specific behaviors began by looking at simple innate behaviors that are directly tied to the animal's nervous system: for example, fighting, mating, and foraging behaviors. However, species-specific behavior may be learned, because members of the species have a biological preparedness to learn certain behaviors easily, whereas other species would never, or only rarely, learn the same behavior: for example, the tendency for raccoons to wash food and other objects in water, or the tendency of songbirds to develop complex vocalizations. The performance of species-specific behaviors is often assumed to be important for welfare (Miller et al., 2020). The reasoning for this is that:

Figure 7.5 Another even more specialist behavior is unique to the acorn woodpecker species, which stashes thousands of acorns into holes drilled into a tree or, in this case, a utility pole. This is the granary pole that will serve a community of approximately a dozen woodpeckers. Photograph taken by Tina Rich.

- Species-specific behaviors contribute to behavior diversity, which reflects how the animal is engaged with enriched or naturalistic environments (where behavioral diversity is the richness of behavior as measured by the number of different behaviors performed and their relative frequency).
- Many of these behaviors will be performed alongside indications of positive subjective states: for example, eating, mating, or play.
- Many of these behaviors are functional and beneficial to the health of the animal: for example, grooming, social cohesion behavior, foraging.

Species-specific behaviors are also important indicators of behavioral genetics, as they can be used to track the role of the evolution of genes in the performance of adaptive behaviors. (Behavioral genetics is the study of how genes and the environment influence behavior.)

Keyword: behavioral diversity, behavioral genetics, species-specific behavior (species typical)

7.9 TELOS

A modern reimagining of the role of the natural life for "higher animals" was proposed by American ethicist Bernard Rollin (1985), describing animals that "have interests and natures, biologically determined, the thwarting of which also matters to them. Using Aristotle's term … the physical and psychological needs and interests … of [the] animal is the telos or nature of that animal". Telos, in Greek philosophy, is an understanding of what gives a person a sense of purpose in their life. Rollin often informally explained the telos of animals as "the pigness of the pig" or "the dogness of a dog" (Rollin, 1992). This concept obviously echoes ideas that have already been discussed as bodily and behavioral integrity and species-specific behaviors. However, the idea of telos is broader and more holistic. It is not a purely scientific idea, but a philosophical principle that is also informed by scientific evidence. In pragmatic terms there are differences between cultures in what we think of as integral to "pigness". For example, some people think pigs should have good hygiene and not be dirty. Others see a dirty pig as a happy pig with access to the outdoors (Lassen, 2006). Either way, the concept of telos invites us to think about how an individual of a particular species experiences their identity and what they need to have a meaningful life. (For more about how we understand what a good life means, see Part V "What Kind of Life". Because how do we support a good state of welfare, holistically, without having some idea of what that is?)

Keyword: telos

READINGS

Natural life

Lindqvist, C., & Jensen, P. (2009). Domestication and stress effects on contrafreeloading and spatial learning performance in red jungle fowl (Gallus gallus) and White Leghorn layers. *Behavioural Processes*, *81*(1), 80–84.

Roocroft, A., & Oosterhuis, J. (2001). Foot care for captive elephants. In B. Csuti, E. L. Sargent, & U. S. Bechert (Eds.), *The elephant's foot: Prevention and care of foot conditions in captive Asian and African elephants* (pp. 21–52). Iowa State University Press.

Natural history

Church, J., & Williams, H. (2001). Another sniffer dog for the clinic? *The Lancet*, *358*(9285), 930.

Delorme, G. (2022). *Deer man: Seven years of living in the wild*. Greystone Books.

Gould, S. J. (1985). *The flamingo's smile: Reflections in natural history*. WW Norton & Company.

Gudger, E. W. (1935). How rats transport eggs. *The Scientific Monthly*, *40*(5), 415–424.

Masson, J. M. (1995). *When elephants weep: The emotional lives of animals*. Delta.

Meijer, E. (2018). *Bird cottage*. Pushkin Press.

Rinaldi, A. (2007). The scent of life. The exquisite complexity of the sense of smell in animals and humans. *EMBO Reports, 8*(7), 629–33. https://doi.org/10.1038/sj.embor.7401029. PMID: 17603536; PMCID: PMC1905909.

Schlanger, Z. (2024). *The light eaters: How the unseen world of plant intelligence offers a new understanding of life on earth*. Harper.

Turin, L. (2002). A method for the calculation of odor character from molecular structure. *Journal of Theoretical Biology*, *216*, 367–385.

Xu, J., Jarocha, L. E., Zollitsch, T., Konowalczyk, M., Henbest, K. B., Richert, S., Golesworthy, M. J., Schmidt, J., Déjean, V., Sowood, D. J. C., Bassetto, M., Luo, J., Walton, J. R., Fleming, J., Wei, Y., Pitcher, T. L., Moise, G., Herrmann, M., Yin, H., Wu, H., Bartölke, R., Käsehagen, S. J., Horst, S., Dautaj, G., Murton, P. D. F., Gehrckens, A. S., Chelliah, Y., Takahashi, J. S., Koch, K. W., Weber, S., Solov'yov, I. A., Xie, C., Mackenzie, S. R., Timmel, C. R., Mouritsen, H., & Hore, P. J. (2021). Magnetic sensitivity of cryptochrome 4 from a migratory songbird. *Nature*, *594*(7864), 535–540. https://doi.org/10.1038/s41586-021-03618-9. Epub 2021 Jun 23. PMID: 34163056.

Bodily integrity

Davison-Vecchione, D., & Pambos, K. (2017). Steven M. Wise and the common law case for animal rights: Full steam ahead. *Canadian Journal of Law & Jurisprudence*, *30*(2), 287–309.

Fraser, D. (2001). The "new perception" of animal agriculture: Legless cows, featherless chickens, and a need for genuine analysis. *Journal of Animal Science, 79*(3), 634–641. https://doi.org/10.2527/2001.793634x. PMID: 11263823.

Sandøe, P., Hocking, P. M., Förkman, B., Haldane, K., Kristensen, H. H., & Palmer, C. (2014). The blind hens' challenge: does it undermine the view that only welfare matters in our dealings with animals? *Environmental Values*, *23*(6), 727–742. https://doi.org/10.3197/096327114X13947900181950.

Yeates, J. (2018). Naturalness and animal welfare. *Animals, 8*(4), 53. https://doi.org/10.3390/ani8040053. PMID: 29621140; PMCID: PMC5946137.

Ethology

Crampton, J., Frère, C. H., & Potvin, D. (2022). Australian Magpies Gymnorhina tibicen cooperate to remove tracking devices. *Australian Field Ornithology*, *39*, 7–11. http://dx.doi.org/10.20938/afo39007011

Johnsen, A., Fiske, P., Amundsen, T., Lifjeld, J. T., & Rohde, P. A. (2000). Colour bands, mate choice and paternity in the bluethroat. *Animal Behaviour*, *59*(1), 111–119.

Olsson, I. A. S., & Keeling, L. J. (2005). Why in earth? Dustbathing behaviour in jungle and domestic fowl reviewed from a Tinbergian and animal welfare perspective. *Applied Animal Behaviour Science*, *93*(3–4), 259–282.

Rewilding

Duncan, I. J. H., Savory, C. J., & Wood-Gush, D. G. M. (1978). Observations on the reproductive behaviour of domestic fowl in the wild. Applied *Animal Ethology*, *4*(1), 29–42.

Gering, E., Johnsson, M., Willis, P., Getty, T., & Wright, D. (2015). Mixed ancestry and admixture in Kauai's feral chickens: invasion of domestic genes into ancient Red Junglefowl reservoirs. *Molecular Ecology*, *24*(9), 2112–2124.

Howard-McCombe, J., Jamieson, A., Carmagnini, A., Russo, I. M., Ghazali, M., Campbell, R., Driscoll, C., Murphy, W. J., Nowak, C., O'Connor, T., Tomsett, L., Lyons, L. A., Muñoz-Fuentes, V., Bruford, M. W., Kitchener, A. C., Larson, G., Frantz, L., Senn, H., Lawson, D. J., & Beaumont, M. A. (2023). Genetic swamping of the critically endangered Scottish wildcat was recent and accelerated by disease. *Current Biology*, *33*(21):4761–4769.e5.

Rosser, N., Seixas, F., Queste, L., Cama, B., Mori-Pezo, R., Kryvokhyzha, D., Nelson, M., Waite-Hudson, R., Goringe, M., Costa, M., Elias, M., Figueiredo, C., Freitas, A., Joron, M., Kozak, K., Lamas, G., Martins, A., McMillan, W., Ready, J., & Dasmahapatra, K. (2024). Hybrid speciation driven by multilocus introgression of ecological traits. *Nature*, *628*, 1–7. https://doi.org/10.1038/s41586-024-07263-w

Savory, C. J., Wood-Gush, D. G. M., & Duncan, I. J. H. (1978). Feeding behaviour in a population of domestic fowls in the wild. *Applied Animal Ethology*, *4*(1), 13–27.

Stolba, A., & Wood-Gush, D. G. M. (1989). The behaviour of pigs in a semi-natural environment. *Animal Science*, *48*(2), 419–425.

Wood-Gush, D. G. M. (1985). The attainment of humane housing for farm livestock. In M. W. Fox & A. N. Rowan (Eds.), *Advances in animal welfare science* (pp. 47–55). Springer Netherlands.

Environmental enrichment

Mellen, J., & Sevenich MacPhee, M. (2001). Philosophy of environmental enrichment: Past, present, and future. *Zoo Biology*, *20*(3), 211–226.

Reinhardt, V., & Reinhardt, A. (1998). *Environmental enrichment for primates* [electronic resource].

The dystopian mouse "utopias"

Calhoun J. B. (1949). A method for self-control of population growth among mammals living in the wild. *Science (New York, N.Y.)*, *109*(2831), 333–335. https://doi.org/10.1126/science.109.2831.333

Calhoun, J. B. (1970). Population density and social pathology. *California Medical*, *113*(5), 54. PMID: 18730425; PMCID: PMC1501789.

Calhoun, J. B. (1971). Space and the strategy of life. In A. H. Esser (Ed.), *Behavior and environment* (pp. 329–387). Plenum Press.

Calhoun, J. B. (1973). Death squared: The explosive growth and demise of a mouse population. *Proceedings of the Royal Society of Medicine*, *66*(1 Pt 2), 80–88. PMID: 4734760; PMCID: PMC1644264.

Ramsden, E., & Adams, J. (2009). Escaping the laboratory: The rodent experiments of John B. Calhoun & their cultural influence. *Journal of Social History*, *42*, 761–792.

Loss of behavioral integrity

Grandin, T. (1994). *Methods to reduce PSE and bloodsplash*. Veterinary Outreach Programs, University of Minnesota.

Species-specific behaviors

Miller, L. J., Vicino, G. A., Sheftel, J., & Lauderdale, L. K. (2020). Behavioral diversity as a potential indicator of positive animal welfare. *Animals*, *10*(7), 1211.

Telos

Lassen, J., Sandøe, P., & Forkman, B. (2006). Happy pigs are dirty!–conflicting perspectives on animal welfare. *Livestock Science*, *103*(3), 221–230.

Rollin, B. E. (1985). The moral status of research animals in psychology. *American Psychologist*, *40*(8), 920.

Rollin, B. E. (1992). *Animal rights and human morality* (2nd ed.). Prometheus Books.

Chapter 8

Subjective Experiences

8.1 SENTIENCE AS SENSITIVITY, CONSCIOUSNESS AS SELF

It is widely considered that individuals must have certain qualities to have welfare. These qualities are often referred to as *sentience*, although sentience does not have a universally agreed-upon meaning. Some definitions circle around themes of being aware of your own experiences, such as thoughts and feelings. For example, Donald Broom (2022) described sentience as "having the capacity, the level of awareness and cognitive ability, necessary to have feelings".

American definitions of sentience are heavily influenced by the English philosopher Jeremy Bentham (1748–1832). Bentham was a social progressive and is considered to be the founder of our modern form of utilitarianism (which he preferred to think of as the "greatest happiness principle"). Bentham probably had little idea of how one sentence he wrote would create the foundation for American animal welfare philosophy and science: "[T]he question is not, Can they reason? nor, Can they talk? but, Can they suffer?" This statement appeared literally as a footnote in his *An Introduction to the Principles of Morals and Legislation* (Bentham, 1823). It put aside Descartes' argument that animals could not suffer due to a lack of rationality, speech, or intellect by declaring those faculties to be irrelevant. This notion is foundational to the current understanding of what is ethically important in mainstream American culture, which is the ability to experience harm or to suffer: the consideration of other traits being a pointless distraction from this ethical imperative.

So, for the purposes of this book, *sentience* is defined as being able to experience feelings such as pain and pleasure, while *consciousness* refers to having awareness, for example a sense of self (and is often elevated to "being aware of being aware" in humans). As discussed for the notion of *drive* in Part I, you will find people using these same words in many different ways. This might be because they are focusing on a specific situation; for example, in anesthesiology "consciousness" is often used to mean "being awake". And at other times,

DOI: 10.1201/9781003375456-10

it is simply because terms are used as synonyms or are distinguished from each other in different ways (see Zalta et al., 2002).

Keywords: consciousness, sentience, Jeremy Bentham

8.2 MARKERS OF SUBJECTIVE EXPERIENCE

There is currently no way to fully share or fully agree upon what another individual is experiencing using the scientific approach of agreement of observers. Subjective experiences are private; however, there are ways that they are expressed, both deliberately and involuntarily, with varying degrees of validity and reliability. For example, there are ways to objectively describe the qualities of material commonly perceived as being the color red. However, how any person experiences the "redness" of any specific object remains private, and of course, some people cannot perceive red at all. This subjective part of a sensory experience is called a *quale* (plural: *qualia*).

Historically, there have been many attempts to develop a better way for people to share their subjective experiences. For example, Wilhelm Wundt, the first person to call themself a *psychologist* (1832–1920), and his student Edward B. Titchener (1867–1927), established an approach to psychology called *Structuralism*. They asserted that through introspection, we could break down the elements of experience into units such as sensations, mental images, and feelings. This approach was eventually abandoned to pursue more objective forms of evidence. It should be noted that approaches that fall out of use are rarely complete dead ends. Titchener is also the person who introduced the word "empathy" to the English language in 1909 (Stueber, 2019) when he found there was no existing English term for what in German is called *Einfühlungsvermögen*. Nevertheless, some forms of evidence are more directly associated with, and so indicative of, internal subjective states. Of these, we will more fully discuss facial expressions, qualitative assessment of body language, cognitive bias, and maladaptive behaviors. But first, we need to talk about sentience and consciousness.

Keyword: qualia

8.3 ANIMAL SENTIENCE

In scientific settings, it is not unusual for some animal welfare scientists to center the entire field around the feelings of animals. Ward et al. (2018) commented: "Animal welfare science is the scientific study of the welfare state of animals that attempts to make inferences about how animals feel". Sentience, from the Latin *sentire*, to feel, is the capacity to experience feelings and emotions (with positive or negative valence, e.g., joy, fear, etc.). In animals seen as less intelligent or developed, you might see a slip in language to *sensing stimuli*, although sentience is much more than this. Sometimes, in settings like science fiction programs, sentience is used to refer to species with human-like intelligence, but this would more accurately be called "sapience", which describes creatures that have wisdom or advanced intellect.

You will also find many definitions of consciousness. They might include the "What is it like to be a Bat?" definition from Nagel's 1974 paper (Nagel, 1974). In other words, if it "feels like something" to be a bat, then the bat is conscious.

(Some argue that Nagel should have used the word *sentient.*) Others will use phenomenal consciousness and sentience interchangeably, while yet another definition would assert that consciousness is richer and denotes introspection. And there's always the medical definition of being awake and aware. The confusion around definitions led Marian Dawkins to note a form of intellectual "bait-and-switch". A publication would promise to resolve the hard problem of consciousness, only to offer a solution to some other, much simpler concept, also labeled consciousness (Dawkins, 2017). Dawkins also raised the point that we can undertake many complex behaviors without "noticing that we do them"; indeed, most of the things that humans do are achieved unconsciously (Dawkins, 2022).

There are standard tests or demonstrations that are made for various facets of sentience, including:

Awareness of self and other: A sentient individual should be aware of themself and their body and how it is separate from others. So, if a mark is placed on their body while they are unconscious, they should react to seeing it in a mirror and examine or touch the marked area (*the mirror test*). They should have systems of communication and the ability to understand the points of view of others, which might be demonstrated by individual recognition, having reputations, and being capable of deliberate deception.

Physical substrates of experience: For a sentient animal to experience pain, they should have a *nociceptor* that detects tissue damage and a signal that can be traced between that sensor and an organ capable of transducing the signal into an experience of pain, e.g., a region of the brain. Similar structures should exist for other feelings and experiences.

Expressing emotions and preferences: A sentient animal should have situations they enjoy or dislike. They should be able to demonstrate preferences and make choices and decisions. For example, they should flee from situations where they are being harmed or caused pain, and they should self-medicate for conditions that cause suffering when given the opportunity. They would be expected to express emotions with their body and behave abnormally when in a state of serious suffering, for example, showing recognizable states of panic or depression. Note that these definitions are skewed toward mobile and visually competent life forms and are essentially anthropocentric.

For a large constituency, a sentient animal is a "somebody" for themselves and to others. They have a life story and other faculties that we should be aware of. Their life should be of value to us because it is to them. Where sentience is difficult to assign or prove, a precautionary principle approach may be warranted whereby we assume that an animal is sentient (Browning & Birch, 2022). We met this principle in the *Brambell Report* in Part I, and it is a risk assessment in the face of uncertainty. Usually, this stops something from happening, like banning a food additive that might be harmful. In this case, we can stop potential harm by declaring sentience for animals. The scope of legal sentience is also important. The United Kingdom's *Animal Welfare (Sentience) Act 2022* recognizes sentience in all vertebrates, cephalopod mollusks, and decapod crustaceans. Notably, orders of certain non-vertebrates are included in a way that leaves the door open for others to be added at a later stage.

Keywords: mirror test, nociceptor

Extraordinary Claims

The science of sentience is taking us into increasingly thought-provoking and contested spaces: spaces traditionally governed by the notion of *ECREE*. The aphorism "extraordinary claims require extraordinary evidence" is attributed to astronomer Carl Sagan (1979). ECREE is often applied to evidence of subjects outside mainstream science (magic, miracles, and the paranormal) and, historically, animal sentience.

As commonly used, or, according to David Deming (2016), misused, ECREE artificially hikes the standard of proof for novel research, when the evidentiary bar should only be set high for findings that buck a massive amount of empirically gathered evidence. The ECREE principle seems to have arisen as Enlightenment scholars tackled the credibility of the incredible, namely, miracles (Deming, 2016). Since then, many have fallen foul of its diktats. One of those victims, before ECREE was even popularized, *may* have been Sagan's ex-wife, Lynn Margulis. Margulis literally deduced the evolutionary backstory of the first eukaryotic cell. This was a prokaryote that had engulfed a free-living bacterium (which became the mitochondrion—our energy factory) in an early act of "domestication" (called endosymbiosis). Her hypothesis was rejected 15 times before becoming a cornerstone of evolutionary biology (Sagan, 1967).

If every new and unusual study was accepted as definite proof, we would believe a lot of nonsense. Of course, if our conceptual framework is wrong, we *are* believing a lot of nonsense, and future generations will marvel at our stupidity. But within our worldviews, reproducibility and technical rigor can be relied on to erase attractive but flawed data from the scientific record. However, irrational bias against a phenomenon, despite evidence that supports its validity, is a more intractable cultural problem rooted in the natural conservatism of science (the bias is that scientists will arbitrarily decide what constitutes an "extraordinary" claim and what does not).

Keyword: ECREE

8.4 THE MIND: IT'S WHAT YOU DO WITH IT

For the philosopher Aristotle, consciousness was about *capacities* (Smit & Hacker, 2020). Smit and Hacker illustrate the usefulness of this approach by asking various questions about the mind. For example, what does it mean if someone "knows their own mind?" The answer—they have an opinion. To "have a mind to do something" is to threaten some action. To "make up your mind" is to decide something—all powers associated with mental activity. So, what you can do with your mind is key. Aristotle's schema was that plants had a *nutritive soul*, concerned with preservation and replication. Up a level were animals with *sensitive souls*. Animals had senses and appetites that they satisfied and were considered sentient. Humans, at the apex of development, had *rational minds*, capable of truth and logic, morality to reign in those animal appetites, plus their nutritive soul.

Most (although not everyone) would assume that something between the nutritive and sensitive soul might be the evolutionary start point for consciousness. Simona Ginsburg and Eva Jablonka set out to find some *new capacity* that marked that transition—the moment at which a minimally conscious lifeform blinked into existence (one that could, according to Ginsburg and Jablonka, "perceive, feel, and think"; discussed in Birch et al., 2020). The advantage of using a functional marker is that it can reveal the most basic model of consciousness, stripped of the distracting bloatware that might get bolted on later. We could then interrogate the tree of life (complex organisms at the highest points, simple beneath) with our new marker and see when it arose. Would we see a wave of consciousness from the top, petering out at the bottom, or would there be sporadic flashes, consciousness arising here and there? It's a compelling question—this is what they came up with (Ginsburg & Jablonka, 2022).

First, the characteristics of minimal consciousness were deduced. For example, while consciously awake, we don't have any annoying "buffering" between sound and action or see objects imperfectly overlapping each other. All perceived information is perfectly merged and continuously refreshes itself. So, this is one of the features of consciousness—unified information. As soon as the "must have" features of consciousness were compiled, the search began for a property that used all of them (Birch et al., 2020). *Unlimited Associative Learning* (UAL), the ability to learn about the world in an open-ended way, seemed to fit the bill. A lifeform capable of UAL is a lifelong motivated learner. Furthermore, performing UAL doesn't just correlate with consciousness; the evidence shows that an animal *must* be conscious to perform UAL. A literature search for the taxa most likely to possess UAL (remembering that this is biased toward the lifeforms most studied) showed evidence of it in vertebrates, some cephalopod mollusks (octopus/squid/cuttlefish), and some arthropods (including honeybees and fruit flies). Consciousness might have evolved separately in invertebrates, cephalopods, and arthropods. So, having UAL likely indicates an at least minimally conscious mind, but not having UAL doesn't rule out the possibility of being a conscious life form.

Another theory, called the cellular basis of consciousness (CBC), is that consciousness is, evolutionarily speaking, a bottom-up and inherent property of life. CBC would argue that mainstream theories of consciousness fall short of demonstrating the *emergent* property of consciousness because it never "arose" in the first place—it simply "was" (Mallatt & Feinberg, 2017; Reber, 2016). Life and consciousness are, in essence, tethered. This would support the position that all life is conscious, which is called zoopsychism when referring to animal life and biopsychism when referring to all biological life forms. Others believe that consciousness is fundamental to the entire universe; this is panpsychism.

Keywords: biopsychism, panpsychism

8.5 "THAT-WHICH-MUST-NOT-BE-NAMED"

Academia is socially organized into "feudal-like" structures with labor offered to prestigious academics. Those scientists held in the most esteem act as gatekeepers to tenure, laboratory space, funding, speaking engagements, and the like—the levers of control are many. Aspiring researchers shy away from rocking the boat with their avant-garde ideas until they can wriggle free of this patronage. A memorable paraphrase of Max Planck is that: "Science progresses one funeral at a time". Until recently, research into consciousness was strictly taboo and potentially career ending. Eminent researcher Christof Koch (2024) wrote: "We encountered considerable resistance when exploring that-which-must-not-be named. Indeed, I was urged by a Caltech mentor to wait until I had either retired or won a Nobel Prize (as had Crick) to pursue a topic best left to philosophers, religious folks, and mystics" (Figure 8.1). As the task of tackling the "hard problem" of consciousness became acceptable, new ideas began to emerge.

Figure 8.1 Christof Koch's dogs, Bella, Trixie, and Nosy (left to right). Koch, a neuroscientist at the Allen Institute, and former president of the Allen Institute for Brain Science, was convinced that his own companion animals had at least some degree of sentience. Many scientists know intuitively that their companion animals are sentient, and for some, that encourages them in their work. Photograph reproduced with the permission of Christof Koch.

Ordinarily, I would say that an animal welfare scientist might not need to worry about the state of play for consciousness research. But for you, and uniquely for you, you should keep an eye on this research, because it might come in handy. A coherent theory of consciousness would make it much more difficult to defend our poor treatment of animals. It would also allow us to tell the difference between the truly comatose individual and someone who retains some consciousness. But answering the question "what is consciousness?" has taken far too long. What's the problem?

Keyword: Christof Koch

8.6 DO CUTTLEFISH DREAM OF INVERTEBRATE SHEEP?

A cuttlefish sleeps, her limbs twitch, lid-covered eyes dart left, right, then back again, and a flush of aqua blue erupts through her mantle. Could she be dreaming?

What remembered experiences might a dreaming cuttlefish mull as she sleeps in something that looks remarkably like REM-like sleep (rapid eye movement)? But wait, is she "allowed" to do that? Isn't dreaming a human-only activity? It turns out that REM-like sleep has now been observed in many animals, including our cuttlefish. Convergent evolution has generated analogous brain structures for humans and cephalopods (Shigeno et al., 2018). In some places, cephalopods even have legal protections because they're so smart. But jumping spiders also have remarkably twitchy REM-like sleep, and their brains must be tiny (Rößler et al., 2022). A better understanding of the animal mind could well be on the cards if only we could crack the question of consciousness in humans. Rather than indulge in guesswork about cognitive capacity or expected behaviors based on brain size, can the "hard" sciences decipher what the conscious mind *is* and how it is created? Every morning, you awake into life as "you", but how is your conscious self created? For those who'd like a deeper dive, follow me to the next section. If you'd rather not, we can regroup at "The immeasurable mind is meaningless" later in the chapter. You would never be tested on the theories we're about to look at, but they offer a fascinating insight into the fuzziness of hard science, its authority, and the fundamental reality of our world.

Accusations of pseudoscience

There are many theories of consciousness, and I will briefly mention a couple here. Some of the best minds in neuroscience and mathematics are tackling this question, and the first two theories that we'll discuss have very different approaches, supported by different concepts. There will be a couple more theories, but we'll get to those later. The first theory models the brain's *functions* needed for consciousness (functionalism; the Global Neuronal Workspace Theory), and the second ponders the brain's physical *architecture* needed to deliver conscious experience (physicalism; the Integrated Information Theory). Both are known by their acronyms: GNWT and IIT.

GNWT is a *computational functionalist* theory, and so it homes in on brain activity (its functions) rather than the phenomenal aspects of how we experience life (Dehaene, 2015). At its core, GNWT argues that our conscious recognition of information is limited by our ability to pay it any attention. This is why we might run a red light in the car when we have something "on our mind". For GNWT, brain function = computation. Many purists of functionalism assert that simulating every computation in the brain of a human, dog, etc. would allow their consciousness to be created as some form of emergent property (Figure 8.2, top). GNWT also predicts an important role for the prefrontal cortex in supporting consciousness and the requisite processing pathways. In "GNWT world", the conscious mind is all about global brain networking and computation.

Next up is IIT. IIT (devised by Giulio Tononi in 2008) took the incredibly radical start point of subjective experience (how we experience life) and then deconstructed it to its essential elements (i.e., a thought is private, with seamlessly integrated sensory information, etc.). IIT then examined the brain's "real estate" that might support the networking requirements needed for conscious thought (Tononi & Koch, 2015). Take the neuron-packed cerebellum, for example: its super parallel highway of neuronal "wiring" lacks the complexity of the circuitry found in the neocortex (Figure 8.2, bottom). According to IIT theory, the cerebellum is a consciousness-free zone and runs our unconscious motor functions (Koch, 2024). This is why we don't consciously think about every step we take and can drive somewhere "on automatic". IIT predicts that a *posterior hot zone* at the back of the brain is the prime seat of consciousness. This is where neuronal connectivity delivers the integrated information needed for *causal* power—that's the difference between the conscious mind with free will and the zombie. Qualia (the quality of our experiences) are determined by the patterns and structures of brain activity (its network), and *maxima of integrated information* deliver the conscious thought over and above any blips of local activity. IIT can be represented mathematically and even provides a value (termed phi) for the "amount" of consciousness (at least theoretically).

IIT was radically different from "science as usual", and its predictions that any integrated information system could have a smidgeon of consciousness led to accusations of pseudoscience. At present, neither IIT nor GNWT seems to

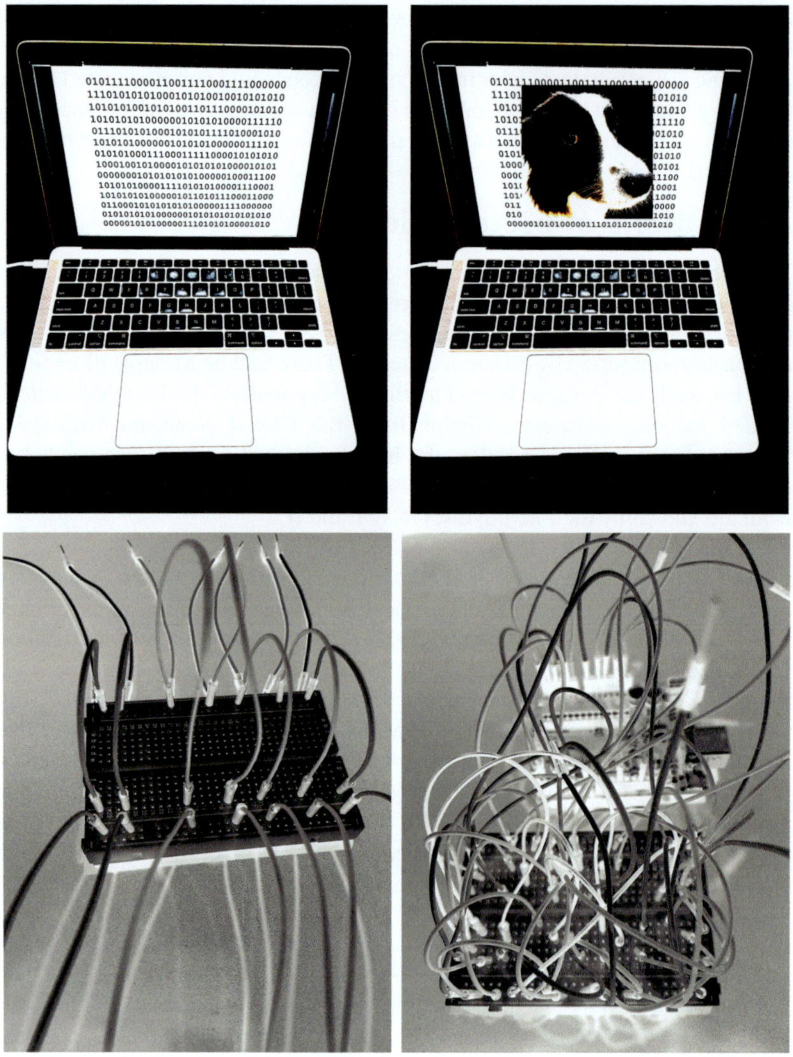

Figure 8.2 Top panel: GNWT predicts that running a whole brain simulation of this border collie will allow her consciousness to emerge. In this model, mind-is-function, and that function can be performed by a computer or a brain. Bottom panel: IIT theory is all about integrated information. In this kid's play science kit, the neater parallel wiring in the left-hand image represents the unconscious brain. The chaotic and complex wiring on the right-hand side represents the more intricate architecture that supports conscious thought and free will through integrated information (from reentrant—back and forth signaling). Image by Tina Rich.

reveal the *essence* of how consciousness is created. We still seem to be assuming that if we work out the brain's computations (as per GNWT), and the physical apparatus that delivers those functions (as per IIT), then a mash-up of the two will allow consciousness to emerge. How does all this relate to having self-awareness, emotions, and feelings? While we wait for consciousness to "emerge", let's see what the scientific fringe has to offer.

Waiting in the wings is the Nobel Prize–winning physicist Roger Penrose. Penrose has the scientific pedigree to be fearless, and he came straight out with "the elephant in the room" in his book *Shadows of the Mind* (1994) when asking: "What is the ultimate scope of science? Is it just the *material* attributes of our universe that are amenable to its methods, whereas our *mental* existence must forever lie outside of its compass? Or might we someday come to a proper understanding of the shadowy mystery of minds?" Penrose was after that "proper understanding", and he was about to introduce something else—quantum physics. Would this provide the spark to ignite consciousness?

Keywords: Global Neuronal Workspace Theory (GNWT), Integrated Information Theory (IIT)

8.7 A NEW FIELD OF QUANTUM BIOLOGY EMERGES

Quantum physics had been a bit of a mixed bag since its discovery in 1900. On the one hand, quantum theory is truly the most successful physics theory ever created, but we don't really know how it works or if it is complete. This was a particular problem for reductionist Western science, as it tends to be *bottom-up* in organization. In other words, we have a habit of drilling down to the smallest components of our world (atoms and subatomic particles and the like) and then attempting to rearrange these to explain our world. The newfound fuzziness of these building blocks (with probability waves instead of particles) felt too precarious for the founding fathers of quantum physics, or at least some of them. To deal with this uncertainty, it was decided that we would use quantum physics as a predictive tool—it was not to be used to poke around into the reality of our world (Becker, 2018). This is akin to having a Ferrari in the garage, and instead of taking it for a drive, we play with the satnav and see where we might go. Furthermore, it was decided that quantum effects only exist at very small scales because they are too fragile to survive in the classical macro world. So, problem solved, quantum craziness couldn't touch our world.

Increasingly, though, and some of this is still speculative, we find quantum effects linked to some incredibly important biologic processes. The best known of these is the ability of birds to navigate using quantum compasses in their heads. Quantum-like effects have also been reported for respiration, photosynthesis, and vision. So, *quantum niches* do exist where quantum effects last long enough to be of use in our world. In "science speak", this quantum jail break is called a *non-trivial* effect and is now driving the creation of a new field of *quantum biology* (Marais et al., 2018; https://www.youtube.com/watch?v=asps5mZ4Kp8). These processes explain many phenomena that were previously inexplicable and probably got called pseudoscience. Could the same processes underlie the unsolved mystery of consciousness? Roger Penrose had a ready-made theory for generating consciousness from quantum effects (Penrose, 1989), if only he could find a neuronal quantum niche. To find out what happens next, we pick up the trail in Arizona with a whiff of gas, some biology, and an anesthesiologist called Stuart Hameroff.

Hameroff was fascinated by how anesthesia works. All living organisms can be anesthetized. The "great vivisector" himself, Claude Bernard (1878), commented: "[W]hat is alive must sense and can be anesthetized, the rest is dead". Anesthesia reversibly severs us from our conscious mind while leaving our unconscious activities intact. What could anesthesia tell us about how consciousness works? One particular molecular target of anesthesia attracted Hameroff's attention. It wasn't a protein on the surface of a neuron, but was inside it, in the cytoplasm. The protein was tubulin, and immense polymers of tubulin create hollow molecular cables called microtubules (MTs).

To cut a long story short, Penrose and Hameroff devised an entirely new theory of consciousness (called Orchestrated Objective Reduction or Orch OR; Hameroff & Penrose, 2014), in which protoconsciousness is generated within MTs within neurons (the mechanism is too complicated to go into here). The interior of MTs serves as a quantum niche *and* acts as a computing architecture to kickstart information processing *up* to the level of the neuronal firing apparatus. It is the collapse of the quantum wave within the MT that is proposed to give rise to a moment of proto-consciousness. Anesthesia is thought to "mangle" the quantum information processed within MTs (Craddock, 2017), and so the patient loses consciousness. MT instability is also a feature of Alzheimer's disease, which might explain the confusion that the disease causes. According to Hameroff (2014), treating the brain as a neuron-only "computer" approximates 10^{16} operations per second. But an MT bolt-on gets us to 10^{27} operations per second—a major boost in brain power. For context, the most powerful supercomputer currently sits at slightly more than 10^{18} operations per second.

Every cell type has tubulin and MTs, as do other animal and even plant cells, but neurons don't divide. (Cell division radically alters MT dynamics.) So, neurons are *the* cell type in which these biologic quantum computing devices are safe and sound. There are also enormous clusters of MTs in neurons, so lots of computing power.

Orch OR is falsifiable (Hameroff, 2020), but that hasn't happened yet. Indeed, evidence for quantum consciousness keeps stacking up (Yamauchi, 2024; Craddock et al., 2017; Babcock et al., 2024). All these consciousness models are testimony to the powers of human ingenuity, and it's intriguing to think that some type of quantum effect may underlie consciousness, but it seems that we're struggling uphill to a theory of consciousness rather than just investigating consciousness directly. None of these theories have provided an answer as to what consciousness is, but they do tell us something about how we do our science.

Keyword: Roger Penrose

8.8 THE IMMEASURABLE MIND IS MEANINGLESS

So, the conclusion is that science as we know it cannot explain consciousness. It's not even close. So, what's going on? To work, science must pair its way of studying and questioning our reality (which we call *ontology*) with a theory of knowledge (*epistemology*) that allows us to know the truth. In other words, philosophically, you need to come to some conclusions about how you view the world before you can successfully study it. This is where things get tricky and might be going wrong.

Earlier, in Chapter 6, we mentioned Cartesian Dualism and the hard problem of how an intangible mind can interact with a physical body. Outside of philosophy, there is very little consideration given to these questions, and yet a moment's thought reveals how impoverished science is without properly considering the mind. Our lives are about subjective experiences. We might be very attached to matter, especially matter organized as an iPhone, but it is the moment of expectation, joy, triumph, upset, etc. that we have when using these objects that is fundamental to us. The object just facilitates. The problem that we have is that mind/subjective experience cannot be measured or copied, and so for many mainstream scientists, subjective experience simply does not exist. It's extraordinary that in much of our current research into the mind, our scientific methodology forces us to ignore the only thing that "objectively" matters to us because our science cannot abstract it to a value. This has led many scientists to engage in worldviews that, at their most extreme, predict that computation alone can create consciousness in a full capitulation to the notion of *man as a machine*. Termed by Penrose (not an advocate) and others, as strong AI or functionalism (Penrose, 1994), many would find this notion to be preposterous, and yet there it is.

The strange thing about this is that all the wonderful theories that we have discussed are created by the *minds* of scientists, but the mind must take a back seat in its own story. As these theories fail to explain consciousness, there is a growing anxiety that their philosophies might explain the world *as we define it*, but not necessarily as it is. This is not a new worry (Wigner, 1960), and others have proposed an alternative worldview in which the mind is fundamental. This is called *Idealism* (Chalmers, 2019; Kastrup, 2021), and idealists would argue that making the mind subordinate to matter is allowing the tail to (try to) wag the dog. This is an extraordinary moment in the study of mind and consciousness, and you'll notice that many of the books and articles referred to are very recent. There is a whiff of a paradigm shift in the air.

Federico Faggin helped to invent the first microprocessor at Intel, the touch pad and touch screen technologies, and knows a thing or two about computer networks. Interestingly, Faggin has now committed to an idealist worldview with a new theory that puts consciousness front and center. Faggin proposes that the unfathomably rich experiences that we enjoy in our minds are quantum in nature, and these are abstracted in the classical world to qualities that we can share with others (Faggin, 2024). So, quantum information is personal (and because it is quantum, it is uncopyable and private), and classic is for sharing. Penrose, with his quantum vision for mind, also put it beyond the reaches of classical computation (so no conscious computers) and suggested no fewer than four worldviews to study consciousness (Penrose, 1994). Whichever it is, we will know the correct worldview when we have it because it will enable us to derive a thorough understanding of mind and matter. So, we embarked on a quest to learn about consciousness, only to discover that we are not sure about the right worldview to study consciousness. Christof Koch, Federico Faggin, Roger Penrose, and Bernardo Kastrup regularly discuss these issues, and many of their chats can be viewed online. You might want to check them out yourself and see how their opinions have shifted. At this point, I should say that one of us (EPK) remains a materialist (i.e., a physicalist), whereas TR is much more of an idealist.

However you think of the brain, mammals have what it takes to be called conscious. Neither IIT nor Orch OR would have any problem with consciousness being generated in living systems without a brain given the right structures or informational networking (and all animals and plants have MTs). Theoretically, GNWT wouldn't even need a brain, as information processing alone should be sufficient for consciousness (Dehaene et al., 2017). And if we still want to take the brain to be a relevant indicator of consciousness, then a recent paper tells us that across mammals, brain wiring is equivalent (Assaf et al., 2020). This section has been included to show you how uncertain science can be about the biggest questions science can ask. We could not be on shakier ground. That said, these theories have done us a favor in showing that there may be no reason why consciousness shouldn't cascade throughout the tree of life, and perhaps into its roots.

Keywords: idealism, epistemology, ontology, Bernardo Kastrup

8.9 SENTIENCE IN NATURE AND PLANTS

The book *The Blind Spot* (Gleiser & Thompson, 2024) has the following subheading: "Living is sensemaking in precarious conditions". Organisms are described as interacting with their environments in such a way as to prioritize their survival. Do we allow plants the option of making "sense" of their environment? Do they even exhibit the behaviors of a sentient life form? It turns out that not only are plants immobilized by anesthesia (Kelz & Mashour, 2019) but they also self-medicate with anesthetics when stressed (as usual, this is all highly contested [Yokawa et al., 2018; Draguhn et al., 2021]). Plants are extraordinary, as they terraformed this planet and are the only living organisms capable of feeding themselves. Putting aside zoocentric notions of the "wetware" needed for sentience (a brain), plant sentience follows the same arguments made for simple animal sentience (Segundo-Ortin & Calvo, 2022).

Researchers tread carefully when suggesting experiments to interrogate these possibilities (Mediano et al., 2021). A rooted life appears largely sessile (without a good time lapse camera) and without localized organs (too risky for an immobile life form). Instead, distributed information and structural redundancy is deployed; the plant can jettison parts under attack or wall them off. Underground, each tree has fantastic connectivity via the mycorrhizal fungal network and, up top, politely gives their neighbors space in that curious feature of "canopy shyness": leaves close but not quite touching. The remarkable capabilities of plants as chemists have led some to suggest "plants as a model system for understanding information sharing and communication among organisms in general" (Kalske et al., 2019).

Plants undertake a cost benefit analysis when signaling danger using volatile organic compounds (VOCs) (Kessler et al., 2023). They have private chats with relatives using heritable chemotypic VOCs and use more generic VOCs to allow unrelated plants to eavesdrop. As noted by Schlanger (2024), it is impossible to ignore the notion of "intent" in some of these interactions, although researchers studiously do. Instead, "fitness outcomes" are discussed. Reputational anxiety is a constant issue. The fledgling field of behavioral botany was torpedoed following the publication of *The secret life of plants* in 1973 by Tompkins and Bird, which was roundly criticized as pseudoscience. A more innate issue is our "Plant Blindness", a cognitive bias first described by Wandersee and Schussler (1999) in which we mentally screen out plants as background decoration. So, given an innate ability to ignore plants, do we also underestimate their capabilities? Why is a "feeling" for an animal downgraded to a "stimulus" for a plant, even when the experimental scenarios look remarkably similar?

In total, you may not be convinced that plants are sentient to the level of having welfare, but they are measurably less insentient than a rock. It's probably fair to say that both authors had, for most of their lives, dismissed out of hand the idea that plants could be sentient. It is for this very reason that we've paid close attention to the study of plant sentience and subsequently broadened our thinking about what sentience is.

Historically, claims for sentience in anything other than animals and humans came mainly from an intuitive sense of spiritual sentience or animism, where animism is a sense that something has a spiritual essence, identity, or presence beyond being simply dull matter. From this perspective, any entity, regardless of its qualities, may be understood spiritually as a "being" that should be respected.

It can also be asserted that the existence of anything that is complex and has a recognizable identity conveys rights, even in the absence of universally recognized sentience: for example, the recognition of the legal personhood of the Magpie River/Mutehekau Shipu in Canada. This eco-sentience encompasses those who believe that the river has a significant form of sentience, as well as those who believe that the river is important and should not be "harmed" even in the absence of sentience. That is, the integrity and health of the river are inextricably linked to the health of the land and its occupants. The river is also seen as aesthetically and spiritually important, which creates a duty to care for its wellbeing. This general approach to protecting significant parts of nature forms the idea of *Earth Jurisprudence* or the *rights of nature movement*, which gained traction early in the 21st century. This movement is represented in the United States by The Community Environmental Legal Defense Fund (CELDF), founded in Pennsylvania in 1995.

By now, you may be wondering why a book on animal welfare spent a whole section talking about plants. Again, the point is to show you that many sources of authority will present facts as "cut and dried" when they are anything but. Likewise, throughout this book and most animal welfare science discussions, we talk very glibly about animals. When discussing animal sentience, we generally only think about mammals, and possibly a few other taxa like the octopus. When it comes to taxa that are alive but genetically and evolutionarily distant from human beings, plants represent a very distant point. But in their qualities, they expose something about the size of the area of ambiguity when assigning sentience.

Keyword: animism

Alex and Irene Pepperberg—relationships can make good science

In the last few sections, we've seen lots of examples of relationships: between ecosystems, animals, scientists, and even their worldviews. Sometimes, it's the relationship with your subject that makes all the difference. It was for Irene Pepperberg when she began her language and cognition studies with an African grey parrot called Alex (his name was an acronym for avian learning or language experiment). Parrots had long been seen as promising species for teaching an animal to make human-like use of language. Not only do they have the capacity to speak human language, but there are anecdotal reports of their use of language in a meaningful context.

Most of the scientific research into parrot language used repeated, recorded language, to produce uniform stimuli for multiple parrots. Unsurprisingly, this approach did not produce parrots that could converse in a human-like way. Irene Pepperberg made a breakthrough in her one-to-one work with Alex (Pepperberg, 1981, 1983). This was a labor-intensive language class within a human–animal relationship; Alex was a pet but also a colleague. Alex became the first scientifically studied parrot that would reliably demonstrate the ability to select objects based on abstract categories such as color or the material that they were made of.

At this time, people were already familiar with the use of American Sign Language (ASL) with chimpanzees and great apes (Gardner & Gardner, 1969). Koko the gorilla (1971–2018) was especially famed for using sign language to communicate with people and other gorillas, and for adopting a kitten *she named* "All Ball" (Gold & Watson, 2018). The kitten was a Manx kitten with no tail, so was distinctly ball-like. The signing apes were part of a general increase in acceptance of primates as close relatives to humans, not only genetically, but socially, emotionally, and intellectually.

Going even further back, relationships with communicating animals were discussed only as undesirable confounds to research. The example often given was "Clever Hans" (circa 1895–1916), a horse that could give correct answers to questions and mathematical problems. It was eventually discovered that he did so by reading the unconscious body language of the people around him, who would relax slightly when he reached the right number of hoof taps for the correct answer or when his nose was over the right card. While disappointing for those who thought a horse could understand language and do complex mathematics, it did reveal Clever Hans' ability to interpret body language, which is in many ways just as impressive.

Alex and other parrots that joined the Pepperberg lab became important celebrity animals because they showed an easily appreciable use of language (for the majority not fluent in ASL) in animals that are genetically very distant from humans. They showed that animals could learn to communicate with us with clarity and sophistication if they learned to do so within the context of a meaningful and positive friendship.

The research with Alex was later supported by research into the avian brain that revealed analogous structures to the "cortico-cerebellar" circuitry of primates (Gutiérrez-Ibáñez, et al., 2018). Parrots became a vanguard species for appreciating how animals with different evolutionary histories can demonstrate cognitive abilities that were more advanced than previously appreciated. But perhaps we could have made more progress if we'd ever tried to meet these animal scholars halfway, and made more of an attempt to learn their languages.

Keywords: Alex the parrot, Clever Hans, Irene Pepperberg

8.10 INTELLIGENCE

Having just explained that intelligence is not relevant to sentience, discussions of animal intelligence are still wrapped up in the issue. When the public learns that animals are clever, have organized societies, or can use language in meaningful ways, we become more interested in protecting them. There is much debate about which individuals are intelligent and how we can prove it. Now, we have the added complication of intelligent AI that can lure us into anthropomorphizing *it*, a mistake famously made by a Google engineer in 2022.

In 1950, a test called the "imitation game" was devised by a British computing genius, Alan Turing. This test, through oral interrogation, was devised to detect whether the tester was interacting with an artificial intelligence or not. Computers have been passing Turing tests (fooling us that they are human) for over a decade. *Eugene Goostman*, a chatbot pretending to be a 13-year-old Ukrainian boy, was the first AI to pass a Turing test, although that test was highly controversial. More recently, Chat-GPT4 passed more rigorous Turing tests with flying colors (Mei et al., 2024). These tests may be telling us more about what it is to be human than what it is to "be" an artificial intelligence. We're clearly not ready to concede that these AIs are "thinking" as we do, and so we devise more and more demanding tests designed to trap "them".

Regardless of the test, you will only find intelligence where and how you look for it (Kilgour, 1981). New Zealander Ronald (Ron) Kilgour (1932–1988) noticed that psychologists very rarely studied learning in livestock (pigs, cattle, poultry). Luckily, Kilgour was particularly forward thinking when devising methods to illustrate the abilities of animals of different species in maze running tasks. The highest-scoring species were, in order: human children, dogs, cows, and goats. The lowest were brush tail possums, who are rather averse to walking long distances on the ground. Kilgour not only challenged the idea of livestock as "big dumb animals" but also championed studying behavior for its own sake rather than only as a management tool. Kilgour was happy to try to explore the animal's point of view to better understand their behavior, an approach that provided some support for the work of animal welfare scientists, including Marian Dawkins and Temple Grandin.

In research tasks, chimpanzees will tend to choose what is best for them personally and will outperform humans in patience for a food reward, given the food insecurity of chimpanzee society. Likewise, if you want to test the full cognitive ability of a dog, put yourself in the canine world of olfaction and design a task around that sense. It goes without saying that animals will outperform humans on spatial memory tasks when those animals need to remember where they hid their food to survive the winter. In short, if you want to know the true cognitive power of an animal, design a test that is relevant to that animal (a *biocentric task* and not anthropocentric) as opposed to asking them to perform some irrelevant human-like skill (Bräuer et al., 2020).

Keyword: Ronald "Ron" Kilgour

8.11 CONSCIOUSNESS AND WELFARE

Changes in consciousness have considerable implications for whether animals can suffer. For example, during a potentially painful surgery, animals should be administered analgesia (preventing pain from being experienced by a conscious animal), anesthesia (preventing an animal from being conscious), or both. However, rather than being very clearly present or absent, consciousnesses can vary on a spectrum or be altered by natural states or by pharmaceuticals, with effects that are not limited to analgesia and anesthesia but also impact anxiety reduction, sedation, and memory loss. *Transitional states of consciousness* are often contentious areas for animal welfare. For example, when killing animals in many contexts, especially when unattended, death might occur slowly (e.g., trapping, pest control) and with a poor idea of the pain or stress that may be experienced by the animal.

Beliefs about the management of conscious experience can change. For a long time, it was thought to be beneficial for animals recovering from injury or surgery to feel pain. The idea was that the pain would stop them from moving around too much, which might worsen their injury or slow healing and recovery. This belief is now less common as research indicates that pain control typically aids recovery, and effective pain control techniques are available, at least for companion animals.

In the past, there was also a tendency to assume that many surgical procedures performed on animals were not very painful. However, prey animals like sheep tend to be stoic and try not to react to painful events to avoid drawing attention to their weaknesses. Cats are both hunters and, in some places, the hunted. These mesopredators can also mask pain. In a recent article on prey species that hide pain, Carbone (2020) noted that "if virtually all species in the close confines of a laboratory setting would innately see humans as threatening potential predators, then all animals are 'prey species' and all should be unreadable in their pain".

There is now a high degree of agreement that interventions like burning or cutting and even ligature banding will almost always cause pain. However, there continue to be systematic biases in how painful we think these procedures are. For example, veterinary nurses perceive veterinary procedures to be more painful than veterinarians, and procedures on dogs were rated as being more painful than similar procedures performed on cats. In human medicine, self-administration has provided an avenue for determining appropriate dosing to prevent pain. The same could potentially be applied to animals, as animals have demonstrated their ability to self-medicate based on need, for example by choosing between medicated and unmedicated water.

Keywords: analgesia, anesthesia

Example: Consciousness and the egg

If an animal is sentient (able to suffer) and conscious (awake and able to integrate and experience sensory inputs), their welfare must be considered. One implication of this is that a fertilized bird egg that is not capable of suffering can be discarded like any other refuse. But if the chick within the egg can suffer at any point, then an approved method of humane euthanasia needs to be used. The standards for euthanasia describe a *good death* as being instantaneous, or unconsciousness followed by death.

There has been considerable research into whether a chick could become conscious in the egg before hatching. To answer this, we need a coordinated understanding of the neural substrates being developed, how consciousness is moderated in the egg, and how behaviors indicate conscious experience. Despite the wide variety of egg-laying species, this process has only really been studied in the domestic chicken (*Gallus domesticus*) because of the sheer scale of the chicken egg and meat industries.

The American Veterinary Medical Association (AVMA) maintains documents on what is acceptable for animal euthanasia, and these represent a legal standard in most US states (Underwood & Anthony, 2020). These guidelines are based on research suggesting that chick consciousness is suppressed until they hatch and breathe, which increases the oxygen content of the blood. However, eggs at 80% of incubation would have the ability to become conscious if exposed to air from outside the egg. Activities occurring in ovo prior to this point are therefore unconscious even when they involve responding to external stimuli. However, as is clear in the guidelines, this is a nuanced and somewhat contested situation, with a need for more research.

Keywords: euthanasia (animal/veterinary) neural substrate

8.12 FACIAL EXPRESSION

Between humans, one of the behaviors used to judge internal states is facial expression. A particular area of focus is the expression called the grimace. A grimace is contortion and tensing of the face associated with various forms of distress, including pain. Grimacing is often involuntary but can also be used deliberately to communicate feelings such as disgust or disapproval. Grimaces have a long history of being used to assess non-verbal humans. For example, newborn babies (at one and five minutes after birth) are assessed using the APGAR scale to see how well they are doing. First developed in 1952, this scale is still used today. APGAR is a mnemonic for Appearance, Pulse, Grimace, Activity, and Respiration. "Grimace" is a test for reflexes by giving the baby a small pinch and seeing whether they grimace. APGAR is not only the mnemonic but also the surname of Virginia Apgar (1909–1974), an obstetrical anesthesiologist at Columbia University who developed and standardized this test for newborns.

There are also methods to rate how intensely an individual is grimacing. Normally, a scale is presented using pictures, where you pick the one that most closely resembles what you are feeling (or see). These scales can be used by people to self-identify their level of suffering or by an assessor for people unable to respond verbally or choose for themselves, for example due to dementia.

It was often considered that many animals do not grimace or do not have expressive faces. However, assessment of mouth, nose, and ear landmarks showed that there were often strong parallels to the human grimace, but we were not particularly adept at reading and interpreting them. A visual scale showing members of that species can help people to identify and code degrees of grimace in other animals, and these variations can be automatically identified by machine learning or an AI interface (Steagall et al., 2023).

The grimace scale is just one example of a tool developed for use with humans that has been repurposed for animals (Langford et al., 2010; Mogil et al., 2020). Human-oriented methods are often a good place to start when developing a concept or tool for animal welfare assessment. Other examples referred to elsewhere in this text include qualitative behavioral assessment and quality of life scales.

Keywords: APGAR test, grimace

8.13 ANTHROPOMORPHISM

In scientific research, you should be able to describe how you collected and interpreted your data in such a way that other interested and impartial parties can, with sufficient resources, replicate your work. For this reason, modern scientific research is hesitant to rely on subjective impressions (empathy, intuition, etc.,) as a sole methodology for drawing conclusions. This sometimes leads to an assumption that any information gained from animals through intuition is not only questionable but always wrong. Specifically, the concern is that the observer is *anthropomorphizing* their animal subjects.

One well-known anecdote about the pitfalls of anthropomorphism was shared by Ronald Kilgour, who described a time when a lion was being transported by air. The transporters thought they could make the lion more comfortable in the same way that they would for a human. They therefore provided the lion with a pillow. During the flight, the lion ate the pillow and as a result had died by the time they arrived at their destination. The point being, you cannot assume that a lion will understand and use a pillow in the same way that a human will.

Table 8.1 Terminologies relating to the subjective assessment of behaviors

Incorrectly assume equivalence to human expression	**Correctly assume equivalence to human expression**
ANTHROPOMORPHISM	EMPATHY
Incorrectly assume difference from human expression	**Correctly interpret a difference in meaning based on species**
SPECIESISM / ANTHROPODENIALISM	ZOOFLUENCY

However, it is important to keep in mind that information gained through intuitive reasoning should not be assumed to be wrong (Lockwood, 1985). Some humans do share a lot of experiences and expressions with a wide variety of other animals, such as the fight or flight reaction. Emory primatologist Frans de Waal introduced the idea of *anthropodenialism*. Anthropodenialism is used to deny that animals express traits regarded as being only human, even when those traits have negative connotations, like the ability to commit suicide (de Waal, 1999). Useful terminologies for the subjective assessment of animal behavior are shown in Table 8.1.

Keywords: anthropodenialism, anthropomorphism

8.14 QUALITATIVE BEHAVIORAL ASSESSMENT

There are methodologies that use subjective data in ways that are scientifically and objectively robust. One particularly interesting approach is called the *Qualitative Behavioral Assessment* (QBA). This is a method developed by Françoise Wemelsfelder at Scotland's Rural College (SRUC) in Edinburgh. QBA assesses animal experience by considering behavior as a form of expressive *body language* that reflects the emotional aspects of their wellbeing. QBA does not measure "what" animals are doing physically (e.g., walking, eating) but "how" they are performing those behaviors (e.g., do they move in a relaxed and playful way, or in a tense and anxious way?).

QBA is the application of three elements: 1) a list of qualitative descriptors for animal emotions (created by the observers through a process called "Free Choice Profiling" or pre-fixed and given to observers by a project leader); 2) a Visual Analogue Scale next to each descriptor that observers use to score the intensity of an observed emotion; and 3) a statistical method for multivariate analysis, which identifies larger patterns of emotional expressivity from the scores. The resulting expressive profiles can be used to compare animals with each other, measure the impacts of welfare challenges, and compare the findings of QBA with other methods of assessing welfare (Figure 8.3).

This combination of techniques had previously been put to good use when analyzing the sensory experiences of people (with no prior training or special expertise) in food tasting and food product development. Wemelsfelder adapted the same methodology to the assessment of the emotional expression of animals (Wemelsfelder, 2001). When observers were shown the same animals, or videos

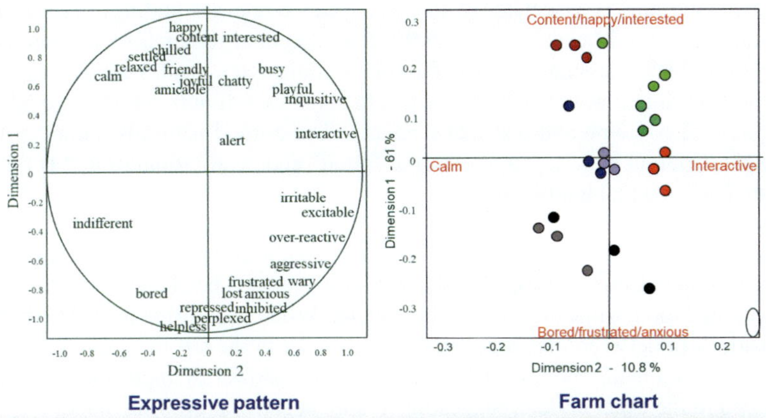

Figure 8.3 QBA analysis, showing the principal components in dimensions 1 and 2, and the distribution of assessment terms across the four quadrants that describe those dimensions. The resulting farm chart shows data points colored according to the strength of the principal components. Reproduced with the permission of Françoise Wemelsfelder.

of animals, a high degree of inter-rater reliability was found (Phythian et al., 2013). QBA ratings have been shown to be sensitive to both naturally occurring differences between animals (e.g., temperament) and variables relevant to welfare (a new and frightening setting or a familiar and pleasant one).

If you compare a QBA with an ethogram, they are almost complete opposites. In an ethogram, the scientist decides what the important behaviors are and when an animal is performing them (i.e., pigs walk, this pig is walking). In a QBA, there is no attention paid to what the animal is doing but only to the quality with which it is done (this pig is anxious). The information from QBAs has often been found to be more accurate than that from ethograms or physiological stress assessments, and reproducible, irrespective of whether the rater is a scientist, student, or farmer.

Keyword: qualitative behavioral assessment

8.15 PESSIMISTIC BIAS

Cognitive bias is a systematic distortion in perception or understanding of a situation shown by an individual due to a flaw in their processing of information. We saw this earlier in the description of "plant blindness". The idea of cognitive bias was proposed in 1972 to encourage research into the causes of this apparently flawed or irrational behavior. It is currently understood that these biases are often adaptive, or have some underlying adaptive basis, even when they appear illogical.

There are many types of cognitive bias, such as *confirmatory bias*, where we are quicker to accept and retain new information that matches our opinions and reject anything contradictory. Or *the gambler's fallacy*, where we think that a randomly occurring event becomes more likely if it has not happened for a while. One cognitive bias that is studied in animal welfare science is the *optimism/pessimism* bias. This is how an individual in a negative mood is more likely to expect that bad things will happen to them, while an optimist will expect good things from the same situation. This bias is studied in animals precisely because you can infer mood from expectation. That is, animals that expect bad outcomes in neutral settings are probably experiencing a negative mood state, implying poor welfare (Bateson et al., 2011).

Pessimistic bias is associated with other welfare-relevant findings, like serotonin depletion. Michael Mendl (2009) led influential research to develop the cognitive bias assessment as part of animal welfare assessment. This is achieved in an experiment where animals are taught about two stimuli, one that provides a pleasing outcome and one that does not. The animal naturally shows or is taught different ways to respond to the "good" and "bad" stimulus and is then tested with an ambiguous stimulus.

An even simpler assessment is to see how fast an animal approaches the stimulus. For example, Casey et al. (2021) showed animals that a bowl on the left always contained food, while one on the right did not. When bowls were placed in three positions in between, the dogs ran to those closer to the left more quickly. But dogs trained using more coercive methods were slower to approach the bowl placed in the exact middle, suggesting a more negative prevailing mood.

Keyword: cognitive bias

8.16 EXTREME MALADAPTIVE BEHAVIORS

Some behaviors are recognized as completely abnormal or *maladaptive*. Examples would include self-mutilation, injuring or killing conspecifics (e.g., "cannibalism"), barbering (the excessive chewing/plucking of hair or whiskers in rodents), horse crib-biting or wind sucking, and consuming vomit. Many of these behaviors visibly harm the body of the animal or another of the same species. Once an animal is showing these behaviors, they may persist throughout life even if moved to a better living situation. Some of these behaviors are not directly harmful, but people find them distressing and will often perceive them as aesthetically disgusting.

These maladaptive behaviors can be described in terms that seem to blame the animal: for example, as a *behavioral vice*, where "vice" is an old term for crimes that are considered immoral or evil. You will also see references to *problem animals*. Animal owners and caretakers certainly tend to find these behaviors distressing and difficult to address, which may explain the emotional tone in which they are described.

Because you can't just sit still and watch animals be harmed, people often resort to directly preventing the distressing behavior. For example, there are many contraptions through history that have tried to prevent horses from sucking air into their stomachs, often performed with an arched neck and loud gulp or grunt ("wind sucking"). While any link between wind sucking and digestive disorders and colic is disputed, people certainly hate to see it, and later efforts to stop the behavior included cutting nerves and/or muscles of the neck.

Maladaptive behaviors were first scientifically classified as having no function or an improper function based on the assumption that any behavior without an obvious and wholesome goal is aberrant. Over time, we've come to realize that most of these behaviors are either a redirected form of functional behavior or an attempt to cope with unsupportable demands from the environment, such as stress or boredom. In a small, barren environment, animals may do unusual things simply to provide some activity and stimulation to their lives, usually selecting a behavior that their species is biologically predisposed to perform (e.g., hens peck, horses chew, mice groom each other). While they might present as a "badly behaved animal", these behaviors are diagnostic of a current or past environment that failed to adequately support their needs.

There is still some dispute as to whether these physically harmful behaviors are an extreme example of being able to cope, or tip over into not coping. The idea of "good" coping comes from research showing that animals exhibiting these behaviors have indications of better welfare, such as lower levels of physiological stress, compared with animals not showing the behavior. However, this is within an environment that is apparently very challenging. This might seem like a distinction without a difference, but nonetheless, is used to debate whether or not the animal's welfare is adequate.

One important category of maladaptive behaviors is *stereotypy*. This refers to any repetitive actions conducted for large periods of the animal's day and with no apparent function. Wind sucking, pacing, and weaving might be nested within this category. The exact form of the stereotypy varies by animal and context, and once established, it can become difficult to reverse. The more rigid and time consuming a behavior is, the more likely it is to be considered a stereotypy.

Given their occurrence in smaller and more barren settings, stereotypies may relate to coping with a lack of engagement with the environment. But they can also be associated with highly anticipated events such as being fed, especially when these occur at regularly scheduled times. Some of the most well-researched stereotypies include bears and other predators pacing in an enclosure, the swaying of tethered elephants, and bar-biting by caged rats (Mason, 2006). The latter is now understood to be a redirected escape behavior, which, over time, becomes fixed, repetitive, and less associated with the original goal. While maladaptive behaviors are often a problem for their owners or caretakers, their role in animal welfare science is to indicate a past or currently compromised welfare. We group these kinds of behaviors not purely due to their shared characteristics but because they elicit distress or disgust. When we see them, we intuitively feel that something has gone horribly wrong for and with this animal. And in most cases, this is probably correct.

Keywords: wind sucking, stereotypy

READINGS

Sentience as sensitivity, consciousness as self

Bentham, J. (1823). *An introduction to the principles of morals and legislation*. Hafner Pub. Co.

Broom, D. M. (2022). Concepts and interrelationships of awareness, consciousness, sentience, and welfare. *Journal of Consciousness Studies*, *29*(3–4), 129–149.

Zalta, E. N., Nodelman, U., Allen, C., & Anderson, R. L. (2002). Stanford encyclopedia of philosophy. http://plato.stanford.edu/

Markers of subjective experience

Stueber, K. (2019). Empathy. In Edward N. Zalta (ed.), The stanford encyclopedia of philosophy (Fall 2019 Edition), https://plato.stanford.edu/archives/fall2019/entries/empathy/

Animal sentience

Browning, H., & Birch, J. (2022). Animal sentience. *Philos Compass*, *17*(5), e12822.

Dawkins, M. S. (2017). Animal welfare with and without consciousness. *Journal of Zoology*, *301*, 1–10. https://doi.org/10.1111/jzo.12434

Dawkins, M. S. (2022). Unconscious humans, autonomous machines and the difficulty of knowing which animals are sentient. *Animal Sentience*, *32*(20). https://doi.org/10.51291/2377-7478.1746

Nagel, T. (1974). What is it like to be a bat? *The Philosophical Review*, *83*(4), 435–450. https://doi.org/10.2307/2183914

Ward, S. J., Sherwen, S., & Clark, F. E. (2018). Advances in applied zoo animal welfare science. *Journal of Applied Animal Welfare Science*, *21*(sup1), 23–33.

Extraordinary claims

Deming, D. (2016). Do extraordinary claims require extraordinary evidence? *Philosophia (Ramat Gan)*, *44*(4), 1319–1331.

Sagan, C. (1979). *Broca's brain, reflections on the romance of science*. Random House.

Sagan, L. (1967). On the origin of mitosing cells. *Journal of Theoretical Biology*, *14*(3), 255–274. https://doi.org/10.1016/0022-5193(67)90079-3. PMID: 11541392.

The mind: It's what you do with it

Birch, J., Ginsburg, S., & Jablonka, E. (2020). Unlimited associative learning and the origins of consciousness: A primer and some predictions. *Biology and Philosophy*, *35*, 56. https://doi.org/10.1007/s10539-020-09772-0

Ginsburg, S., & Jablonka, E. (2022). *Picturing the mind: Consciousness through the lens of evolution*. The MIT Press.

Mallatt, J., & Feinberg, T. E. (2017). Consciousness is not inherent in but emergent from life. *Animal Sentience*, *1*(11), 15.

Reber, A. S. (2016). Caterpillars, consciousness and the origins of mind. *Animal Sentience*, *11*(1). http://dx.doi.org/10.51291/2377-7478.1124

Smit, H., & Hacker, P. (2020). Two conceptions of consciousness and why only the neo-Aristotelian one enables us to construct evolutionary explanations. *Humanities and Social Sciences Communications*, *7*, 93.

That-which-must-not-be-named

Koch, C. (2024). *Then I am myself the world: What consciousness is and how to expand it*. Basic Books.

Do cuttlefish dream of Invertebrate sheep?

Rößler, D. C., Kim, K., De, Agrò, M., Jordan, A., Galizia, C. G., & Shamble, P. S. (2022). Regularly occurring bouts of retinal movements suggest an REM sleep-like state in jumping spiders. *Proceedings of the National Academy of Sciences of the United States of America*, *119*(33), e2204754119. https://doi.org/10.1073/pnas.2204754119

Shigeno, S., Andrews, P. L. R., Ponte, G., & Fiorito, G. (2018). Cephalopod brains: An overview of current knowledge to facilitate comparison with vertebrates. *Frontiers in Physiology*, *20*(9), 952.

Accusations of pseudoscience

Dehaene, S. (2015). *Consciousness and the brain* (pp. 161, 177). Viking.

Koch, C. (2024). *Then I am myself the world: What consciousness is and how to expand it*. Basic Books.

Penrose, R. (1994). *Shadows of the mind: A search for the missing science of consciousness*. Oxford University Press.

Tononi, G., & Koch, C. (2015). Consciousness: Here, there, but not everywhere. *Philosophical Transactions of the Royal Society B*, *370*, 2014.0167.

A new field of quantum biology emerges

Babcock, N. S., Montes-Cabrera, G., Oberhofer, K. E., Chergui, M., Celardo, G. L., & Kurian, P. (2024). Ultraviolet superradiance from mega-networks of tryptophan in biological architectures. *The Journal of Physical Chemistry B*, *128*(17), 4035–4046. https://doi.org/10.1021/acs.jpcb.3c07936

Becker, A. (2018). *What is real? The unfinished quest for the meaning of quantum physics*. Basic Books.

Bernard, C. (1878). *Lectures on phenomena of life common to animals and plants*. Ballliere, and Son.

Craddock, T. J. A., Kurian, P., Preto, J., Sahu, K., Hameroff, S. R., Klobukowski, M., & Tuszynski, J. A. (2017). Anesthetic alterations of collective terahertz oscillations in tubulin correlate with clinical potency: Implications for anesthetic action and post-operative cognitive dysfunction. *Scientific Reports*, 7(1), 9877.

Hameroff, S. (2014). Consciousness, microtubules, & 'Orch OR': A 'space-time odyssey'. *Journal of Consciousness Studies*, *21*(3–4), 126–153.

Hameroff, S. (2020). 'Orch OR' is the most complete, and most easily falsifiable theory of consciousness. *Cognitive Neuroscience*, *12*(2), 74–76.

Hameroff, S., & Penrose, R. (2014). Consciousness in the universe: A review of the 'Orch OR' theory. *Physics of Life Reviews*, *11*(1), 39–78.

Marais, A., Adams, B., Ringsmuth, A. K., Ferretti, M., Gruber, J. M., Hendrikx, R., Schuld, M., Smith, S. L., Sinayskiy, I., Krüger, T. P. J., Petruccione, F., & van Grondelle, R. (2018). The future of quantum biology. *Journal of the Royal Society, Interface*, *15*(148), 20180640. https://doi.org/10.1098/rsif.2018.0640

Penrose, R. (1989). *The emperor's new mind: Concerning computers, minds and the laws of physics*. Oxford University Press.

Yamauchi, A., Tanaka, K., Fuki, M., Fujiwara, S., Kimizuka, N., Ryu, T., Saigo, M., Onda, K., Kusumoto, R., Ueno, N., Sato, H., Kobori, Y., Miyata, K., & Yanai, N. (2024). Room-temperature quantum coherence of entangled multiexcitons in a metal-organic framework. *Science Advances*, *10*(1), eadi3147. https://doi.org/10.1126/sciadv.adi3147. Epub 2024 Jan 3. PMID: 38170775; PMCID: PMC10775993.

The immeasurable mind is meaningless

Assaf, Y., Bouznach, A., Zomet, O., Marom, A., & Yovel, Y. (2020). Conservation of brain connectivity and wiring across the mammalian class. *Nature Neuroscience*, *23*(7), 805–808. https://doi.org/10.1038/s41593-020-0641-7. Epub 2020 Jun 8. PMID: 32514137.

Chalmers, D. (2019). Idealism and the mind-body problem. In W. Seager (Ed.), *The Routledge handbook of panpsychism* (pp. 353–373). Routledge.

Dehaene, S., Lau, H., & Kouider, S. (2017). What is consciousness, and could machines have it? *Science (New York, N.Y.)*, *358*(6362), 486–492. https://doi.org/10.1126/science.aan8871

Faggin, F. (2024). *Irreducible: Consciousness, life, computers and human nature*. Essentia Books.

Kastrup, B. (2021). Science ideated. In B. Kastrup (Ed.), *The fall of matter and the contours of the next mainstream scientific worldview*. IFF Books.

Penrose, R. (1994). *Shadows of the mind: A search for the missing science of consciousness*. Oxford University Press.

Wigner, E. P. (1960), The unreasonable effectiveness of mathematics in the natural sciences. Richard Courant lecture in mathematical sciences delivered at New York University, May 11, 1959. *Communications on Pure and Applied Mathematics*, *13*, 1–14.

Sentience in nature and plants

Draguhn, A., Mallatt, J. M., & Robinson, D. G. (2021). Anesthetics and plants: No pain, no brain, and therefore no consciousness. *Protoplasma*, 258(2), 239–248.

Frank, A., Gleiser, M., & Thompson, E. (2024). *The blind spot: Why science cannot ignore human experience*. The MIT Press.

Kalske, A., Shiojiri, K., Uesugi, A., Sakata, Y., Morrell, K., & Kessler, A. (2019). Insect herbivory selects for volatile-mediated plant-plant communication. *Current Biology*, *29*(18), 3128–3133.e3.

Kelz, M. B., & Mashour, G. A. (2019). The biology of general anesthesia from paramecium to primate. *Current Biology*, *29*(22), R1199–R1210. https://doi.org/10.1016/j.cub.2019.09.071. PMID: 31743680; PMCID: PMC6902878.

Kessler, A., Mueller, M. B., Kalske, A., & Chautá, A. (2023). Volatile-mediated plant-plant communication and higher-level ecological dynamics. *Current Biology*, *33*(11), R519–R529.

Mediano, P. A. M., Trewavas, A., & Calvo, P. (2021). Information and integration in plants: Towards a quantitative search for plant sentience. *Journal of Consciousness Studies*, *28*, 80–105; *Interdisciplinary Reviews: Cognitive Science*, *13*(2), e1578.

Schlanger, Z. (2024). *The light eaters: How the unseen world of plant intelligence offers a new understanding of life on earth*. Harper.

Segundo-Ortin, M., & Calvo, P. (2022). Consciousness and cognition in plants. Wiley *Interdisciplinary Reviews: Cognitive Science*, *13*(2), e1578.

Tompkins, P., & Bird, C. (1973). *The secret life of plants*. Harper and Row.

Wandersee, J. H., & Schussler, E. E. (1999). Preventing plant blindness. *The American Biology Teacher*, *61*, 82–86.

Yokawa, K., Kagenishi, T., Pavlovic, A., Gall, S., Weiland, M., Mancuso, S., & Baluška, F. (2018). Anaesthetics stop diverse plant organ movements, affect endocytic vesicle recycling and ROS homeostasis, and block action potentials in Venus flytraps. *Annals of Botany*, *122*(5), 747–756.

Alex and Irene Pepperberg—relationships can make good science

Gardner, R. A., & Gardner, B. T. (1969). Teaching sign language to a chimpanzee: A standardized system of gestures provides a means of two-way communication with a chimpanzee. *Science*, *165*(3894), 664–672.

Gold, K. C., & Watson, L. M. (2018). In memoriam: Koko, a remarkable gorilla. *American Journal of Primatology*, *80*(12), e22930.

Gutiérrez-Ibáñez, C., Iwaniuk, A. N., & Wylie, D. R. (2018). Parrots have evolved a primate-like telencephalic-midbrain-cerebellar circuit. *Scientific Reports*, *8*, 9960. https://doi.org/10.1038/s41598-018-28301-4

Pepperberg, I. M. (1981). Functional vocalizations by an African Grey parrot (Psittacus erithacus). *Zeitschrift für Tierpsychologie*, *55*(2), 139–160.

Pepperberg, I. M. (1983). Cognition in the African Grey parrot: Preliminary evidence for auditory/vocal comprehension of the class concept. *Animal Learning & Behavior*, *11*, 179–185.

Intelligence

Bräuer, J., Hanus, D., Pika, S., Gray, R., & Uomini, N. (2020). Old and new approaches to animal cognition: There is not "one cognition". *Journal of Intelligence*, *8*(3), 28. https://doi.org/10.3390/jintelligence8030028. PMID: 32630788; PMCID: PMC7555673.

Kilgour, R. (1981). Use of the Hebb-Williams closed-field test to study the learning ability of Jersey cows. *Animal Behaviour*, *29*(3), 850–860.

Mei, Q., Xie, Y., Yuan, W., & Jackson, M. O. (2024). A Turing test of whether AI chatbots are behaviorally similar to humans. *Proceedings of the National Academy of Sciences of the United States of America*, *121*(9), e2313925121.

Consciousness and welfare

Carbone, L. (2020). Do "prey species" hide their pain? Implications for ethical care and use of laboratory animals. *Journal of Applied Animal Ethics Research*, *2*(2), 216–236.

Consciousness and the egg

Underwood, W., & Anthony, R. (2020). AVMA guidelines for the euthanasia of animals: 2020 edition. *Retrieved on March, 2013*(30), 2020–2021.

Facial expression

Langford, D. J., Bailey, A. L., Chanda, M. L., Clarke, S. E., Drummond, T. E., Echols, S., Glick, S., Ingrao, J., Klassen-Ross, T., LaCroix-Fralish, M. L., & Matsumiya, L. (2010). Coding of facial expressions of pain in the laboratory mouse. *Nature Methods*, *7*(6), 447–449.

Mogil, J. S., Pang, D. S., Dutra, G. G. S., & Chambers, C. T. (2020). The development and use of facial grimace scales for pain measurement in animals. *Neuroscience & Biobehavioral Reviews*, *116*, 480–493.

Steagall, P. V., Monteiro, B. P., Marangoni, S., Moussa, M., & Sautié, M. (2023). Fully automated deep learning models with smartphone applicability for prediction of pain using the Feline Grimace Scale. *Scientific Reports*, *13*(1), 21584. https://doi.org/10.1038/s41598-023-49031-2

Anthropomorphism

de Waal, Frans B. M. (1999). Anthropomorphism and anthropodenial. *Philosophical Topics*, *27*(1), 255–280.

Lockwood, R. (1985). Anthropomorphism is not a four-letter word. In M. W. Fox & L. D. Mickley (Eds.), *Advances in animal welfare science* 1985/86 (pp. 185–199). The Humane Society of the United States.

Qualitative behavioral assessment

Phythian, C., Michalopoulou, E., Duncan, J., & Wemelsfelder, F. (2013). Inter-observer reliability of qualitative behavioural assessments of sheep. *Applied Animal Behaviour Science*, *144*(1), 73–79.

Wemelsfelder, F., Hunter, T. E., Mendl, M. T., & Lawrence, A. B. (2001). Assessing the 'whole animal': A free choice profiling approach. *Animal Behaviour*, *62*(2), 209–220.

Pessimistic bias

Bateson, M., Desire, S., Gartside, S. E., & Wright, G. A. (2011). Agitated honeybees exhibit pessimistic cognitive biases. *Current Biology*, *21*(12), 1070–1073.

Casey, R., Naj-Oleari, M., Campbell, S., & Mendl, M., & Blackwell, E. (2021). Dogs are more pessimistic if their owners use two or more aversive training methods. *Scientific Reports*, *11*(1), 19023. https://doi.org/10.1038/s41598-021-97743-0

Mendl, M., Burman, O. H., Parker, R. M., & Paul, E. S. (2009). Cognitive bias as an indicator of animal emotion and welfare: Emerging evidence and underlying mechanisms. *Applied Animal Behaviour Science*, *118*(3–4), 161–181.

Extreme maladaptive behaviors

Mason, G. (2006). *Stereotypic behaviour in captive animals: fundamentals and implications for welfare and beyond*. 10.1079/9780851990040.0325.

Part III

Evaluation

For many animals, there is safety in numbers. Welfare assessments for individuals can be complicated by their reticence to be separated from the herd. This is true for wild horses but can also manifest as intense separation anxiety in domesticated horses. We will see how this complicates welfare assessments in Chapter 10, one of several chapters that tackle practical aspects of animal welfare assessment. Herd of horses, vector illustration (#152915177) by Kvocek/Shutterstock.com.

DOI: 10.1201/9781003375456-11

Chapter 9

Animal Welfare Assessment

9.1. DEVELOPING AN ASSESSMENT

When tasked with assessing animal welfare in a particular setting, you need to select, adapt, or create an assessment tool. An *animal welfare assessment* is a structured method used to determine or estimate the welfare of an animal according to categories or a scale. Assessment systems are developed by a single expert, a team, or a consultative process. One approach to creating a robust assessment is to use a *Delphi survey*, which involves gathering diverse experts and facilitating their discussion until a consensus is reached (Zapf et al., 2023).

Botreau (2007) suggests that any assessment scale for farm animals should use the smallest number of easily understood and independent criteria necessary to be exhaustive and supported by stakeholders. Browning (2022) asserts that the metrics making up an assessment must be correct, valid, accurate, complete, reliable, and useful. In the next few chapters, many different assessment tools will be discussed. Those that are most effective do tend to follow these guidelines in being clear, simple, and sufficient. However, there is a hidden ingredient that goes into achieving this goal—your powers of observation (Figure 9.1).

Keywords: Animal welfare assessment, Delphi survey

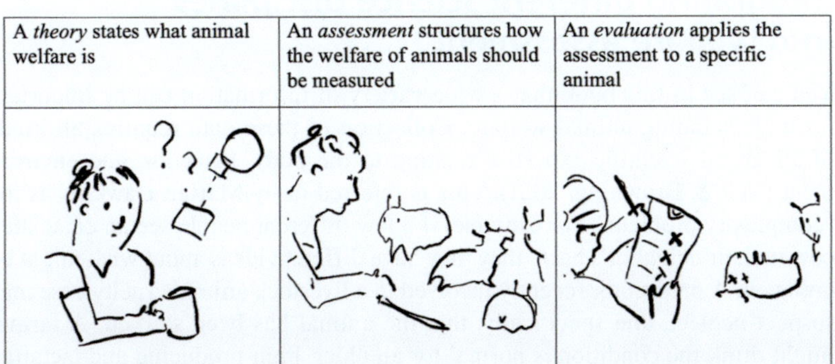

Figure 9.1 The stages of developing an animal welfare assessment. Image by Emily Patterson-Kane.

DOI: 10.1201/9781003375456-12

9.2 FIVE DOMAINS

One prominent example of an assessment method is the *Five Domains*. (Many people treat the Five Domains as "theory", but the intention was that it be used as a guide for how to carry out an assessment, so we describe it here.) The Five Domains model, first outlined in 1994 (Mellor & Reid, 1994), shows welfare as a subjective state (the mental domain) that is influenced by the domains of nutrition, environment, health, and behavior. This model evolved from the Five Freedoms framework but differs from it in several important ways. First, as mentioned already, the function of the Five Domains is to guide assessment in applied settings, not to serve as a theoretical framework. Second, the Five Domains was reformatted in 2015 to accommodate an assessment of both negative and positive contributors to welfare states, whereas the Five Freedoms was oriented toward negative experiences (Mellor & Beausoleil, 2015). The Five Domains model has been iteratively developed over many years by David Mellor and his colleagues (e.g., Mellor et al., 2020) as befits an applied "work in progress" rather than a strict theory/definition. Therefore, when reading articles about this model, it is important to note their year of publication so that you are up to date.

Finally, while the Five Freedoms includes consideration of subjective states as a part of welfare, an evaluation of the Five Domains comprises an audit of the three *internal* states of nutrition, environment, and health, as well as behaviors that are influenced by *external circumstances*. Each domain gives rise to positive or negative experiences (affects) that feed through and combine in the mental domain. See Mellor (2017) for a breakdown of how each domain can be assessed in different evaluation settings. You can imagine how being hungry, cold, or injured would contribute negative experiences to the mental domain. So, in this model, it is how the animal experiences life that directs welfare (Figure 9.2).

This structural relationship between the domains is absent from the Five Freedoms, which would allow interpretations such as a physical health deficit that impacts welfare without registering in the mental domain (e.g., an asymptomatic infection). As such, the Five Domains are designed to guide assessments on the ground and are aligned with welfare definitions such as this one offered by John Webster (2016): "The welfare of any sentient animal is determined by its individual perception of its own physical and emotional state".

Keywords: The Five Domains, David Mellor

Integrating different science disciplines into welfare assessment

You can see in this book that a wide variety of information can be important for understanding animal welfare. Collection of these data requires all kinds of advanced scientific expertise relating to the body, behavior, and environment (Veit & Browning, 2021). This is referred to by Marian Dawkins as the "complexity problem". For example, if a few different people see an emaciated cow in their neighbor's barn, they may have different ideas about what might be going on. A prosecutor recently involved in a livestock animal cruelty case may suspect neglect, and specifically, that the animal has been starved. A farmer might think the condition is normal for an older, high-producing and lactating

Figure 9.2 The Five Domains model illustrates how the domains of nutrition, health, environment, and behavior all influence the mental domain of a dog called Pai (meaning "good" in Māori). The captions are all obvious apart from his "being a walking trip hazard", which, given his robust good health, is the likeliest way that he will get injured. These images were collated from a collection of cartoons drawn by Emily Patterson-Kane about her own dogs.

Holstein milk cow. A veterinarian who noticed broken-down farm machinery near their animal might wonder if the cow has ingested a foreign body and is sick with a condition such as reticulitis, which caused them to stop eating and become ill.

So, based on their observations and experience, these individuals' initial impression of whether there is a welfare problem, and if so, what needs to be done, are sharply different. In each case, closer observation, collecting more information, and discussion should lead them to a more consistent point of view and agreement about what should happen next. For example, if the lactating cow has ample access to food, is quite young, has diarrhea, and is not producing much milk, then the veterinarian's point of view is likely the best one to follow. The affected animal would be treated, other animals assessed, and foreign objects removed from the enclosure.

In research settings, it is sometimes not clear what is going on, or what expertise is needed to solve an issue. Research questions may circulate between disciplines as different hypotheses are developed and the requisite information to test them is collated. At other times, groups of people from different scientific domains might collaborate on the same research project and successfully combine their skills—this is an example of a *pluralistic* approach (Figure 9.3). Openness to different ideas and perspectives can be particularly valuable when grappling with an unknown or complex cause, as different sources of expertise (and authority) may be needed that might not be obvious at first sight.

Keywords: pluralism (re: work approach)

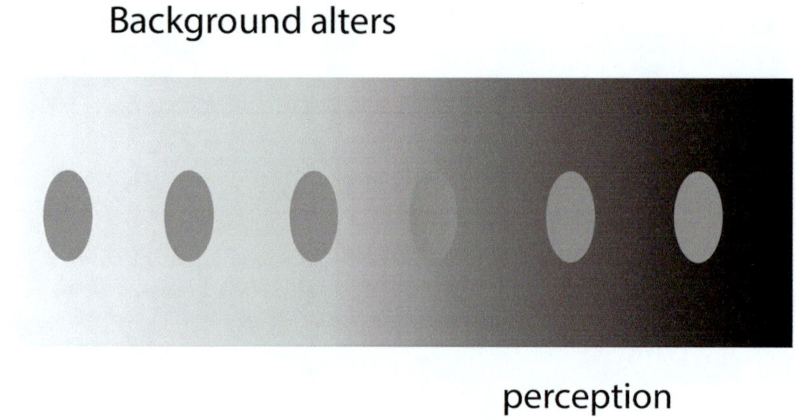

Figure 9.3 Background influences interpretation. In this optical illusion, each oval is exactly the same hue of gray; we just perceive them to be lighter or darker because of the background. Likewise, the backgrounds that we bring to our everyday work tasks can influence how we see things. Image by Tina Rich.

9.3 INDIVIDUAL ASSESSMENT

In this section, we focus on situations where we are managing individual animals. This tends to occur in settings where the animal is highly valued, such as a pet, a research primate, or a large exotic zoo animal. With these animals, there is often a desire, or regulatory pressure, to assess the welfare state of each individual animal and to track it on an ongoing basis.

Sometimes, welfare assessments are carried out as an investigation by an expert. This typically involves: 1) prior research by consulting the brief, relevant research, and discussions with important stakeholders; 2) information being gathered by observation of the animals, with examination and testing as required; 3) the expert opinion being provided as a narrative supported by evidence and analysis, with a conclusion, either verbally or in a written report. These investigations may or may not use a structured evaluation tool but will always go far beyond this approach to produce a critique and recommendations for welfare improvement.

Where repeated evaluations are required, a more structured assessment will usually be developed. Most individual assessments will then typically take the form of a *multi-item survey* with an observer using a tool with preset questions, each with a rating scale. There may be one or multiple questions for each domain being considered. As an illustrative example, one tool that has been developed for this is the *Animal Welfare Assessment Grid* (AWAG), which tracks data under the categories of Physical, Behavioral/Psychological, Environmental, and Procedural (e.g., how the animal deals with husbandry/handling, etc). This can be tracked with proprietary software made available at awag.org.uk. This product was initially applied to animals in research but has since been adapted for horses, cattle, dogs, and many kinds of exotic animals, such as those at zoological gardens (Malkani et al., 2022).

AWAG works by making a table for each of the domains. Along the top are key indicators as determined by experts familiar with the species. The cells below each indicator give descriptions of what this rating might look like in the target species. The animal's status for each individual indicator can be ranked from 1 (normal) to 10 (extremely compromised). This leads to a total score in each domain and an overall score. Any welfare deficit being experienced by the animal can be plotted as an area using these four scores and tracked over time. See https://awag.org.uk/. Other broadly comparable welfare tracking programs include ZooMonitor, https://zoomonitor.org/; the Chicago Zoological Gardens WelfareTrak https://www.welfaretrak.com/; and the Cetacean Welfare Assessment Process (C-Well), https://www.animalwelfareexpertise.com/the-cetacean-welfare-assessment, which is specifically for captive bottlenose dolphins. More specific schemas will continue to be developed, and they tend to follow the same basic model of identifying key indicators, rating each one, and possibly adding these values up to make sub-scores and an overall score.

Any scale is a combination of different items, usually covering several different domains. Other important factors might be:

- *Animal-based*, where some aspect of the animal is measured, or *engineering-based*, where you look at what is provided by their environment
- *Objective,* where they can be precisely measured, or *subjective*, where some degree of judgment is required by the rater
- *Observational*, where the animal is not disturbed, or a *procedural* observation, where the animal is responding to a normal husbandry activity or a special test

An assessor may choose to take a particular approach, for example using all animal-based measurements, or they may select a wide range of different types of measurements. This will be driven by 1) the assessor's understanding of animal welfare, 2) their understanding of what drives welfare for *that* animal, and 3) situational factors such as what is accessible to measurement. What unites most individual animal assessments is that they are a curation of multiple specific variables that can be measured. The goal is that this selection of measurements is sufficient to form a valid impression of that animal's welfare. The final evaluation will be completed based on this information, either intuitively or by adding together the sub-scores, possibly with weightings, into a specific number that is compared with a target range.

Just as an animal welfare theory aims to include every consideration that is truly important to understand welfare in an abstract or generalized sense, an assessment must include everything that is truly important to understand the welfare of that particular animal at that time and place. The main reason valid animal welfare tools should vary is because what is important for welfare will differ between individual-based factors (species), environments, and life stages. Welfare factors will also vary in more debatable ways depending on human priorities, such as what is necessary for an animal to produce product, breed successfully, look attractive, or be behaviorally compliant—i.e., whatever makes that animal fit for its role from our perspective.

Keyword: multi-item survey

9.4 QUALITY-OF-LIFE ASSESSMENT

Multi-item surveys of animal welfare can be grouped further into general types. One of these is the *quality-of-life* (QOL) assessment, which focuses on trying to assess how the animal is experiencing their life, from their point of view (Taylor & Mills, 2007). Quality of life assessments for animals were developed from tools initially created for people. In that setting, they would help a team of people such as medical professionals to understand how well a person is functioning in different aspects of their life, and the impact on their overall wellbeing. These assessments often look at the domains of physical, social, and psychological functioning. The overall concept of human quality of life is defined by the World Health Organization, https://www.who.int/tools/whoqol, as follows: "Quality of Life: an individual's perception of their position in life in the context of the culture and value systems in which they live and in relation to their goals, expectations, standards and concerns".

Animal quality of life assessments retain the QOL assessment approach of interrogating something about how an individual is experiencing their life overall. However, they have developed questions representing different categories drawn from animal welfare theories or definitions. Where the motivation is to assess whether a life is subjectively experienced mainly in suffering, or with a balance of positive and negative experiences, then a quality-of-life framework is the valid assessment to use. A QOL assessment is accompanied by a philosophy that people should understand the animal's experience of their life and reach a consensus about how to protect their interests. This may or may not be part of other types of animal welfare assessment, although philosophy-driven "subcultures" are quite common (e.g., organic or regenerative agriculture within the general activity of agriculture).

Keyword: quality of life assessment

Example: End-of-life QOL tools

One setting is assessing whether a companion animal has a life worth living when a hospice protocol or euthanasia is being contemplated. Several assessment tools are used and shared by veterinarians to help an animal guardian determine whether their companion animal has a life worth living. QOL is especially important to understand and communicate clearly when an animal is at the end of life. In a veterinary context, *end of life* is a period that begins when a life-limiting condition is deemed not curable. Given the extremely high potential for managing chronic conditions such as diabetes or feline immunodeficiency virus (FIV), this can be a very extended period where the animals' welfare is not significantly impacted by their condition.

When an animal is experiencing suffering or a significantly reduced quality of life associated with one or more life-limiting conditions, consideration should be given to entering a *palliative care* phase similar to human hospice, and where euthanasia may become an option. An end-of-life QOL assessment can help an animal caretaker carefully consider their animal's best interests and align their needs and expectations with their veterinary provider. When this period is poorly managed, it can lead to considerable distress for caretakers and veterinary professionals relating to premature or unnecessary euthanasia, on the one hand, or an animal experiencing protracted poorly controlled suffering, on the other. It is important to note that the assessment scale does not directly answer the question of what to do about that animal's care. However, working through the questions guides the family and the veterinarian to discuss the animal's quality of life and reach a consensus that serves the animal's interests.

There are several different scales available. They involve grading the animal according to a set of questions that relate to different aspects of their life. Using the same scale over time can help track an animal's deterioration or improvement. Links to some of the more popular scales are provided in Table 9.1.

Table 9.1 QOL assessment instruments

There are many different quality-of-life assessments intended to help manage a companion animal's end-of-life period.			
Assessment Scale	Link	Format	Domains
HHHHHMM (or H5M2) Quality of Life Scale	The HHHHHMM (H5M2) Quality of Life Scale – Sentience Research (https://sentience-research.org/the-hhhhhmm-h5m2-quality-of-life-scale/)	1 Domain 7 questions 0–10 scale	Hurt Hunger Hydration Hygiene Happiness Mobility More good days than bad
Lap of Love Quality of Life Scale	lap-of-love-quality-of-life-scale.pdf(https://www.lapoflove.com/how-will-i-know-it-is-time/lap-of-love-quality-of-life-scale.pdf)	4 Domains 16 questions 0–2 scale	Social Functions Health Mental Health Natural Functions
OSU (Ohio State University) Quality of Life Scale	https://vet.osu.edu/vmc/sites/default/files/import/assets/pdf/hospital/companionAnimals/HonoringtheBond/HowDoIKnowWhen.pdf	1 Domain 25 questions 1–5 scale	Quality of Life

Quality-of-life scales are not exclusively used for end-of-life care but can be adapted for any stage of life. There are also QOL scales that relate specifically to tracking and management of a particular disorder, especially those associated with chronic pain, such as cancers, arthritis, or skin conditions. The important thing is that the scale supports a conversation about how the animal experiences their own life. This can help track their condition, assess different treatments, create a supportive environment, and make any other adjustments to support good welfare. Many scales are focused on dogs and cats, and tools for other species are limited.

Keywords: end-of-life, palliative care (hospice)

9.5 OTHER APPROACHES

While I have focused most of this chapter on the current most common approach to repeated assessment, there is really no limit as to how an assessment might be performed. For this reason, I am giving a few representative examples of how assessments might differ based on their philosophical purpose or function. The examples given below include the use of free reports, single scores, and single-domain assessments.

Free report: Even the best assessment tool may fail to include important factors. For this reason, a flexible approach that facilitates reporting from people who observe the animals may be encouraged, perhaps verbally or with free text reports: for example, a "lab book" in an animal research facility where anyone present can note down unusual events, such as construction noise, a rat with a rough-looking coat, or a hen that didn't lay an egg that day. These books can help track subclinical findings to see if they persist and may also explain unusual variations in the data. (*Subclinical* refers to signalment [an indicator or description] that is not clear or serious but could possibly relate to a disease or disorder.)

Free reports can capture considerations that are new, unexpected, and only accessible to certain observers (e.g., those highly familiar with the animal or at the facility at unusual hours). It is not uncommon for investigations into an animal's welfare to move between deep qualitative understandings of the animal and more structured assessment tools that can be iteratively modified. For example, if an animal started showing stereotypical behaviors, this might not be part of the existing assessment. But if the cause cannot be immediately identified and remediated, it should be added to the assessment as a new indicator of impaired welfare.

Single score assessments: To make frequently repeated assessments easier, sometimes a single score may be used. This might be having an experienced caregiver like a pet guardian or zookeeper report each day whether welfare seems good or compromised, or even just whether the animal had a "good day" or a "bad day". A more objective approach is to track a measurement that is quite sensitive to an animal's overall welfare. Qualitative behavioral assessment (Chapter 8, Part II) can be repeated in a more simplified form with pre-selected terms. This creates a good single score assessment because the overall way an animal is behaving will tend to reflect how good or bad their welfare is, even if it is not specific as to what might be upsetting welfare. Another example is how some fish show color fading or body striping when they are sick or stressed, and so these changes can be a general indication that they are under pressure. Other possible indicators of poor animal welfare might include a reduction in behavioral diversity (Miller et al., 2020), vigilance behavior, huddling, hiding, ungroomed appearance, or a loss of appetite.

One single behavior thought to be broadly indicative of welfare is nest building in mice (Jirkof, 2014). Nest building is a species-specific behavior that indicates a level of physical fitness and motivated activity. Once created, the nest provides preferred conditions such as darkness and warmth, and so, opportunities to control their own environment. As such, nest building can be seen as intersecting with many different aspects of welfare. Failure to nest build is associated with many other indicators of poor welfare. And if nest material is provided, it is extremely easy to visually check whether a nest has been built at all, and if so, how well it was built. So, while it does not replace a comprehensive animal welfare assessment, it does provide a very easy tracking system that will detect many threats to welfare.

Another much more obvious single score assessment is death, or other very severe outcome such as serious injury or immobilization. Even if you do not know the cause, these outcomes are unambiguous indicators of the presence of a serious threat to welfare. There is a developing understanding that evidence of catastrophic subjective suffering, even without organic causes, should be given similar consideration: for example, complete panic, disorientation (e.g., collapse, confusion, dissociation, depression), or *psychological trauma* (where psychological trauma is defined as a long-term debilitating response to a negative experience potentially including conditioned reactions, volatile emotions, and poor physical health).

Focused assessments: Some assessments may be focused on a single domain but still be considered a welfare assessment because that domain is assumed to be the most pressing issue for welfare at the time. For example, after an operation, pain might be assumed to be the most important threat to welfare. A tool that focuses tightly on indicators of wound healing, sickness behavior, and pain indicators might still be considered a welfare test despite its narrow focus.

Similarly, with a research animal under end-of-life monitoring, their assessment will focus on forms of suffering associated with the expected cause of death. There is an expectation that even in a research setting, where a disease or toxin is expected to cause death, *humane endpoints* should be used. A humane endpoint is the requirement to kill an animal who is suffering at the point where their death is imminent. At the point where death can be reasonably predicted, the suffering of the animal is not necessary to draw a conclusion about the lethal outcome.

Another example of a risk-based focus is the welfare assessment used to ensure that a horse can continue in a competition. The rules for this type of assessment focus on the welfare risks most prevalent for the event. For example, what is colloquially referred to as "the blood rule" means that a competitor is disqualified for the use of tools like the bit, spurs, or whip in a way that draws blood on the horse and is important in show jumping and dressage. In contrast, long-distance races focus more on the overall evidence of the horse being physically exhausted.

Keywords: humane endpoint, subclinical, psychological trauma

The Anna Karenina Principle

An oft-quoted passage from the novel *Anna Karenina* (published in installments from 1875 to 1877) by Leo Tolstoy is: "All happy families are alike; each unhappy family is unhappy in its own way." This statement has been used as a metaphor by diverse scientists, including ecologist Dwayne Moore and scientist-historian Jared Diamond.

It is taken as a tidy way of expressing that if any of the many parts of a complex situation is bad, the whole situation is bad. As applied to animal welfare, it means that if any aspect of welfare is extremely bad, this will tend to compromise the animals' overall welfare. Therefore, one of the main ways that a welfare assessment will fail is by not assessing the one thing that is, or becomes, extremely deficient. And the things that go wrong could be anything impacting the animal: inherited disorders, gut microbiome, conflict with a conspecific, acquired allergy to a food … and so much more.

For this reason, no animal welfare assessment should be considered completely sufficient or eternal. Over time, animals change, environments change, even things that are meant to be uniform like feed ingredients will change without you even knowing (because the crop growing conditions change or the formulations change, or something goes wrong in storage, processing, or transport). For these reasons, one of the implicit roles of an animal welfare scientist is to be vigilant—watch out for any indications that something is going wrong, and be willing to be tenacious in tracking down the underlying cause, no matter what it is.

Keyword: Anna Karenina principle

READINGS

Developing an assessment

Botreau, R., Veissier, I., Butterworth, A., Bracke, M. B., & Keeling, L. J. (2007). Definition of criteria for overall assessment of animal welfare. *Animal Welfare, 16*(2), 225–228.

Browning, H. (2022). Assessing measures of animal welfare. *Biology & Philosophy, 37*(4), 36.

Zapf, R., Schultheiß, U., Schubbert, A., Gieseke, D., Cimer, K., March, S., Brinkmann, J., & Knierim, U. (2023). Evaluation of animal welfare in on-farm self-assessments–development of a reference framework with target and alarm values based on a Delphi survey. *LANDTECHNIK, 78*(3), 125–138.

Five domains

Mellor, D. J. (2017). Operational details of the five domains model and its key applications to the assessment and management of animal welfare. *Animals: An Open Access Journal from MDPI, 7*(8), 60. https://doi.org/10.3390/ani7080060

Mellor, D. J., & Beausoleil, N. J. (2015). Extending the 'five domains' model for animal welfare assessment to incorporate positive welfare states. *Animal Welfare, 24*(3), 241.

Mellor, D. J., Beausoleil, N. J., Littlewood, K. E., McLean, A. N., McGreevy, P. D., Jones, B., & Wilkins, C. (2020). The 2020 five domains model: Including human–animal interactions in assessments of animal welfare. *Animals, 10*(10), 1870.

Mellor, D. J., & Reid, C. S. W. (1994). Concepts of animal well-being and predicting the impact of procedures on experimental animals. In R. M. Baker, G. Jenkin, & D. J. Mellor (Eds.), *Improving the well-being of animals in the research environment*. Australian and New Zealand Council for the Care of Animals in Research and Teaching (pp. 3–18) (ANZCCART), Glen Osmond, Australia.

Webster, J. (2016). Animal welfare: Freedoms, dominions and "a life worth living". *Animals, 6*(6), 35.

Integrating different science disciplines into welfare assessment

Veit, W., & Browning, H. (2021). Perspectival pluralism for animal welfare. *European Journal for Philosophy of Science, 11*, 1–14.

Individual assessment

Malkani, R., Paramasivam, S., & Wolfensohn, S. (2022). Preliminary validation of a novel tool to assess dog welfare: The Animal Welfare assessment grid. *Frontiers in Veterinary Science, 9*, 940017.

Quality of life assessment

Taylor, K. D., & Mills, D. S. (2007). Is quality of life a useful concept for companion animals? *Animal Welfare*, *16*(S1), 55–65.

Other approaches

Jirkof, P. (2014). Burrowing and nest building behavior as indicators of well-being in mice. *Journal of Neuroscience Methods*, *234*, 139–146.

Miller, L. J., Vicino, G. A., Sheftel, J., & Lauderdale, L. K. (2020). Behavioral diversity as a potential indicator of positive animal welfare. *Animals*, *10*(7), 1211.

Chapter 10

Assessment—Group

10.1 ASSESSING WELFARE AT THE HERD OR GROUP LEVEL

Within any group, some individuals will tend to have poorer experiences than others. Therefore, a group assessment can be used to show some kind of average, variability, and whether any of the animals are falling below an acceptable standard. When deciding what to include in a group assessment, it is a good idea to diagram important events that drive or hinder good welfare over the scope of the assessment or situation (e.g., time periods/life stages and locations).

Be sure to include consideration of the following and interactions between them:

- Events that impact all animals, or only some
- Events that will definitely occur and those that might foreseeably occur
- Events that will have mild effects and those that would have severe effects
- Events that impair welfare, are good for welfare, or protect against harm

Keeping this in mind, select items to measure that give broad coverage but ensure consideration of critical and highly probable threats to welfare, as well as protective factors.

Visibility stress

When assessing a large group, a common problem is that you can't see the animals. There is an impulse to do things that make the animals more visible, so that we can feel sure that they are safe and sound. However, these actions may be a source of harm to them.

A clear example of this came from the human realm in considering how to care for prematurely born infants. For some time, these infants were routinely kept in brightly lit environments with quite loud settings and frequent blood testing (heel sticks). However, consideration was then given to what you might call the

DOI: 10.1201/9781003375456-13

"natural history" of what would optimally still be a fetus, that is, being in the womb. Light and sound stresses were discovered to relate to worse outcomes for the brain, body, and emotional development, compounded by preterm babies often having other vulnerabilities. So, modern approaches are very mindful about providing darker and quieter settings with more judicial and less stressful testing.

A similar case from an animal setting is the housing of captive dolphins (Clegg et al., 2017). It had been assumed that the best conditions had to be with very clean, clear water and open enclosures, with the benefit that the dolphins could be easily seen and monitored. However captive dolphins cannot dive very deep, and so unobstructed water clarity can expose them to sunburn and even provoke blindness. Nor can captive dolphins choose whom to associate with, with the result that social conflict may arise. Finally, the *microbiome* (population of microorganisms in a location) on and inside a cetacean's body interacts with the air that they breathe (its quality/filtered) and water, so that, for example, overly sterile water (artificial seawater) might not support a healthy microbiome (Cardona et al, 2018).

Some animals are stressed or challenged simply by being looked at. This might include those who are subject to predation, naturally live in dark or enclosed settings, or defend their group or territory. One initiative to address this is the use of camouflage nets in front of the viewing public (Blaney et al., 2004). This allows visitors to see, for example, gorillas, through gaps in the net but makes the human visitors harder for the gorillas to see, reducing their stress.

In summary, while we may want to have as much information as possible about any group of animals whose welfare we are monitoring, making them easy to see, and over-testing them, can be counterproductive. Human beings tend to address a problem by doing something, when sometimes the best approach is precisely the opposite. Always carefully consider that doing nothing might be the best available option for the wellbeing of the animal.

Keyword: microbiome

10.2 PROBLEMS WITH REMOVING AN ANIMAL FROM A GROUP TO ASSESS IT

Welfare measurements of social animals are often performed while the animal is in the group, without removing them for separate testing. This may be criticized, as you are not getting a "pure" assessment of that animal. However, social animals often perform poorly when separated from their group, especially if this is not common for them (Lundmark et al., 2015). As we've already discussed, Ron Kilgour took an early interest in assessing the learning ability of various animals, especially those used in agriculture. Kilgour found that many animals proved willing to run a particular maze (Hebb–Williams Maze) designed to test the intelligence or cognitive ability of rodents, as long as it was sized appropriately.

Sheep were something of an exception, which might initially have supported the general notion that sheep are not very intelligent. However, follow-up studies showed that sheep can learn rapidly but will not display good learning when it is contrary to good flocking behavior. So, to reveal the sheep's true ability in running complex mazes, it was necessary to test them in small groups.

Several options can be taken if a large proportion of animals need to have their welfare assessed in a group context. These are:

> *Group-level measures without individual-level implications:* There are some measures of compromised welfare that are impossible to attribute to any particular individual: for example, alarm chemicals in the water. As early as 1938, ethologist Karl von Frisch showed that a chemical alarm signal was emitted by minnows, which he called *Schreckstoff* (fright substance; Stensmyr, 2024). Chemicals are released by fish that have been injured, so their detection indicates that at least one fish in that body of water recently experienced poor welfare. This kind of assessment can tell you when there might be a reason to investigate further. Other examples of this type might include numbers or volume of vocalizations, or blood in the environment.
>
> *Measuring individuals to assess groups:* It is common to use measurements taken from a representative sample of individuals to draw conclusions about an entire group. So, in this case, you know which animals have indicators, like scratches on their skin. But if you want to be able to determine what proportion of a group have these injuries, you need to assess a sample that is large enough and representative enough to generalize to the population. You also need some basis for saying what proportion with scratches might be considered acceptable (as, for example, injuries from normal exploration or play). It would be optimal that animals showing an undesirable assessment, such as severe lacerations, could be located and brought up to the acceptable level through veterinary care. But again, in high volumes and in inaccessible populations, these data may be treated more as an indicator that corrections need to be made rather than attending to the needs of individuals.

The most common type of group assessment carried out in the United States relates to agricultural species such as cattle, sheep, and poultry kept together in herds or flocks, and fish in schools. An idea of the welfare of the group can be arrived at by carrying out assessments of all or some of the animals. Depending on group size, this assessment may need to be relatively simple, based on a small number of items that can be recorded by an observer with a limited amount of training. Ordinarily, the questions can be answered as present or absent or by using other fixed options as described in the previous chapter.

Keyword: fright substance (Schreckstoff)

How to get a representative sample

There are different ways to choose a sample from a larger population where the goal is to generalize the result to the entire population. A *representative sample* is one where every individual has an equal chance of being selected, and so measurement of them can be used to estimate a characteristic in the wider population with a known level of certainty. One method for doing this is random sampling. In this case, you can determine how many animals to test based on how big the population is, your margin of error (the range within which the true number will be), and your confidence level (how confident you can be that the number is within that range). You can determine the sample size you will need using a dedicated calculator hosted at https://www.calculator.net/sample-size-calculator.html.

However, a key element of the exercise is that the animals to be assessed must be chosen randomly; for example, by assigning each animal a number and then randomly selecting numbers. This is not too difficult if you want to run an animal welfare assessment of a representative number of rats in a research colony, as each rat will already have a unique identifier and a known accessible location. What if you want to assess a population of sheep out on rangeland or fish in a large sea net (Torgerson-White & Sánchez-Suárez, 2022; Wisenden et al., 2004), or a completely wild population? You might think that you could sample the animals based on when you happen to see them, or at any event when you might collect them together, such as during round ups or musters. However, neither of these are random in nature, as the likelihood and order of seeing animals at these events is affected by their location, heath, behavior, and even size or color.

If you use the muster to examine animals, then identifying them by a unique identifier like an ear tag is a way around the problem. But if the animals do not have unique identifiers, it is better to choose a different method of sampling, such as *quota sampling*, where you identify important subtypes in the population and ensure that you assess examples of each subtype; for example, individuals of each sex, various ages, sizes, etc. However, you may still be capturing biased observations within each of these categories.

Sometimes, neither of these methods for representative sampling is possible. For example, when you are observing grizzly bears in the wild, it might be difficult to know how many bears there are, which ones you are seeing, or even if you are seeing the same bear repeatedly. This does not mean that your observations are without value. Instead, rather than quantitatively generalizing your findings, you should focus on why certain animals are showing signs of good or poor welfare, and what might be driving the welfare experiences of animals in that group. For example, consider these two studies that assessed how grizzly bears are negatively affected by roads (Proctor et al., 2020; Bourbonnais et al., 2014). Review their methodologies and see how the conditions of different bears were associated with their proximity to different features in the landscape.

Keyword: representative sample

10.3 ASSESSMENT AT SLAUGHTER

Animals raised for meat often experience the greatest challenges to their welfare during transportation and slaughter, and many slaughterhouses assess welfare indicators of animals moving though their facilities. In the United States, Temple Grandin has been a highly influential advocate in improving facility design and handling prior to and during slaughter, and all through the remarkable prism of her neurodiversity (Grandin, 2006; Grandin & Johnson, 2006). This included setting key metrics that slaughterhouses should monitor, and values that they should stay within. Grandin identified a relatively small number of assessment metrics to indicate how humane the overall process is, with each animal being scored positive or negative on each metric. The assessment is simple enough to include every animal going through the system.

Variables that should be evaluated to assess animal welfare during handling, stunning, and slaughter included: stunning efficacy, insensibility of stunned animals hanging on the bleeding rail, vocalization, electric prod use, slipping, and falling down. For a full description, see Grandin (1998). Audits of slaughterhouses showed that deficient performance was often associated with simple failures, such as not properly maintaining the equipment used for stunning and not installing non-slip flooring in stun boxes. A common recommendation was also to make fairly trivial changes in the facility to encourage the forward motion of the animals. There was also a persistent need for personnel training that has not abated. In the early 1970s, the meatpacking industry had a low turnover rate, but with union busting and slashed salaries and benefits, the turnover rate skyrocketed (Hansen, 2023). That, combined with deregulation, has had knock-on effects for safety, both for the workers and for animal welfare. Of course, the poor working conditions came to the fore during the COVID-19 pandemic, when cramped conditions in meatpacking plants led them to become major viral hotspots.

Grandin particularly emphasized that some outcomes have *no* acceptable level and should never happen. So, while these assessments are conducted at the group level, the goal is that no individual has an experience indicating an unacceptable degree of suffering, such as falling to the ground or being hoisted by the leg while still conscious.

Keyword: Temple Grandin

10.4 AUTOMATED ASSESSMENT

Most agricultural production in the United States takes place on farms that are large and remote from population centers. In these cases, there may be limited contact between animals and caretakers, either because the animals are scattered over a large range or because very large numbers are kept in more concentrated settings. This has caused considerable interest in the use of automated systems to collect information about animals for various purposes, including welfare assessment (Winckler, 2019; Buller et al., 2020). One important application of automation is to detect early signs of potential problems, which buys time to intervene, protects the welfare of individual animals, and detects emerging threats to the welfare of the whole group.

For example, GPS ear tags can alert farmers that cattle might have strayed off the property, have slowed down, or have stopped moving altogether, any of which indicates that help needs to be directed to the last reported location of the animal. This kind of real time tracking of animals is sometimes referred to as *precision livestock farming* (PLF). Precision livestock farming is when individual livestock animals are constantly monitored in real time for factors relevant to their health, welfare, productivity, or impact on the environment.

In a more intensive setting, automated systems for feeding and milking allow animals to use them as needed and can detect changes such as loss of appetite or increased cell counts in milk (that might signify developing illness). These can be combined with grimace scale information (Part II) and facial recognition technology, with cameras used to scan groups of animals for expressions of discomfort (Hansen et al., 2021). Given that many vocalizations also carry emotional content, automated bioacoustics has been discussed as an area open to development (Mcloughlin et al., 2019). Vocalizations can have very specific meanings, such as rat "laughter" associated with play, or the bleat of a separated lamb and ewe.

However, the use of automation can also have net negative effects on animals. Real-time monitoring of animals is a powerful tool, but it is based primarily on easily quantified aspects of welfare. Over-reliance on these systems might give users an oversimplified view of animal welfare. Also, while these technologies represent ways for motivated caregivers to be more informed and attentive, they can also allow producers to become more detached and technology reliant, and have a false sense of security about their animals' welfare. There are other dangers associated with using automation as a replacement for routine monitoring like riding the herd and walking the pens. Livestock unused to people may become more frightened when they are later transported and handled, which increases stress levels and the risks of injuries to everyone concerned.

Keyword: precision livestock farming

10.5 FRAME SHIFTING WELFARE ASSESSMENTS

Purchasers of beef products increasingly require some level of animal welfare assessment or assurance. However, their methods for understanding the welfare of animals are often biased toward familiar farming styles rather than understanding the welfare of the animal in a context-agnostic way. For example, the UC Davis Cattle Assessment is aimed at cow/calf ranches that use chutes (https://www.ucdcowcalfassessment.com/). Bayvel et al. (2010) contrasted the intensive production systems in Europe, Asia, and North America with "extensive systems of production … in New Zealand, Australia, South America, and Africa for grazing species", noting that the "public perception [was] that more extensive management systems are synonymous with better welfare".

In the United States, welfare assessment focuses on whether the cattle are injured, thin, or stumble when moved, and if electric cattle prods are used. By contrast, in New Zealand, the assessment was adapted to a fully pastoral system, without feedlots, in a temperate climate. This is semi-intensive in that pasture quality is carefully managed, cattle are contained, and their diet supplemented. The New Zealand assessment looks at nutrition, access to pasture shade, and stockpersonship. In essence, the focus is on items at risk of being a problem in *that* setting.

In Namibia, the assessment needed further adaptation. First, the farming is both extensive and in an arid climate. Cattle breeds in these areas are adapted to hot and dry conditions that breeds from more temperate climates find challenging. Therefore, engineering and environmental criteria derived from temperate settings would be misleading. On the other hand, periodic droughts represent a significant and predictable threat to animal welfare.

Second, beef from animals in northern parts of Namibia is not usually exported. Instead, animals are raised in a communal manner for use by the family and neighbors. Some of these animals will not be confined by fences, and so there is an increased risk of them ingesting poisonous plants or being subject to predation. Horns are typically left on, which may assist with fending off predators but leads to cattle injuring each other. Handling is also impaired by less well-developed facilities, but there are no electric prods used. Overall, these comparisons show that the different settings vary in their benefits and risks to the welfare of animals.

In order to fairly compare the performance of a particular farm, an assessment would need to accurately identify the risks to the animals and how they can be measured. The varied range conditions that we've just outlined can also be applied back to the United States. The North American continent is very diverse, including sub-arctic, semi-arid, and subtropical zones, and farming occurs across these areas on different scales and with different cultural and economic conditions. There are also genetically distinct types of cattle in the United States with different preferences and tolerances in terms of nutrition and climate. For example, *American Brahman cattle* have been bred in the United States since the mid-19th century for their ability to tolerate heat and insects. They descend from south Asian Zebu cattle, and most retain the distinctive humped back and loose skin. Therefore, as initial welfare systems were developed in small and less diverse territories (e.g., the United Kingdom, the Netherlands, and New Zealand), extensive adaptation may be required to accurately understand and fairly assess welfare concerns on farms within the United States and economically linked areas such as Canada and Mexico.

Keyword: Brahman cattle (American Brahman)

READINGS

Visibility stress

Blaney, E. C., & Wells, D. L. (2004). The influence of a camouflage net barrier on the behaviour, welfare and public perceptions of zoo-housed gorillas. *Animal Welfare*, *13*(2), 111–118.

Cardona, C., Lax, S., Larsen, P., Stephens, B., Hampton-Marcell, J., Edwardson, C. F., Henry, C., Van Bonn, B., & Gilbert, J. A. (2018). Environmental sources of bacteria differentially influence host-associated microbial dynamics. *mSystems*, *3*(3), e00052-18. https://doi.org/10.1128/mSystems.00052-18

Clegg, I. L. K., Van Elk, C. E., & Delfour, F. (2017). Applying welfare science to bottlenose dolphins (Tursiops truncatus). *Animal Welfare*, *26*(2), 165–176.

Problems with removing an animal from a group to assess it

Lundmark, F., Berg, C., Wahlberg, B., & Röcklinsberg, H. (2015). 'One animal is no animal'–consequences of measuring animal welfare at herd level. In D. E. Dumitras, I. M. Jitea, & S. Aerts (Eds.), *Know your food: Food ethics and innovation* (pp. 254–271). Wageningen Academic Publishers.

Stensmyr M. C. (2024). Schreckstoff: It takes two to panic. *Current Biology: CB*, *34*(7), R278–R281. https://doi.org/10.1016/j.cub.2024.02.069

How to get a representative sample

Bourbonnais, M. L., Nelson, T. A., Cattet, M. R., Darimont, C. T., Stenhouse, G. B., & Janz, D. M. (2014). Environmental factors and habitat use influence body condition of individuals in a species at risk, the grizzly bear. *Conservation Physiology*, *2*(1), cou043.

Proctor, M. F., McLellan, B. N., Stenhouse, G. B., Mowat, G., Lamb, C. T., & Boyce, M. S. (2020). Effects of roads and motorized human access on grizzly bear populations in British Columbia and Alberta, Canada. *Ursus, 2019*(30e2), 1639.–

Torgerson-White, L., & Sánchez-Suárez, W. (2022). Looking beyond the Shoal: fish welfare as an individual attribute. *Animals*, *12*(19), 2592.

Wisenden, B. D., Vollbrecht, K. A., & Brown, J. L. (2004). Is there a fish alarm cue? Affirming evidence from a wild study. *Animal Behaviour*, *67*(1), 59–67.

Assessment at slaughter

Grandin, T. (1998). Objective scoring of animal handling and stunning practices at slaughter plants. *Journal-American Veterinary Medical Association*, *212*, 3640.

Grandin, T. (2006). Progress and challenges in animal handling and slaughter in the US. *Applied Animal Behaviour Science*, *100*(1–2), 129–139.

Grandin, T., & Johnson, C. (2006). *Animals in translation: Using the mysteries of autism to decode animal behavior.* Houghton Mifflin Harcourt.

Hansen, R. (2023). 'What we eat I: The rise and fall of meatpacking unions', *War, Work, and Want: How the OPEC Oil Crisis Caused Mass Migration and Revolution* (New York, 2023; online edn, Oxford Academic, 24 Aug. 2023). https://doi.org/10.1093/oso/9780197657690.003.0015

Automated assessment

Buller, H., Blokhuis, H., Lokhorst, K., Silberberg, M., & Veissier, I. (2020). Animal welfare management in a digital world. *Animals*, *10*(10), 1779.

Hansen, M. F., Baxter, E. M., Rutherford, K. M., Futro, A., Smith, M. L., & Smith, L. N. (2021). Towards facial expression recognition for on-farm welfare assessment in pigs. *Agriculture*, *11*(9), 847.

Mcloughlin, M. P., Stewart, R., & McElligott, A. G. (2019). Automated bioacoustics: Methods in ecology and conservation and their potential for animal welfare monitoring. *Journal of the Royal Society Interface*, *16*(155), 20190225.

Winckler, C. (2019). Assessing animal welfare at the farm level: Do we care sufficiently about the individual? *Animal Welfare, 28*(1), 77–82.

Frame shifting welfare assessments

Bayvel, A. C. D., & Cross, N. (2010). Animal welfare: A complex domestic and international public-policy issue—who are the key players? *Journal of Veterinary Medical Education*, *37*(1), 3–12.

UC Davis University of California. Davis Cow-Calf Health and Handling Assessment. Retrieved February 16, 2023, from https://www. ucdcowcalfassessment.com/

Chapter 11

Implications for Conservation

11.1 WILDLIFE CONSERVATION AND ANIMAL WELFARE

For a long time, the paradigm of animal welfare science was rarely applied to free-roaming wildlife. This is because wildlife species are often managed as populations, with *conservation* referring to protecting or restoring wildlife and their environments, including the prevention of species extinction. In each case, the focus was on having the right species in the right place and in the right numbers, rather than on their wellbeing.

Occasionally, species preservation may clash with animal suffering. Common examples of this would include killing animals that are seen as invasive pests or threats to endangered species: for example, rats that predate on birds and their nests, or the cane toad, which outcompetes native amphibians (Figure 11.1). In an unbalanced ecosystem, even endemic species can become harmfully over-abundant. Animal welfare is also impacted by "control" practices such as round ups, netting, tracking, marking, reproductive control, or attempts to avoid animal–human conflict.

Most wildlife will be stressed by any period of restraint or captivity, but the true lethality of these events only became evident with post-capture monitoring. One known effect is called *capture myopathy*, in which severe and prolonged stress results in muscle damage, including the heart (Paterson, 2014). Capture myopathy can be fatal, meaning that capture for research or relocation should only be undertaken when absolutely necessary and with mitigating measures, especially for the most susceptible species, such as terrestrial ungulates (Breed et al., 2019).

Since the 1980s, there have been active efforts to acknowledge animal welfare concerns in conservation practices. Increased regulation of predator eradication activities (e.g., fox hunts) and the development of more moderate and targeted

DOI: 10.1201/9781003375456-14

Figure 11.1 Local crows in Australia carefully handle cane toads (*Bufo marinus*) to avoid poison-secreting glands at the shoulders. These toads were introduced in 1935 to control beetles that infested sugarcane. The cane toads failed in pest control (Shine et al., 2020) but did enjoy their new environment and are busy populating the continent. Photograph reproduced with the permission of Janet Patterson-Kane.

methods of control are evidence of this. Leg hold traps could maim or cause the death of any animal passing by, and so these have been modified with softer edges to allow reliable and less harmful capture of the target species (Muth et al., 2006). Poisons applied to sheep collars to target coyotes that attack livestock by biting the throat are another effort to reduce harm to non-target species and individuals (Savarie & Sterner, 1979).

Efforts have also been made to develop guidelines for ethical animal control when coexistence breaks down, such as those workshopped by Sara Dubois and her colleagues (Dubois et al., 2017). The principles that they outlined were: "modify human practices when possible, justify the need for control, have clear and achievable outcome-based objectives, cause the least harm to animals, consider community values and scientific information, include long-term systematic management, and base control on specifics of the situation".

There are two main efforts to reconcile animal conservation and animal welfare. Around 2010, there were moves to create a synthesis of mainstream conservation and animal welfare as two allied mission-driven multidisciplinary fields. A field that includes both has been suggested, which might be called *Conservation Welfare*. Their core missions, to preserve the wilderness and prevent animal suffering, might not always completely align, but where differences emerge, a utilitarian framework could be used to inform ethical decision making.

An alternative approach from the perspective of critical animal studies (animal rights) is called *compassionate conservation.* A key player in the field, Marc Bekoff, edited a collection of essays in support of the idea titled *Ignoring nature no more: The case for compassionate conservation* (Bekoff, 2013). The central ideas are to do no harm, that individuals matter, that we should treat all animals, regardless of species or status, with empathy and compassion, and that we should aim for peaceful coexistence. Interestingly, a recent publication found no difference in the ethics, morality, or motivations between traditional and compassionate conservationists, noting that the main area of difference was in the effectiveness of wildlife management practice in minimizing animal harm (Bobier & Allen, 2022).

Keywords: compassionate conservation, capture myopathy

11.2 PERSPECTIVES FROM OTHER KNOWLEDGE TRADITIONS

In the last decade, there have been increased efforts to recognize the value of *Indigenous Traditional Ecological Knowledge* (ITEK) in federal scientific and policy circles. ITEK has various names, but for simplicity's sake, I will use this term. There are also numerous ITEKs, as knowledge is place and people specific. However, broadly speaking, ITEK is a holistic cosmology (worldview) based on a profound knowledge of place, culture, all living things, and ancestors. That knowledge is acquired over many generations and supports responsible land stewardship. The elevation of ITEK may have been accelerated by notable blunders in fire control policy (where Indigenous leaders have tremendous knowledge around prescribed burns), issues of climate change, and the appointment of Deb Haaland, a member of Pueblo of Laguna, as Interior Secretary for the Biden-Harris administration. Haaland was the first Native American to serve as a Cabinet Secretary. Fast forward to the Trump-Vance administration, and we see a refusal to recognize certain Indigenous place names combined with a desire to open up Native lands to fossil fuel extraction. These actions suggest that any enlightened approach to ITEK may now be under threat.

We also see a recognition of how different perspectives on land and animal management have been exacerbated by clumsy language or an insufficient appreciation of local cultures. For example, use of the word "harvest" when killing animals is insensitive, as it suggests dominion, ownership, and entitlement (Goldman et al., 2011). Habits can take time to change, although one new initiative is underway. Any survey conducted under the Endangered Species Act (ESA) must now consider ITEK, if any is available. This is how the US Fish and Wildlife Service, working against the clock to get an ESA survey completed for the Alexander Archipelago Wolf in SE Alaska, reached out to the Tlingit (people of the tides; Brooks et al., 2024). Luckily, the collaboration suited the Indigenous knowledge holders, as they needed an accurate wolf count to ensure sufficient numbers of Sitka black-tailed deer and moose for their food security. However, there is a concern that sacred Indigenous knowledge, shared with federal authorities under the premise of its remaining confidential, is then revealed through Freedom of Information Act requests to third parties. This scenario has already been reported.

Deer and Murphy's excellent law review article "Animals may take pity on us: Using traditional tribal beliefs to address animal abuse and family violence within tribal nations" points out many truisms about the impact of Western society on Indigenous peoples. One of the many cultural differences is the treatment of animals as sentient teachers of adults and children in Native American myths and stories. The Euro-American worldview is that we have little to nothing to learn from animals and that animal stories are for children, i.e., they are childish (Deer and Murphy, 2017).

In terms of science, we've moved beyond seeing the practitioners and authorities of science as value free and neutral. ITEK has long been marginalized and sometimes openly denigrated by Western scientists. Richard Dawkins, a well-known British celebrity-academic, famously criticized a New Zealand initiative to more fully recognize *Mātauranga Māori* (Māori knowledge). Dawkins described Māori knowledge as myth and opined that science needed no such "cultural" identifier. Perhaps Dawkins missed the point. Western science is the dominant epistemology, and most consider Western science *to be* science. Dawkins' argument is the archetypal complaint that denies "a seat at the table" to others and is "the wrong" that the New Zealanders were trying to right.

A more useful start point is to ask what our cultures demand of science. Let's look at clinical research because of its use of experimental animals. The truth is that much of Western biomedical research is a promissory note—an indication of some future benefit. The article "Why most clinical research is not useful" was a devastating critique and concluded that "not only are most research findings false, but, furthermore, most of the true findings are not useful" (Ioannidis, 2016). Conversely, ITEK is seen as worthless if it does not serve the community. ITEK is community focused and participatory, and exists within a worldview of connectivity. For an interesting viewpoint on this, you might watch Dion Pou's YouTube video (https://www.youtube.com/watch?v=cHqyj5zoufI)—"applying *Mātauranga Māori* principles to science, environmental engagement, and education".

Western scientists often communicate using acronym-laden and confusing language, leading many to think that the basic concepts must be much more complicated than they are. In this way, science is legitimized by its incoherence. When practitioners of ITEK explain themselves in a common-sense fashion, Westerners distrust their science by virtue of its making sense. I've even seen academics say that Indigenous viewpoints are suitable for children but must make way for "real science" for older students. The shame of this is that ITEK, among other things, would appear to be a culturally embedded systems biology (Trewavas, 2006). Systems biology in Western science is a data-driven exercise to understand complexity. ITEK does the same through detailed observation and a granular understanding of the whole. ITEK's preference for a holistic and top-down organization contrasts with the bottom-up organization favored by reductionist Western science. The latter tries to use the smallest components of a system to reconstruct the whole, and perhaps, some emergent property; this largely succeeds in studying the parts but provides an incoherent vision of the whole. Instead, emergent properties of the whole system should inform the parts. For a deep dive into complexity, I would study Alicia Juarrero's 2023 book *Context changes everything: How constraints create coherence.*

To date, Indigenous lands have largely been used as a resource, and there is a long history of extraction, both of materials and of knowledge. Biopiracy and the bioprospecting of plants used in traditional medicine by Western ethnopharmacologists is just one example. Compensation for resource sharing is often paltry, and the longstanding Western practice of *parachute science* (Odeny & Bosurgi, 2022) is still exploitative. (Parachute science is a transient practice of engaging in field studies in third-party countries with little to no meaningful engagement with local participants.)

As Indigenous communities interact more with conservationists, it is important to manage expectations, and especially, the economic realities of local participants (Kohler & Brondizio, 2017). The tradition of ITEK is to dynamically learn and adapt. Indigenous people should not feel compelled to stay on the land to manage it for anyone's benefit, nor should Indigenous cultures be treated as "a relic". Likewise, there should be a sense of reciprocity and capacity building for participating communities; otherwise, there is no reason for Indigenous groups to engage.

It might seem daunting to try and incorporate different cultural perspectives into your work, but much of the heavy lifting has been done. Cross-cultural working can be supported using a rubric called *Two-Eyed Seeing* (or *Etuaptmumk*), devised by Mi'kmaw Elder Dr Albert Marshall. This is described as "learning to see from one eye with the strengths of Indigenous knowledge and ways of knowing, and from the other eye with the strengths of mainstream knowledge and ways of knowing, and to use both these eyes together, for the benefit of all". Useful references that describe how Two-Eyed Seeing came about and might be applied are Bartlett et al. (2012) and Reid et al. (2021). Even easier, you can watch *Signal Fire*, a film based on a 2020 article "Towards reconciliation: 10 Calls to Action to natural scientists working in Canada" (Wong et al., 2020). You can watch the film here: https://www.signalfirefilm.ca/watch.

Researchers should also be aware of legal protections like the UN's Convention of Biological Diversity (1993) and its Nagoya Protocol (which regulates the equitable sharing of genetic materials, associated traditional knowledge, and the benefits from their use). It is no longer trivial for genetic material to cross international borders, and failures to comply can result in stiff penalties. Problems remain with these legal arrangements (Lajaunie & Morand, 2020; Piano, 2023), and some complain that it is too difficult to share pathogens and microbiota for research. This came to the fore during the recent SARS-CoV-2 outbreak, when the hunt was on for the virus reservoir and its intermediate host. Just as a practicality, anyone intending to work across international borders should be mindful of any compliance risks (e.g., sanctions lists for people/organizations/countries available at the US Department of the Treasury) and that materials can be legally exported.

Lastly, when researching Indigenous materials, you will find many spurious or fabricated stories and data online. These texts are almost entirely created, amplified, quoted, misquoted, and then debunked by Western sources. For example, Fernández-Llamazares et al. (2024) recently debunked one claim about the richness of biodiversity on Indigenous land that no one took particularly seriously (other than the bureaucrat who probably misquoted a misquote to create it in the first place). Many of these stories create stereotypes of Indigenous people that serve Western interests, or some policy agenda, but little else (Abruzzi, 2000). If you don't find the material quoted by Indigenous authorities, treat it as suspect until verified.

Keywords: Indigenous Traditional Ecological Knowledge (ITEK), Two-Eyed-Seeing

11.3 WHEN SHOULD PEOPLE RESCUE WILDLIFE OR VULNERABLE HUMANS?

Animal welfare scientists took little interest in wildlife because suffering caused by nature did not seem to be part of the remit, but is this really such a simple distinction? Take, for example, the ethics of documentary film creation. It is a strong part of this culture that the documentarian should not intervene to prevent the suffering of animals being photographed or filmed. However, for a long time, there was little hesitation about *causing* harm to create an entertaining image: for example, using captured or partially disabled prey animals to film predatory behavior. Over time, these methods have fallen from favor, not so much for animal welfare reasons but because they were seen as misleading the audience. BBC filmmaker Jeffery Boswall became a keen advocate for ethics in natural history broadcasting and was influential in creating standards of transparency and disclaimers for nature documentaries (Boswall, 1962; Boswall, 1986; Boswall, 1998).

The current BBC standards outlaw:

- "capture of any birds for filming purposes"
- "feeding live mammals, birds and reptiles to any other animal"
- "tethering or restricting a vertebrate by any means to attract a predator"
- "cruel goading of an animal to fury"

These standards exist at least partially because these actions are now illegal. However, the result is a set of standards that continue the norm of not intervening with wildlife for its benefit, while retaining the prerogative to intervene (in some ways) to make a better film. There is a tacit acknowledgment that filmmakers will tend to harass the animals that they film, and it must be tempting to manufacture scenes to capture compelling footage. The situation in filmmaking is just one example of how we are inclined to bend the rules for our own benefit but not for animals.

However, some spectacles place the documentarians' version of the "prime directive" under stress. For example, in some areas, such as the Zimbabwe Mana Pools, large areas of slick mud can trap animals even as large as elephants. These animals are often harassed and eventually killed by hyenas in what can be a protracted and grueling process. Photographer Jens Cullmann recorded watching an elephant calf being killed in this way while the mother was stuck nearby, observing, dying days later after park officers made an unsuccessful rescue attempt. I [EPK] recall watching a documentary of a similar event as a child where the film maker ultimately intervened to save a wildebeest calf stuck in mud, simply unable to continue to watch this torturous but entirely natural process.

There is a good case for having a blanket rule that humans should not interfere with nature, as it is a system with its own processes and integrity in need of protection, and any disruption of these processes is almost certain to be unbeneficial to the greater whole. However, welfare-motivated counter arguments are made and fall into three main areas: 1) that when the cause of the suffering is or might be of human origin, preventing and responding to that harm is ethical; 2) that when a person encounters an individual wild animal in need, there is no clear reason not to assist; and 3) that intervention that causes a net benefit to animals and the ecosystem is ethical.

Not intervening to assist those who are suffering can have its own costs. Inner demons haunted photojournalist Kevin Carter after taking the infamous *vulture and girl* photograph of the 1993 famine in Sudan. Carter won a coveted Pulitzer Prize the next year; a few months later, he would be dead by suicide. In this instance, a starving child was being stalked by a vulture, watched by Carter as he selected the best light and composition for his shot—his first instinct was to take the photograph. Later, Carter reported chasing off the vulture. Carter's life had been spent filming horror, and when a close friend was killed filming a gun battle, the horror overtook him. On publication of the picture, there was uproar in the United States about the fate of the child. It turned out the young child was a boy, not a girl, and, remarkably, survived through the famine.

Human intervention has already disrupted almost every wild ecosystem, and so withholding it only when it might benefit animals is somewhat disingenuous. Much animal suffering in the wild can be tracked to anthropogenic causes: for example, from the effects of human structures, from being repelled from cropland, from protecting livestock from predators, or from pollution. As just one small example, human activities create many highly toxic bodies of water, like gold mine retention ponds contaminated with cyanide, manure and sewage treatment ponds, oil field skim pits, and many other stagnant bodies of water that accumulate toxic material like heavy metals and botulism. The number of waterfowl and migratory birds killed annually by these deadly bodies of water is incalculable because to the flying bird, they look like a normal pond or lake where they can take refuge. The deaths are certainly in the many millions.

A second argument underlies a lot of wildlife rehabilitation. It goes like this—is it valid to act on empathy to help an animal found in need of help? (Note that "empathy" is an attempt to share and understand another's emotional experience, whereas sympathy is a form of pity that retains an aspect of separation.) Within this notion, there is controversy when lay people needlessly intervene, for example, with deer fawns and other juveniles left in hiding by their parents (Figure 11.2). Likewise, many fledgling birds seem helpless and inept during

Figure 11.2 Hiding fawn—black-tailed deer. Photo taken by Loren Lane of the US National Park Service at Mount Rainier National Park, Washington. No protection is claimed in original US Government works.

the period of their life where they have not developed effective flight. So, while they can seem a bit pathetic, there is nothing actually wrong with them.

Also controversial is the rehabilitation of invasive species such as young robins, which are found early in the season and swamp rescues before the native species reach the same stage. And rehabilitation that is effective and leads to reintroduction is neither easy nor straightforward. Animals that become habituated to humans may do poorly in the wild and become dangerous to people. Released animals may also spread diseases that decimate wild populations. For these reasons, rehabilitation is generally an activity that requires some training and a license. On an animal-by-animal basis, there is a very uneasy balance between acknowledging that people will help wildlife out of empathy and trying to limit this activity to competent and careful individuals.

The tension between wildlife officers, who are more cautious, and the public, who want to save animals in need, can be seen in news stories about rescued wildlife being seized and euthanized (e.g., elk that have become comfortable around people and solicit food from them, which can make them dangerous to people when they are adults). Indeed, some animal experts have described how even some hand-raised domestic animals can become dangerous from mis-imprinting on people, a syndrome referred to as *berserk male syndrome* or "aberrant behavior syndrome" in animals like llamas and other camelids (Ball et al., 2015).

A number of theories have also directly opposed the notion of non-intervention, arguing that as long as the net effect benefits animals, then intervention to assist wildlife is ethical. This point of view might necessitate a nuanced understanding of how natural systems are affected, or it may choose to wholly prioritize the rights of the animals involved. In the debate around assisting wildlife, the implicit question is: Could unrestrainedly attempting to help and rescue ultimately do more harm than non-intervention?

Keywords: berserk male syndrome, empathy

Anthropogenic harm—the DDT story and *Silent Spring*

It is often the most unintended impacts that are the most catastrophic. One famous example is the invention of the insecticide DDT (dichlorodiphenyltrichloroethane). The DDT controversy is a good example of how science could link human activity to animal welfare and then help to advocate for change. DDT became commercially available in 1945. In 1962, Rachel Carson published *Silent Spring*, detailing the devastating impacts of DDT on entire ecosystems and especially on some of America's most beloved species, including the bald eagle (Carson, 1962). DDT was clearly impacting the reproductive success of many birds, and research would ultimately point to eggshell thinning as one of the problems, along with endocrine disruption.

Carson's message leaned into topics that her human audience found to be compelling, namely, that DDT threatened our own health, that we exploit wild animals, and that our thoughtlessness would ultimately come back to haunt us. But she also explicitly showed that the harm from DDT was cross cutting and was fundamentally the same for people as it was for other animals. Carson's warning about the toxicity of DDT to the dividing cell was made clear: "In this elemental drama all life is revealed as one. … Anything that disturbs mitosis is therefore a grave threat to the welfare of the organism affected and to its descendants". DDT was *the* issue that provoked the creation of the Environmental Defense Fund in 1967, and the insecticide was ultimately banned in 1972. However, this close call with calamity was soon forgotten, as was our near miss with atmospheric ozone depletion and especially, the "ozone hole" at the polar regions. Even now, as soon as the immediate danger passes, so does our memory of it.

11.4 DE-EXTINCTION

The passenger pigeon (*Ectopistes migratorius*) represents an iconic example of the need to understand welfare at the population level: that is, to see how the needs of the individual animal may include the need for certain group sizes, locations, and freedom of movement (i.e., the welfare of the individual is indivisible from that of the group). The passenger pigeon was endemic to America and existed in immense migratory flocks that bred in huge colonies. At their peak, passenger pigeons may have been the most populous bird species on the planet. Human hunting, habitat loss, and disease were a potent trifecta that doomed the species. Total extinction was officially recorded with the passing of Martha (29 years old) on September 1, 1914, at the Cincinnati Zoo. For the passenger pigeon, breeding success was closely tied to the size of the nesting group, so as the group size shrank, so did breeding. Without extremely large groups of their own kind, the passenger pigeon could not experience good welfare or even survive (Halliday, 1980).

This explanation for passenger pigeon extinction was applied retrospectively and is contested. It has also been proposed that natural selection in passenger pigeons went awry when flock sizes shrank (Murray, et al., 2017); without a colossal flock, the passenger pigeon genetically crashed. Whether the death knell was behavioral, genetic, or a mixture of both, what is clear is that it was difficult for people to appreciate just how vulnerable the passenger pigeon was, and so they kept on killing them as a convenient food source. The pigeon was a cheap kitchen staple. A few years after Martha died, scientists began work on developing the broiler chicken.

The issue of animal welfare appears on the horizon again as biotech attempts to resurrect the passenger pigeon by *de-extinction*. Apparently, the massive flocks of passenger pigeons drove regenerative growth of hardwood tree species, and the aim is to reintroduce this ecological engineering process. To achieve this, molecular biologists will tweak the genes of its nearest living relative, the band-tailed pigeon, to create a functional proxy of the passenger pigeon. So, strictly speaking, we're not reviving the passenger pigeon, but we're "editing" another pigeon to look and behave like it. Recreating the desired flocking behavior might be a tough ask, as band-tailed pigeon nesting and flocking habits don't appear to be anything like those of the passenger pigeon. Technically, identifying the underlying genes responsible for behavior is challenging at the best of times, without working with non-model species, or even worse, species that are extinct. Where these passenger pigeon proxies would live is another issue, as the continuous eastern deciduous forest is long gone. Even if you could magically re-animate a genuine passenger pigeon and succeed in creating massive flocks, where is the environment in which they could thrive or even be welcome?

Another planned de-extinction is of the Woolly mammoth, this time using the endangered Asian elephant as the genetic template (https://colossal.com). The idea is that the pseudo-mammoth will act as an environmental engineer to convert the frozen tundra to steppe grasslands. This will help to lock in sequestered carbon by keeping the permafrost frozen. Some of the issues include: 1) the highly social Asian elephant is endangered and susceptible to deadly viral infections in captivity; 2) any viable proxy could be rejected by adult elephants; 3) the newborns would be "culturally blank" and unprepared for life on the tundra; 4) proxies may not be able to digest the forage, as their microbiome will need to be engineered as well; 5) the proxies would also have an Asian elephant epigenome that may be entirely unsuitable. As an update, the "woolly mouse" has just been unveiled to great fanfare. They look cute, but what really is the point of all of this?

While we await the appearance of a proxy mammoth, beavers are already moving into northwest Alaska and are happily turning tundra into wetlands. The biggest issue with all of this is us. If nothing else, these projects treat animals as machines to be delivered to various ecosystems to do our bidding. The notion of de-extinction may even exacerbate the problem by suggesting a reversibility that cannot yet be delivered (Campagna et al., 2017). There are worthy tech-heavy challenges for extant species, such as engineering immunologic resilience or supporting animals to mitigate the effects of climate change. Or just advocating for them. For example, a great win has recently been achieved for Horseshoe crabs. Up to a million of these crabs are bled annually by the pharmaceutical industry, which uses their bright blue blood to test for bacterial contamination in their products. Now, a synthetic alternative has been approved, which should mean that we no longer need to bleed these remarkable animals. More good news is that a new mRNA vaccine against elephant endotheliotropic herpes virus, which kills about half of all captive elephant calves, was just given to its first recipient, Tess, a 40-year-old Asian elephant at Houston Zoo (June, 2024). So, the power of tech can be formidable, especially when we use it to benefit animals that still exist. For an interesting debate on these issues with some of the main protagonists, watch https://www.youtube.com/watch?v=N-liqmKITs8. Which side would you pick?

Keyword: de-extinction

11.5 CONSERVING DOMESTICATED BREEDS

It is interesting to consider how the concept of conservation is also applied to groups of domesticated animals, specifically *breeds*, where a breed is a stock of plants or animals with specific characteristics created by artificial selection. So, breeds are to artificial selection what species are to natural selection. The concept of a pedigree breed, where the exact heritage of the animal must be known and recorded by a third party who issues "papers", is relatively modern. During the Victorian period, certain individuals decided to preserve or "improve" breeds of dogs and gravitated to creating pure-bred lines with no ancestors outside their breed.

This can lead to conflict between those who see dogs as defined by function versus show pedigree, and this divide led to opposition to the Border Collie being recognized by the American Kennel Club (AKC) in 1995. It was argued that an emphasis on a limited range of physical characteristics needed in the show ring might undermine the true core identity of the Border Collie as an intelligent herder (McCaig, 2007). It is noticeable that dogs with working lines and show lines do differ. For example, show greyhounds, which have been recognized by the AKC since 1885, are noticeably different in form from those used in racing. The show dogs tend to be larger, more slender, and with less muscle.

There are many movements and groups with different motivations that act to preserve one or more breeds of domesticated animals. It may be from personal affection for the breed, for traditional reasons, or to preserve valuable genetics. For example, keeping old forms of turkeys (heritage breeds) maintains genetic diversity that would be lost if we retained only commercial breed varieties. We could use those genetic data to develop turkeys that are able to run or fly. These breeds also retain many welfare-relevant features, like slower growth, longer life, camouflaging pigment, and the ability to breed without intervention. The Wildlife Conservancy states: "Rare farm animals represent an irreplaceable piece of earth's biodiversity and offer incredible variety that may be needed for future farms – robust health, mothering instincts, foraging, and the ability to thrive in a changing climate" (https://livestockconservancy.org/, 2023).

By contrast, some commentators who oppose the use of animals for human purposes propose that all domesticated species should be allowed to become extinct, as they have been created to depend on us. For example, domesticated sheep need to be sheared as, unlike their wild type, their hair coat does not stop growing. This worldview would turn back the clock on domestication and sanction the existence of only those animals that can competently live within their natural environments.

Keyword: wild type

11.6 SPECIESISM

This chapter has dealt largely with how the wellbeing of individuals is, to varying extents, inseparable from welfare across the group, stock, species, or ecosystem. In addition to this, we need to understand that our concepts of what species are, and their importance, are variable and changeable. For example, it used to be commonplace to apply the idea of selective breeding to humans, which provided scientific support to the many atrocities carried out in the name of eugenics in the late 19th and early 20th centuries. With updated human pangenome maps (from 47 individuals that better represent our diverse global communities https://www.genome.gov/about-genomics/new-human-pangenome-reference), we find that human genomes are 99.9% identical (in terms of single nucleotide changes and 99.6 if you include multiple changes); drop another percentage point and that is our similarity with chimpanzees and bonobos, with whom we, and the other great ape species, now share the taxonomic family of Hominidae.

"Hominid" used to be a much more exclusive club, containing only modern humans and their immediate ancestors, but as the genetic data rolled in, we could no longer justify our separatist status. Across the animal kingdom, our understanding of which populations segregate to separate species, and how far apart they are, only matters to the extent that the category of "species" itself is important (if it is at all). The various branches of biology suggest to us that the loss of a species is significant for maintaining a balanced ecology and a genetic reservoir. But it is also a loss of potential resources for human benefit, a link to our past, a part of our culture, and something else that is perhaps a bit less tangible. For many, the species is a repository of something unique, and its loss can evoke a sense of pathos and a feeling of guilt or failure, as humans are often a main or contributing cause for extinction.

By contrast, if your approach to animal welfare is more focused on addressing or preventing suffering, it is possible to suggest that species as a concept does not matter at all. If you follow an animal rights–style approach, then an animal should not be sacrificed or used by humans, or be used to benefit other non-human animals. This viewpoint would encompass actions like controlling invasive pests or killing animals to feed other animals in settings like wildlife rehabilitation or captive breeding.

One idea that became pivotal in this discussion is *speciesism*. The idea that distinctions between species are often made improperly has been around for some time, but the idea became supercharged in 1970. This was when Richard Ryder resurrected the word in objections to the use of animals in research. Disagreeing with those who would advocate for using animal research for human benefit, Ryder argued that "it is still just 'speciesism', and as such it is a selfish emotional argument rather than a reasoned one". This term was picked up by philosopher Peter Singer, in his book *Animal liberation*, who gave it a clearer definition of "prejudice or attitude of bias in favor of the interests of members of one's own species and against those of members of other species" (reviewed by Ryder, 2011).

Singer's highly accessible writing style helped to spread the term across many schools of animal activism. However, both conceptions tend to leave out that we also sacrifice the lives or wellbeing of some animals for the benefit of other animal species that we particularly like, such as *charismatic megafauna* (large animal species with symbolic value or widespread popular appeal), or our pets. We would suggest an even simpler definition of "an inaccurate or unethical discrimination based on species".

Keywords: speciesism, charismatic megafauna

Rewilding tainted land

Sometimes land is unwanted, and sometimes that unwanted land is a poisoned chalice. This is the land that might be freely given to animals, as happened in Northern Ukraine. Let's meet a fox in a Ukrainian wood.

> *The young fox tensed, she hadn't been this far west before, but the wolves were moving—so should she. The track seemed safe, but no time to get distracted. Before the wolves came she might have dozed, scattering dewy spider webs for fun. But not today, she had to keep her wits about her—she'd seen no spider webs anyway. Jumping the ditch she melted into the forest, sinking up to her tummy, pine needles everywhere. Her lustrous red coat was just visible, and the forest was quiet, eerily quiet. She'd treat herself to a bank vole later, and yes, a mushroom or two if the boar didn't beat her to it. She'd passed up on a bird earlier, something about the milky eyes had put her off. No, this had been a good choice after all, she was pleased, this new home might work out.*

If that fictional fox could read, she would have seen the Cyrillic script on the railroad signal box that she passed—it read "Chernobyl". We'll come back to our fox in a minute. Chernobyl's nuclear reactor disaster was an opportunity to study the effects of long-term radiation exposure on animals in their natural setting. The Chernobyl Exclusion Zone (CEZ), a 30-km swath of land around the nuclear reactor that exploded in 1986, was cleared of people and later expanded to create the third largest unscheduled rewilding project in Europe (equivalent in size to Rhode Island). This land is some of the most contaminated in the world, with the radiation released equivalent to 400 Hiroshima bombs.

So, what happened to our fox looking for a new home? Research aimed at uncovering the fate of animals in the CEZ has been a bit of a mess, and the scientific community has, by its own admission, not done well. The mission of studying a radioactive rewilding project was a tall order, and there have been problems with data reproducibility and confusion over the actual doses of radiation received by animals. The International Union for Radioecology tried to reach some kind of consensus after a 2015 meeting in Miami that focused on methodological issues and the source of discrepancies between data sets. Their findings, and seven consensus statements about environmental protection from radiation, were published in 2016 (Bréchignac et al., 2016). Recent reports like the "Dogs of Chernobyl" (Dillon et al, 2023) and an unpublished wolf study have garnered more media attention than anything else. As much of the earlier research is contested, the little narrative of our fox, with her observations of the wood, are designed to cross reference to some of the topics of investigation rather than agreed-upon data.

So, back to our fox. She moved through radioactive pine needles as she entered the red forest after the nuclear accident. The coniferous trees near the power plant got a dusting of radioactivity that was later washed by rain into the soil, killing juvenile insects. The dead and dying trees turned brown/red—that's how the red forest got its name. Biodegradation was reported to have slowed because of the death of various microbial species, and so the leaf litter and pine needles were deeper than usual and would have contaminated her belly (Mousseau et al., 2014 and disputed in Bonzom et al., 2016). Numbers of predatory spiders were (supposedly) depressed by the radiation, so fewer spider webs. Eating mushrooms would not have been a good idea, as they are excellent "sponges" for radioactivity, so much so that they've been considered as agents for bioremediation (Dighton et al., 2008). Bank voles that burrow through radioactive soil would also have been a radioactive dinner for our fox. Her red fur contains pheomelanin, which was found to be less protective against damaging free radicals than eumelanin, as also shown for birds (Galván et al., 2011). And cataracts were a huge problem (Mousseau & Møller, 2013); the bird that our fox declined to eat had cataracts.

Many news outlets reported abundant wildlife in the CEZ. Years later, the allegation has been made that reports of apparently thriving wildlife might have been used by bureaucrats to downplay the effects of radiation (Mousseau, 2021). Kate Brown's 2019 book *Manual for survival: A Chernobyl guide to the future* makes a similar claim for the human population (Brown, 2019). Animals may have been continually migrating into the CEZ and then dying, with incomers taking their place, but we simply don't know. PubMed searches using "animal", "welfare", and "Chernobyl" or "Chornobyl" (the Ukrainian spelling), gave a single hit, a study assessing the risks posed by semi-feral dogs to human workers and tourists in the CEZ.

Let's go now to the United States, and Colorado's "infinity rooms". These buildings were so contaminated by radiation that the levels were off scale—infinity. They were part of the Rocky Flats Nuclear Weapons Plant, where triggers were made for atomic weapons. Let's just say that safety was not going well, and come the joint FBI/EPA raid of 1989, things had become as bad as they could get. The buildings, air conditioning, plumbing, it was a parade of plutonium. This is where the US *Superfund* comes in. Superfund is the informal name for a US government act that sanctioned the cleanup of toxic work sites. These are the worst of the worst, and some have been converted into wildlife refuges. This makes the cleanup cheaper, but the public tends to be a little more cautious about the end result.

In 2018, after cleanup and remediation, 5000 acres (plus) of Rocky Flats National Wildlife Refuge opened to the public. Of its current residents, it's the black-tailed prairie dogs that have people worried (Figure 11.3). Prairie dogs are a keystone species and ecosystem engineer (Duchardt et al., 2021). They burrow deep, and it's the burrowing that might get them into trouble. What if they dig up the buried plutonium? Rocky Flats has been one of the last possible homes for prairie dogs displaced by development, ranchers, and cattlemen. Failing that, they end up as dinner at a raptor rescue or an endangered species program (likely a black footed ferret breeding station). So, for the prairie dogs, it's tainted land or bust, but in the future, even Rocky Flats might be out of bounds.

Keyword: superfund site

Figure 11.3 Top: Prairie dogs are displaced by development and ranchers throughout the Denver Metropolitan area's Front Range and are running out of suitable relocation sites (which have included Rocky Flats). Bottom: Connected wildlife areas form corridors for animals, like these deer, to move through the mixed prairie grassland. Connectivity isn't always welcome, though, and plans to connect Rocky Flats to other wildlife refuges through a trail system have stalled, as locals worry about disturbing buried plutonium. Photographs reproduced with the permission of Janet Patterson-Kane.

READINGS

Wildlife Conservation and Animal Welfare

Bekoff, M. (Ed.). (2013). *Ignoring nature no more: The case for compassionate conservation.* University of Chicago Press. https://doi.org/10.7208/chicago/9780226925363.001.0001

Bobier, C. A., & Allen, B. L. (2022). Compassionate conservation is indistinguishable from traditional forms of conservation in practice. *Frontiers in Psychology, 13*, 750313. https://doi.org/10.3389/fpsyg.2022.750313. PMID: 36262450; PMCID: PMC9574382.

Breed, D., Meyer, L. C. R., Steyl, J. C. A., Goddard, A., Burroughs, R., & Kohn, T. A. (2019). Conserving wildlife in a changing world: Understanding capture myopathy-a malignant outcome of stress during capture and translocation. *Conservation Physiology, 7*(1), coz027. https://doi.org/10.1093/conphys/coz027

Dubois, S., Fenwick, N., Ryan, E. A., Baker, L., Baker, S. E., Beausoleil, N. J., Carter, S., Cartwright, B., Costa, F., Draper, C., Griffin, J., Grogan, A., Howald, G., Jones, B., Littin, K. E., Lombard, A. T., Mellor, D. J., Ramp, D., Schuppli, C. A., & Fraser, D. (2017). International consensus principles for ethical wildlife control. *Conservation Biology, 31*(4), 753–760. https://doi.org/10.1111/cobi.12896. Epub 2017 Feb 9. PMID: 28092422.

Muth, R. M., Zwick, R. R., Mather, M. E., Organ, J. F., Daigle, J. J., & Jonker, S. A. (2006). Unnecessary source of pain and suffering or necessary management tool: Attitudes of conservation professionals toward outlawing leghold traps. *Wildlife Society Bulletin, 34*(3), 706–715.

Paterson, J. (2014). Capture myopathy. In G. West, D. Heard, & N. Caulkett (Eds.), *Zoo animal and wildlife immobilization and anesthesia* (pp. 171–179). John Wiley & Sons, Inc.

Savarie, P. J., & Sterner, R. T. (1979). Evaluation of toxic collars for selective control of coyotes that attack sheep. *The Journal of Wildlife Management, 43*(3), 780–783.

Shine, R., Ward-Fear, G., & Brown, G. (2020). A famous failure: Why were cane toads an ineffective biocontrol in Australia? *Conservation Science and Practice, 2*. https://doi.org/10.1111/csp2.296

Perspectives from other Knowledge Traditions

Abruzzi, W. (2000). The myth of chief Seattle. *Human Ecology Review, 7*, 72–75.

Bartlett, C., Marshall, M., & Marshall, A. (2012). Two-eyed seeing and other lessons learned within a co-learning journey of bringing together Indigenous and mainstream knowledges and ways of knowing. *Journal of Environmental Studies and Sciences, 2*(4), 331–340.

Brooks, J. J., Markegard, S. I., Langdon, S. J., Anderstrom, D. S. E., Douville, M. G., George, T. A., Jackson, M. K., Jackson, S. G., Mills, T. K., Ramos, J. D., et al. (2024). Indigenous knowledge and species assessment for the Alexander Archipelago wolf: Successes, challenges, and lessons learned. *Journal of Wildlife Management, 88*, e22563. https://doi.org/10.1002/jwmg.22563

Deer, S., & Murphy, L. (2017). "Animals may take pity on us": Using traditional tribal beliefs to address animal abuse and family violence within tribal nations. *Mitchell Hamline Law Review, 43*(4), Article 1. https://open.mitchellhamline.edu/mhlr/vol43/iss4/1

Fernández-Llamazares, Á., Fa, J. E., Brockington, D., Brondízio, E. S., Cariño, J., Corbera, E., Farhan Ferrari, M., Kobei, D., V, P., Márquez, G. Y. H., Molnár, Z., Tugendhat, H., & Garnett, S. T. (2024). No basis for claim that 80% of biodiversity is found in Indigenous territories. *Nature, 633*(8028), 32–35. https://doi.org/10.1038/d41586-024-02811-w

Goldman, M., Nadasdy, P., & Turner, M. (2011). *Knowing nature. Conversations at the intersection of political ecology and science studies.* University of Chicago Press. https://doi.org/10.7208/chicago/9780226301440.001.0001

Ioannidis J. P. (2016). Why most clinical research is not useful. *PLoS Medicine, 13*(6), e1002049. https://doi.org/10.1371/journal.pmed.1002049

Juarrero, A. (2023). *Context changes everything: How constraints create coherence*. The MIT Press.

Kohler, F., & Brondizio, E. S. (2017). Considering the needs of indigenous and local populations in conservation programs. *Conservation Biology: The Journal of the Society for Conservation Biology, 31*(2), 245–251. https://doi.org/10.1111/cobi.12843

Lajaunie, C., & Morand, S. (2020). Nagoya protocol and infectious diseases: Hindrance or opportunity? *Front Public Health, 16*(8), 238.

Odeny, B., & Bosurgi, R. (2022). Time to end parachute science. *PLoS Medicine, 19*(9), e1004099. https://doi.org/10.1371/journal.pmed.1004099. PMID: 36067221; PMCID: PMC9488789.

Piano, F., International Bar Association (2023). *Where is my profit share? The Nagoya Protocol and how companies must compensate Indigenous traditional knowledge.* Report on the joint session of the Indigenous Peoples Committee, the Latin American Regional Forum and the Professional Ethics Committee at the IBA Annual Conference in Miami. https://www.ibanet.org/where-is-my-profit-share-report

Reid, A. J., Eckert, L. E., Lane, J.-F., Young, N., Hinch, S., Cooke, S., Ban, N., Darimont, C., & Marshall, A. (2021). "Two-eyed seeing": An Indigenous framework to transform fisheries research and management. *Fish and Fisheries, 22*(2), 243–261.

Trewavas, A. (2006). A brief history of systems biology. "Every object that biology studies is a system of systems." Francois Jacob (1974). *The Plant Cell*, *18*(10), 2420–2430. https://doi.org/10.1105/tpc.106.042267

Wong, C., Ballegooyen, K., Ignace, L., (Gùdia) Johnson, M. J., & Swanson, H. (2020). Towards reconciliation. 10 Calls to Action to Natural Scientists Working in Canada. *FACETS*, *5*(1), 769–783.

When Should people Rescue Wildlife or Vulnerable Humans?

Ball, S. R., Way, K., Schleining, J. A., & Millman, S. T. (2015). Survey-based examination of demographics, potential causes and treatments of aberrant behavior syndrome (berserk male syndrome) in camelids. *Iowa State University Animal Industry Report*, *12*(1). https://doi.org/10.1016/j.smallrumres.2023.106918

Boswall, J. (1962). Filming wild nature: Fair means or foul? *Scientific Film*, *3*, 110–113.

Boswall, J. (1986). The ethics and aesthetics of slow motion in wildlife films. *Image Technol*, *68*(11), 560–561.

Boswall, J. (1998). Wildlife film ethics: Time for screen disclaimers? *Image Technology*, *80*(9), 10–11.

Anthropogenic Harm: the DDT story and Silent Spring

Carson, R. (1962). *Silent spring*. Houghton Mifflin.

De-extinction

Campagna, C., Guevara, D., & Le Boeuf, B. (2017). De-scenting extinction: The promise of de-extinction may hasten continuing extinctions. *The Hastings Center Report*, *47*(2), S48–S53. https://doi.org/10.1002/hast.752

Halliday, T. R. (1980). The extinction of the passenger pigeon Ectopistes migratorius and its relevance to contemporary conservation. *Biological Conservation*, *17*(2), 157–162.

Murray, G. G. R., Soares, A. E. R., Novak, B. J., Schaefer, N. K., Cahill, J. A., Baker, A. J., Demboski, J. R., Doll, A., Da Fonseca, R. R., Fulton, T. L., Gilbert, M. T. P., Heintzman, P. D., Letts, B., McIntosh, G., O'Connell, B. L., Peck, M., Pipes, M. L., Rice, E. S., Santos, K. M., Sohrweide, A. G., … Shapiro, B. (2017). Natural selection shaped the rise and fall of passenger pigeon genomic diversity. *Science (New York, N.Y.)*, *358*(6365), 951–954. https://doi.org/10.1126/science.aao0960

Conserving Domesticated Breeds

McCaig, D. (2007). *The dog wars: How the Border Collie Battled the American Kennel club*. Outrun Press.

Speciesism

Ryder, R. (2011). *Speciesism, painism and happiness: A morality for the 21st century*. Imprint Academic.

Rewilding Tainted Land

Bonzom, J. M., Hättenschwiler, S., Lecomte-Pradines, C., Chauvet, E., Gaschak, S., Beaugelin-Seiller, K., Della-Vedova, C., Dubourg, N., Maksimenko, A., Garnier-Laplace, J., & Adam-Guillermin, C. (2016). Effects of radionuclide contamination on leaf litter decomposition in the Chernobyl exclusion zone. *Science of the Total Environment*, *562*, 596–603. https://doi.org/10.1016/j.scitotenv.2016.04.006. Epub 2016 Apr 22. PMID: 27110974.

Bréchignac, F., F., Oughton, D., Mays, C., Barnthouse, L., Beasley, J. C., Bonisoli-Alquati, A., Bradshaw, C., Brown, J., Dray, S., Geras'kin, S., Glenn, T., Higley, K., Ishida, K., Kapustka, L., Kautsky, U., Kuhne, W., Lynch, M., Mappes, T., Mihok, S., … Tsukada, H. (2016). Addressing ecological effects of radiation on populations and ecosystems to improve protection of the environment against radiation: Agreed statements from a Consensus Symposium. *Journal of Environmental Radioactivity*, *158–159*, 21–29. https://doi.org/10.1016/j.jenvrad.2016.03.021

Brown, K. (2019) *Manual for survival: A chernobyl guide to the future*. W.W. Norton & Company.

Dighton, J., Tugay, T., & Zhdanova, N. (2008). Fungi and ionizing radiation from radionuclides. *FEMS Microbiology Letters*, *281*(2), 109–120.

Dillon, M. N., Thomas, R., Mousseau, T. A. et al. (2023). Population dynamics and genome-wide selection scan for dogs in Chernobyl. *Canine Medicine and Genetics*, *10*, 1.

Duchardt, C. J., Porensky, L. M., & Pearse, I. S. (2021). Direct and indirect effects of a keystone engineer on a shrubland-prairie food web. *Ecology*, *102*(1), e03195. https://doi.org/10.1002/ecy.3195

Galván, I., Mousseau, T. A., & Møller, A. P. (2011). Bird population declines due to radiation exposure at Chernobyl are stronger in species with pheomelanin-based coloration. *Oecologia*, *165*, 827–835.

Mousseau, T. A. (2021). The biology of Chernobyl. *Ann Rev Ecology, Evolution and Systematics*, *52*, 87–109.

Mousseau, T. A., & Møller, A. P. (2013). Elevated frequency of cataracts in birds from Chernobyl. *PLoS One*, *8*(7), e66939.

Mousseau, T. A., Milinevsky, G., Kenney-Hunt, J., & Møller, A. P. (2014). Highly reduced mass loss rates and increased litter layer in radioactively contaminated areas. *Oecologia*, *175*(1), 429–37. https://doi.org/10.1007/s00442-014-2908-8. Epub 2014 Mar 4. PMID: 24590204.

Part IV

Interactions—Humans and Animals

Even the pared down social interactions of solitary animals, like this snow leopard, are of importance and have a value. Part IV will address various types of interaction, some involving animals of the same species, different species, and even humans. Vector snow leopard, irbis wild cats, graphic illustration (#1934106671) by Transferaza/Shutterstock.com.

DOI: 10.1201/9781003375456-15

Chapter 12

Interactions—Conspecific

12.1 THE UBIQUITY OF INTERACTIONS

These lines from the metaphysical poet John Donne (1572–1631), "no man is an island, entire of itself", can be said for all animals. We are not wholly self-sufficient; we need others in our lives. This is problematic, as animal welfare is a quality of the individual and does not contain a simple or obvious way to account for the welfare impacts imposed by others. Animals often have under-appreciated relationships between members of their own species and other non-human animal species (in which human influence still plays a major, albeit indirect, role).

This section of the book covers a broad variety of topics, but what they have in common is a focus on interactions between individuals. We begin by considering animal-to-animal and human-to-animal engagement at the individual and group levels. Later, in Part V, we move to considering how people interact with animals according to informal codes and more explicit rules. Specific attention is given to ethics, economics, law, and religion (Morris, 1990; Budiansky, 1999; Palmer, 2011). Non-human animals also have systems that govern interactions for individuals of the same species or across species, which remains an under-developed area of study. For that reason, this section relies more heavily on field observations, traditional practices, and anecdotes.

The following chapters will discuss the animal welfare science implications of interactions in three categories: conspecific (between individuals of the same species), heterospecific (between individuals of different species), and a sub-category of heterospecific: human–animal interactions.

Throughout this section, you will see examples of how we pivot between considering people and animals, and then considering their interactions and relationships. This reframing can bring neglected aspects of a situation into focus. Many animal welfare issues are contested, entrenched, or difficult to understand, and so having a method to change our perspective is a valuable tool

DOI: 10.1201/9781003375456-16

in creating constructive change. Let's begin with an example of how *all* these interactions can occur simultaneously to provoke dramatic compromises in animal welfare.

Keywords: conspecific, heterospecific

Example: Transport-related elephant rhinoceros conflict

African elephants and rhinoceros evolved under conditions where they could freely roam, with elephants migrating large distances to access different resources, and rhinos equally wide ranging but without a clear migratory pattern. These days, both species are confined within parks and reserves that, while often very large, are still a fraction of their natural range. This interferes with multiple natural patterns of behavior, with detrimental knock-on effects for how elephants and rhinoceros interact (Berger & Cunningham, 1998).

The potential for conflict between these species became an acute and obvious problem in 1992–1997, when young orphaned male elephants were relocated to the Pilanesberg National Park, South Africa, for conservation reasons. Post relocation, the orphaned elephants developed a pattern of killing white rhinoceros, eventually killing as many as 49 rhinos. It was ultimately determined that in the absence of older, dominant bulls, the young males had prematurely entered the mating conditions of *musth* and stayed in that condition for far too long. Male elephants enter musth when they experience a sharp increase in sex hormones and are ready to mate. This state is shown by the secretion of a liquid called temporin from glands on the sides of their head (as shown in Figure 12.1), and a constant trickle of urine. They spend more time with females, who can use these signs to detect a readiness to mate. Elephants in musth become unpredictable and can be highly aggressive with each other, although actual fights are quite rare, as they can also judge fitness and the likelihood of losing a fight. Encountering larger bulls ordinarily knocks the juveniles out of musth (Poole & Moss, 1981; Slotow & Dyk, 2001; Slotow et al., 2000), and so their antisocial behavior was resolved after the introduction of six older male elephants from Kruger National Park.

Figure 12.1 Savanna Elephant (*Loxodonta africana*) eye close-up showing the Musth gland opening. Photo credit Bernard Dupont licensed under CC BY-SA 2.0 https://creativecommons.org/licenses/by-sa/2.0/ https://www.flickr.com/photos/berniedup/16689303126/.

The older males imported and enforced a social hierarchy and its norms, thereby eliminating the rhino killing—in effect, they became *role models* for the rampaging juveniles. This disturbing event highlighted how animals are not interchangeable (young males could not substitute for an intergenerational group), and that their physical and mental states, and so ultimately their behavior, are strongly influenced by social roles. This breakdown in social structure damaged the welfare of the elephants by causing them to experience states of aggression and maladaptive violent behavior, and damaged the welfare of the critically endangered white rhinos by exposing them to violence, fear, pain, injury, and ultimately, death.

Keywords: musth, role model

12.2 SOCIAL ISOLATION VERSUS CROSS-HARM

Conspecific simply means "of the same species". Any animals that are naturally social are likely to require the company of conspecifics to experience good welfare. However, a supportive environment is needed for social groups to interact harmoniously. When some element is disrupted, however minor, conspecifics can quickly become harmful to each other. For example, when it was required in the 1990s to socially house nonhuman primates in laboratories, some facilities moved to pair housing (DiVincenti & Wyatt, 2011). The struggle was in finding harmonious pairings for animals who had lived alone in small cages for long periods of time. The social housing requirement was based on these animals having social needs for good welfare and was widely beneficial despite fears about the risks. However, being housed in close quarters with an incompatible partner could lead to welfare becoming significantly worse as animals found themselves trapped with an aggressor.

In intensive settings, it is natural to want to avoid animals harming each other through activities like chasing, mounting, and fighting. This is in addition to some of the risks that drove the development of indoor animal housing in the first place (e.g., reducing the transmission of diseases and parasites, and ensuring that all animals have access to resources like food and water). It is easier to understand how controlled single housing of animals developed when you consider what came before. For example, in many parts of the country, pigs would die from predation and disease "plagues", and laboratory rats would be pulled from large trash-filled rooms by researchers wearing reinforced gloves. These conditions were often not the bucolic natural conditions that might be imagined. Indoor isolated housing avoided a lot of premature death and harm to animals. But the question soon arose … at what cost (Rushen & Passillé, 2014)?

Animal welfare science had a role in exposing the more passive but still serious harm caused by isolating animals when they are biologically and psychologically adapted to live in social groups. Nevertheless, balancing the harm from social deprivation against that from a violent conspecific is a frequent and challenging problem. One example of this dilemma is the *cross-sucking* neonate. When newborn kittens are raised without their mother, they will often express their motivation to nurse by sucking on each other, focusing on any protruding body part (Delgado et al., 2020). They do so with enough persistence that they can injure projecting areas, such as the genitals of male kittens. Foster carers have developed ingenious methods to protect kittens from each other, like putting them in socks or using dividers. However, if the recipient kitten is in danger of being seriously injured, the kittens will have to be physically separated. This same phenomenon occurs with many animals born in sibling groups and raised without the mother, such as piglets (Chou et al., 2022) and calves. Adults kept in groups can also harm each other by *feather pecking* (where poultry pluck feathers from each other, wounding or even killing each other, followed by cannibalism; Rodenburg et al., 2013), tail biting in pigs, barbering in mice (a form of systematic fur and whisker plucking), mounting, and other aggressive behaviors.

Some levels of these harmful behaviors are entirely natural. For example, when piglets are nursing, their outwardly projecting teeth (called milk teeth or needle teeth, which evolved to help them to drive siblings away from "their" teat) might scratch a "competitor piglet" and the mother. It was common for this to be managed by clipping or grinding the piglets' teeth, but as this practice is painful, it is actively discouraged and regulated. Other methods are needed to prevent these injuries, such as keeping litters small and supporting good milk production to reduce conflict.

Many conspecific harms will have a natural underlying cause made worse by constrained housing, open areas, crowding, large group sizes, grouping animals of the same age, genetic effects, and other factors. These have outcomes such as animals prevented from retreating from the line of sight of an aggressor; animals unable to recognize all the members in a group, which impacts reputation building; and a lack of older animals to stabilize social norms. Another example of harmful social grouping is found in fish, such as zebrafish (*Danio rerio*), which are widely used in toxicology testing (Lee et al., 2022). These fish are social and experience isolation stress. In a natural setting, they would associate loosely, leaving or joining shoals at will, and can easily escape *agonistic* behavior (relating to conflict or aggressive combat). In the laboratory, these fish are kept in smaller stable groups with no opportunity to hide. This can lead to serious aggression, especially at lower stocking densities. Under laboratory conditions, pair housing, which is common for breeding and testing, appears to be considerably more stressful to these fish than isolation, as they will tend to fight or harass each other.

Another example of this tradeoff is the movement toward group housing of pigs. Much of the industry across the United States needs access to California as a market, given that California produces little pork but comprises 13–15% of the national market for it. After the passage of Proposition 12 (passed November, 2018 and fully implemented January, 2024), swine raised for breeding must be given 24 square feet (about twice the area of a bathtub), which in practice typically means a move to group housing. Research has associated this group housing with an increased euthanasia rate for lameness and other injuries related to fighting or mounting. So at least during the transition period, some pigs, such as subordinate sows, seem to be on the losing end of this welfare outcome. Sometimes, the most dangerous time for an animal is the condition of flux between a mandated improvement and its implementation.

Cross-harming can be difficult to resolve because it is driven by multiple causes across domains. These damaging behaviors tend to wax and wane and are often described as "outbreaks" when they flare up. Research can struggle to find clear causes for the behavior or might produce contradictory results. For example, rates and severity of pecking between hens have been found to relate, variously, to their early environment, their genetics, pigmentation, being more fearful, being more socially dominant, learned behavior, and nutritional deficiency. However, reviews suggest that it is important to recognize that the pecking itself is often a mis-directed foraging behavior, not aggression, and that this intensifies with stress and close confinement (like the maladaptive behaviors described previously). Difficulties in reducing these behaviors to the lower levels found in wild birds relate not only to their complex causality but to difficulties in addressing their root cause in large-scale captivity.

Keywords: agonistic behavior, barbering (especially referencing mice), cross-sucking, feather pecking

The social lives of "unsocial" animals—snow leopard

Even animals that are typically not described as social may have social needs that are critical to their welfare (Macri & Patterson-Kane, 2011). Take, for example, snow leopards (*Panthera uncia*), which lead a predominantly solitary life but still value some degree of social contact in captivity. Captive snow leopards miss out on many of the natural activities of this very wide-ranging, territorial, and predatory animal (Learmonth, 2019). Therefore, living in harmonious social groups, such as a female with her offspring, may be more enriching and enjoyable than a solitary life.

In addition to this, the interactions snow leopards have in areas where their territories meet may also enhance welfare and be an important natural behavior. These "borders" can be simulated in captive settings using scent enrichment. Keepers use scents from other snow leopards or large felines, or chemically similar substances. They may also use scents from prey animal species, or scents that are simply novel and interesting. For example, some zoos accept donations of unwanted human perfume products to share with their animals. (Not having a great sense of smell, humans often neglect the importance of scent to other animals.)

Some zoos will even rotate solitary predator species (that cannot be co-housed) between enclosures. They can then encounter the scent, claw marks, and other signs of their "neighbors" and interact with the environment in a more natural way that promotes a more diverse behavioral repertoire. Over time, these *rotational exhibits* are being used in increasingly creative ways for both predators and prey species in multi-species groups. This allows animals to have an expanded experience of conspecifics and other species while managing the risk of harmful interactions. So, while many species of animals do not live in constant social groups, it is acknowledged that the pared-down social interactions that they do have are valuable to them and should be accommodated in some form. (Check out the Zoo360 project to see captive animals, including big cats, use aerial walkways to explore their environment [https://www.philadelphiazoo.org/zoo360/].)

Keyword: rotation exhibit

12.3 SOCIAL STRUCTURE

One of the earliest concepts developed in natural history and ethology was the idea of *dominance*. If you place food near two hungry animals from the same group, the one that claims the food will tend to be seen as dominant, and the subordinate will avoid conflict by allowing itself to be displaced. Dominance (the ability to claim limited resources) was initially seen as a linear hierarchy, with the animal at the top seen as the *alpha*. This led to many simplistic ideas about "being the alpha". Over time, the roles of dominance and leadership have been found to be considerably more flexible and malleable.

When this author (EPK) was a master's student, I was sent with some grain to map the dominance hierarchy of a group of goats in the pasture. This was a bit of a trick exercise, as our instructors knew that the biggest and most confident goat was friends with the smallest and most timid. So, a good observer would notice that the small goat got access to the grain when his big friend was nearby but not when he was elsewhere in the paddock. This provided a simple example of how a linear hierarchy, even when it applies, can be an incomplete description of what is happening in a group. Certain individuals are also often exempt from social conflict; for example, infant rats may snatch food from a large adult male's mouth without resistance. An adult male trying the same thing would not fare so well! In many settings, a linear hierarchy does not apply at all, such as in groups that have fluid and interchangeable roles, groups with no need to compete for resources, or groups with a large and variable membership.

The tendency for some animals to live in groups created unique strengths and weaknesses for the individual members. Animals that tend to move in herds were easier to domesticate (e.g., horses), whereas those that tend to scatter when threatened resist these efforts (e.g., zebras). And individuals may take different roles in the group. For example, chicken hens are less fearful and aggressive to each other in the presence of a rooster, who has a vigilance and defense role (Odén et al., 2005). The reassuring role of the rooster is just one example of how conspecifics support each other's welfare by providing basic safety needs and reinforcing meaningful relationships between individuals, such as friendships, parenting, and pair and family connections.

Keyword: dominance

12.4 REPRODUCTIVE INTERACTIONS AND HUMAN MANAGEMENT

When we view pets in a yard or a farm animal in a field, it may be easy to overlook parts of their life cycle that are not immediately apparent at that point in time or that have been eliminated by our management choices. Reproductive behaviors represent a large segment of natural animal behavior, but in captive settings, management choices relating to sexual maturity, mating, and reproduction can have significant impacts on animal welfare.

When reproduction is not controlled, this can lead to the proliferation of animals in unsuitable housing, contested issues with feral animals on public rangelands (Figure 12.2), or out-of-control populations of hoarded animals. Once spay and neuter surgery became available for pets, it was rapidly adopted as part of the effort to reduce the killing of unwanted dogs and cats (see Part II).

Figure 12.2 Wild horses of northwest Colorado on range managed by the Bureau of Land Management (BLM). Wild horses are routinely captured on public rangelands, sometimes driven by helicopters, prior to fertility control, adoption, or removal to holding pens. These actions are taken to preserve the rangelands from overgrazing. Needless to say, families are split, animals are distressed, and some die. The Wild Free Roaming Horses and Burros Act of 1971 is supposed to protect these animals as recognition of their status as a national symbol of the West and its pioneer spirit. Photographs reproduced with the permission of Janet Patterson-Kane.

In many settings, animal reproduction is prevented by keeping animals alone or in single-sex groups. But regardless of whether mating is possible, the normal characteristics and behaviors of sexually intact animals may be considered a problem. For example, the castration of male chickens, called *caponizing*, reduces aggression and improves the amount and quality of meat produced. Mature male pigs have organic compounds such as skatole and androstenone in their fat, which gives the meat an unpleasant gamey taste referred to as "boar taint". This is avoided by piglet castration (surgical or by immunocastration), or by slaughter before sexual maturity.

Deliberate breeding of animals may be performed through selective introduction, restraint, or artificial insemination. The influence of the sire (male parent) used in artificial insemination can be rapid and profound. In dogs, this is referred to as the *popular sire effect*, where a favored male is bred prolifically, narrowing the gene pool (Quartuccio et al., 2020). This constriction of genetic diversity can spread disease or pathology that goes unnoticed until it is too late and has become established across the population. The spread of *rage syndrome* (a genetic seizure-like aggression disorder) in English Springer Spaniels is thought to be partly attributable to the popular sire effect, as are a number of other diseases in dog and horse breeds.

Gestation is a time when animals are more vulnerable. Body configurations considered desirable in adults may make natural birth dangerous, difficult, or impossible. This is seen in the routine use of cesarean section in some breeds, such as the Belgian Blue, a cattle breed that carries a myostatin mutation and is heavily muscled, meaning that even calves are so large that natural birth is likely to result in *dystocia* (difficult labor or delivery; Coopman et al., 2004). There has been a growing understanding that maximizing reproductive output can unacceptably impact the welfare of the mother and offspring. With dogs, this is seen in increasing regulation at the state level of the minimum age at which puppies can be separated from the dam and litter for adoption, the permitted frequency of breeding in commercial kennels, and veterinary approval that animals are healthy enough to breed (Andersen et al., 2021).

Finally, weaning often becomes a difficult process in captivity. Laboratory rats find it difficult to wean their young, as they cannot leave them in the burrow and go elsewhere, and so pups will harry the mother and extend the nursing period. In other settings, like dairy farming, weaning of calves tends to be early and sudden, with the calf being removed for sale as bob veal, which means they are raised on controlled amounts of waste milk or bulk milk until they can eat solid food. Separation of calves shortly after birth has poor acceptance from the public, as it is seen as unnatural. However, leaving the calves for longer with their mothers means a stronger attachment and more distress when separation does occur (Lorenz, 2021).

Keywords: caponizing, rage syndrome (dogs), dystocia

12.5 SOCIAL STANDING

Complex social interactions are based on individuals being able to recognize each other and remember their shared histories (Semmann et al., 2004; Riehl et al., 2016). That is, animals have *reputations* (e.g., a widespread belief that an individual has certain characteristics, such as being cooperative or grumpy), although some species are less worried about their own reputation (Engelmann et al., 2012). In general, proactive, non-obligate-group cooperation is seen as being partly driven by direct and indirect reciprocity. Group members may help others in the hope of seeing their largess reciprocated. For this reason, individuals must be able to recognize each other to make accurate value judgments and remember who owes them a favor! An individual might exclude those who "cheat" and break the rules of group membership: for example, a chimpanzee that tries to take a larger portion of meat than they deserved because they only played a minor role in the hunt.

Taking part in joint activities and maintaining relationships and standing may take a lot of time and effort. Take *allogrooming*, for example (maintaining or cleaning a conspecific; Russell & Phelps, 2013). This behavior is common in primates, where it helps with body maintenance and parasite removal, stimulates endorphins, and is mutually enjoyable. Evidence shows that allogrooming is part of developing relationships and group standing. Therefore, witnessing allogrooming can provide a lot of information about the welfare of individuals and their functioning within the social group. Reputation building can be very difficult to simulate or compensate for in captive settings where natural activities like migrations are impossible and foraging or defense against predators becomes redundant. Captive settings may also disrupt normal methods of reputation management, for example, by keeping dairy cows in herds that are too large to recognize and remember individuals, or housing animals of the same age, sex, and body size. Humans, through processes like imprinting (where a newborn animal fixates on their human carer), may also find themselves filling an unanticipated conspecific social role without understanding the rules for navigating this harmoniously. This issue can be very serious with primates raised as pets when they reach sexual maturity and begin to show sexual and aggressive behavior toward humans.

Keywords: allogrooming, reputation

12.6 STOCK HANDLING

One important *applied* aspect of group welfare comes from farmers and stock people with experience of working with herd animals and other socially cohesive groups. However, the proportion of people with this experience is vanishingly small, as fewer than 2% of Americans work on farms, and even fewer work directly with groups of animals in expansive areas such as open pasture. The loss of this common knowledge is glaringly obvious in situations when law enforcement officers need to wrangle escaped cattle, which sometimes leads to shooting the animal rather than using effective driving or tethering techniques. Even people going into livestock-handling jobs may lack appropriate mentor-based or direct knowledge, or have picked up problematic habits, such as using electric prods or tail twisting. It's not surprising that multiple exposés have shown handlers physically beating animals to move them.

There have been attempts to make livestock handling more standardized and skilled using formal standards and training programs. An important early example was the book *Livestock behavior: A practical guide* by Ronald Kilgour and Clive Dalton (1984). Stockperson training programs include Merck's Animal Health Cattle Care365 training and initiatives run by extension services at many universities. These resources aim not only to improve animal welfare but also to protect the health of workers and the reputations of employers and the industry. Facility design is also critical in aiding livestock movement and safety (Grandin, 1980).

One of the first principles taught about moving large animals such as cattle is how to work with the animal's *flight zone*. A flight zone is conceptualized as a roughly oval shaped area around an animal. If a person enters this area, the animal will move away from them. With cattle, it is recommended to generally avoid entering the flight zone to prevent triggering stress and potentially dangerous behavior. Cattle accustomed to gentle human handling might have a flight zone as small as a few meters, and those that are wild or range reared might have one as large as 50 meters.

When cattle need to be moved, the handler stands at the edge of the flight zone, behind and somewhat to the side of the animal (outside the blind spot immediately behind them). Animals that are very tame will not be driven in this way, as they approach instead of fleeing from humans. These animals may need to be led. It is becoming more popular to suggest leading even flighty cattle when they are in groups, taking advantage of group cohesion: for example, by showing the leading cattle the way to resources like food and water, and allowing the rest to follow.

Using a leader animal can dramatically reduce stress and improve cooperation. This tendency was traditionally exploited using a *Judas goat* (later including a Judas sheep or steer), which was a trained goat that would be added to a herd of sheep and then used to lead them to slaughter. The Judas goat's life would be spared so that they could be used again. Stories of these animals were common in the early 1900s, and the practice is probably an ancient one. Handlers are also taught never to separate and isolate herd animals like cattle, sheep, or pigs, as it will provoke distress and potentially dangerous and vigorous escape behaviors. When herd animals do need to be separated, they should be allowed to maintain visual contact with the group by using a "see through" fence.

Keywords: flight zone, Judas goat

READINGS

The ubiquity of interactions

Budiansky, S. (1999). *The covenant of the wild: Why animals chose domestication*. Yale University Press.

Morris, D. (1990). *The animal contract: Sharing the planet*. Virgin Books.

Palmer, C. (2011). The moral relevance of the distinction between domesticated and wild animals. In T. Beauchamp & R. G. Frey (Eds.), *The Oxford handbook of animal ethics* (pp. 701–725). Oxford University Press.

Example: Transport-related elephant rhinoceros conflict

Berger, J., & Cunningham, C. (1998). Behavioral ecology in managed reserves: Gender-based asymmetries in interspecific dominance in African elephants and rhinos. *Animal Conservation*, *1*(1), 33–38.

Poole, J. H., & Moss, C. J. (1981). Musth in the African elephant, Loxodonta africana. *Nature*, *292*, 830–831.

Slotow, R., & Dyk, G. (2001). Role of delinquent young "orphan" male elephants in high mortality of white rhinoceros in Pilanesberg National Park, South Africa. *Koedoe: African Protected Area Conservation and Science, 44*(1), 85–94. https://doi.org/10.4102/koedoe.v44i1.188

Slotow, R., Van Dyk, G., Poole, J., Page, B., & Klocke, A. (2000). Older bull elephants control young males. *Nature*, *408*(6811), 425–426.

Social isolation versus cross-harm

Chou, J. Y., Marchant, J. N., Nalon, E., Huynh, T. T., Van de Weerd, H. A., Boyle, L. A., & Ison, S. H. (2022). Investigating risk factors behind piglet facial and sow teat lesions through a literature review and a survey on teeth reduction. *Frontiers in Veterinary Science*, *9*, 909401.

Delgado, M. M., Walcher, I., & Buffington, C. T. (2020). A survey-based assessment of risk factors for cross-sucking behaviors in neonatal kittens, Felis catus. *Applied Animal Behaviour Science*, *230*, 105069.

DiVincenti Jr, L., & Wyatt, J. D. (2011). Pair housing of macaques in research facilities: A science-based review of benefits and risks. *Journal of the American Association for Laboratory Animal Science*, *50*(6), 856–863.

Lee, C. J., Paull, G. C., & Tyler, C. R. (2022). Improving zebrafish laboratory welfare and scientific research through understanding their natural history. *Biological Reviews of the Cambridge Philosophical Society*, *97*(3), 1038–1056.

Rodenburg, T. B., Van Krimpen, M. M., De Jong, I. C., De Haas, E. N., Kops, M. S., Riedstra, B. J., & Nicol, C. J. (2013). The prevention and control of feather pecking in laying hens: Identifying the underlying principles. *World's Poultry Science Journal*, *69*(2), 361–374.

Rushen, J., & Passillé, A. D. (2014). Alone or together: A risk assessment approach to group housing. In M. C. Appleby, D. M. Weary, & P. Sandøe (Eds.), *Dilemmas in animal welfare* (pp. 169–187). CABI.

The social lives of "unsocial" animals—snow leopard

Learmonth, M. J. (2019). Dilemmas for natural living concepts of zoo animal welfare. *Animals*, *9*(6), 318.

Macri, A. M., & Patterson-Kane, E. (2011). Behavioural analysis of solitary versus socially housed snow leopards (Panthera uncia), with the provision of simulated social contact. *Applied Animal Behaviour Science*, *130*(3–4), 115–123.

Social structure

Odén, K., Gunnarsson, S., Berg, C., & Algers, B. (2005). Effects of sex composition on fear measured as tonic immobility and vigilance behaviour in large flocks of laying hens. *Applied Animal Behaviour Science*, *95*(1–2), 89–102.

Reproductive interactions and human management

Andersen, S. S., Meyer, I., Forkman, B., Nielsen, S. S., & Sandøe, P. (2021). Regulating companion dog welfare: A comparative study of legal frameworks in western countries. *Animals, 11*(6), 1660.

Coopman, F., Gengler, N., Groen, A. F., De Smet, S., & Van Zeveren, A. (2004). Comparison of external morphological traits of newborns to inner morphological traits of the dam in the double-muscled Belgian Blue Beef breed. *Journal of Animal Breeding and Genetics, 121*(2), 128–134.

Lorenz, I. (2021). Calf health from birth to weaning-an update. *Irish Veterinary Journal, 74*, 1–8.

Quartuccio, M., Biondi, V., Liotta, L., & Passantino, A. (2020). Legislative and ethical aspects on use of canine artificial insemination in the 21st century. *Italian Journal of Animal Science, 19*(1), 630–643.

Social standing

Engelmann, J. M., Herrmann, E., & Tomasello, M. (2012). Five-year olds, but not chimpanzees, attempt to manage their reputations. *PLoS One, 7*(10), e48433.

Riehl, C., & Frederickson, M. E. (2016). Cheating and punishment in cooperative animal societies. *Philosophical Transactions of the Royal Society B: Biological Sciences, 371*(1687), 20150090.

Russell, Y. I., & Phelps, S. (2013). How do you measure pleasure? A discussion about intrinsic costs and benefits in primate allogrooming. *Biology & Philosophy, 28*, 1005–1020.

Semmann, D., Krambeck, H. J., & Milinski, M. (2004). Strategic investment in reputation. *Behavioral Ecology and Sociobiology, 56*, 248–252.

Stock handling

Grandin, T. (1980). Observations of cattle behavior applied to the design of cattle-handling facilities. *Applied Animal Ethology, 6*(1), 19–31.

Kilgour, R., & Dalton, C. (1984). *Livestock behaviour: A practical guide*. Granada.

Chapter 13

Interactions—Heterospecific

13.1 INTERACTIONS BETWEEN NON-HUMAN ANIMAL SPECIES

Many scientists are comfortable with the term conspecific but rarely use its counterpart *heterospecific*, meaning interactions between individuals of different species. In this section, heterospecific means interactions between two or more non-human animals of different species. In the natural environment, animals of different species have strategies to avoid conflict, confusion, or competition. For example, frog species have different types of call, and lightning bugs (fireflies) use different types of flashing signal to attract the right mate.

The presence of heterospecifics can also provide important information. For example, aquarists know that their prized specimens are less timid (and more likely to be visible and appreciated) if they add a school of *dither fish*. This is a general name for small schooling fish such as tetras, barbs, or danios. The presence of these small schooling fish reassures other species like cichlids that there are no predatory fish about and so it's probably safe to move around. The role of other species is often neglected in welfare, even though they represent an intrinsic part of how animals relate to and understand their environment. Other species might also provide valuable opportunities to assess how animal welfare might be supported or challenged. Examples of how these interactions might be relevant to animal welfare are too numerous to cover in any depth, but a few examples will be discussed that relate to predation and the use of other species as a proxy for a conspecific.

Keyword: dither fish

DOI: 10.1201/9781003375456-17

13.2 PREDATOR–PREY

While heterospecific interactions can be benign or even beneficial, in many cases, animals exploit each other, with the most obvious example being predation. Hunting behavior is an important purposeful activity for most predatory species. When idle for long periods of time, wide-ranging predators, such as bears and tigers, often develop *stereotypies* like pacing in captivity (Part II). While they can be provided with live prey, this tends to raise welfare issues for the prey animal and has limited acceptability with the public. In these situations, there are often conflicts between groups that will identify or advocate with one group of animals versus the other. This author (EPK) recalls watching a wildlife rehabilitator carefully return owls to health and think nothing of taking a mouse and breaking its back to make it easier for a recovering owl to hunt and kill.

Across many settings, the feeding of live prey is becoming less acceptable, and as a result, researchers have endeavored to recreate authentic hunting experiences artificially. For example, Hal Markowitz was a pioneer in environment enrichment and would develop devices that encourage natural behaviors, such as chasing prey-like stimuli, dragging through clear tubes, or leaping to catch meatballs thrown into the enclosure. Meanwhile, on the ranch or farm, a major welfare problem is the predation of livestock such as sheep by coyotes. Chickens are a "menu item" for a long list of predators, including foxes, cats, coyotes, hawks, racoons, and snakes. In predator–livestock conflict, animals are often seriously mutilated and do not die quickly. Welfare issues can then be compounded when lethal control methods involve poisons.

One interesting aspect of the predator–prey relationship is that predator evasion behavior may be beneficial for the welfare of many prey species—it may break routine, stimulate natural behaviors, and promote social cohesion. There are also ways that prey species communicate directly to predators. For example, many bird species will engage in *tail flicking*. The tail flick is a message to the predator: "I'm healthy and agile, and you probably can't catch me—so don't bother". Tail flicking may be a form of *honest signaling*, because only a genuinely fit and alert animal can produce the flick and notice its predator audience. As such, it might also be used as a reliable part of an animal welfare assessment. For animals with conservation value, it is important that they maintain their predator avoidance behavior and the capacity to perform it, so that they may eventually be released into the natural environment with a hope of survival. For these animals, *predator avoidance training* typically involves being shown predator cues followed by an unpleasant stimulus, and is most widely used on fish released from hatcheries into open waterways (Edwards et al., 2021).

Keywords: honest signaling, predator avoidance training, Hal Markowitz

Example: Spontaneous predation in zoos and aquaria

In captivity, animals may be kept in monocultures away from natural cohabiting species, or they may be placed close to novel species. These heterospecific interactions may be important for animal welfare and are easily overlooked or neglected, or are complex to assess (Veasey, 2022; Veasey & Hammer, 2012). Zoos are environments that typically have many spontaneous residents, such as rodents, squirrels and racoons, small birds like sparrows, and large birds like geese and crows. These animals introduce many risks to welfare, such as zoonotic disease. Many of the methods usually used to discourage or remove these animals would be dangerous or distressing to zoo animals. As a result, some level of incursion is typically tolerated. For example, in 1985, a killer whale (*Orcinus orca*) in the Vancouver Public Aquarium was seen sharing her food with seagulls (Nordquist & Hutchins, 1985). In 1998, orcas were recorded using fish, sometimes regurgitated fish, to lure and hunt seagulls. This newly observed behavior could be interpreted in multiple ways through an animal welfare lens.

First, it was suggested that gull baiting was a form of play or entertainment for the orcas rather than hunting. If true, this behavior might be seen as a kind of orca-invented environmental enrichment and taken as evidence that their environment was overly barren (Kuczaj et al., 1998). Another suggestion was that the gull baiting was intelligent planned behavior with cultural transmission (Kuczaj et al., 2009). Other orcas would watch and reproduce the behavior, with younger female orcas being the quickest to catch on. Seabird hunting by orcas in the wild is rare, but it does occur, normally by juveniles. Studies of interactions between walruses in the wild and seabirds suggest that the "bird treatment" ranges from *object play* (any play with a physical object) to typical predation with consumption of the bird (Giljov et al., 2017). The same may be true for orcas. Understanding the extent to which a behavior is natural (Capella et al., 2014) informs whether its performance in captivity is an indicator of good welfare (applied intelligence, enrichment) or bad welfare (boredom, maladaptive behavior).

What you are not likely to see is any discussion of the welfare of animals endemic to the areas, like the gulls being harmed by zoo animals. Animals entering the zoo under their own prerogative are not a priority, and their suffering or death is often ignored. Particularly vulnerable animals are those with statuses such as being small and less charismatic, being unowned, being common, or being classed as a pest or vermin species (remember the possum ejected from an enclosure by the orangutan in Part I). Extra meaning was given to the captive orca seagull story when journalists suggested that the behavior was an ominous prequel to lethal interactions with humans, most notably the 2010 death of SeaWorld trainer Dawn Brancheau. Orca Tilikum (captured from the seas around Iceland in 1983) grabbed Brancheau and played with her "like a toy" (Kirby, 2012; Zimmermann, 2014). We should note that at this point Tilikum had already been involved with two other deaths: Daniel Dukes, who gained access to Tilikum's tank in 1999, and trainer Keltie Byrne in 1991 (Anderson et al., 2016).

Keywords: object play, predation

13.3 CONSPECIFIC PROXY

There are also cases where the spontaneous entry of an animal into a zoo enclosure has unpredicted positive effects. Take, for example, the story of a small calico cat that, in 2007, entered the enclosure of a Eurasian lynx (*Lynx lynx*) in St Petersburg/Leningrad Zoo. The lynx is considered a solitary predator that does not bond with others of its species. However, videos of the lynx and cat show very typical social behaviors, such as allogrooming, and other evidence of a bonded relationship. It is notable that the cat entered the enclosure when the lynx was very young, potentially enabling a sibling-like connection. In 2015, in another Russian zoo (Novosibirsk Zoo), staff noted that a "runt" lynx was doing poorly and removed her to a nursing mother cat, who accepted her.

The 2004 European Association of Zoos and Aquaria (EAZA) guidelines say: "Established pairs and groups can be kept together continuously (for enrichment), although incidents may occur", which is a tacit indication that accidents might happen (Krelekamp, 2004). For example, a previous study had shown that co-housed lynx will form close social bonds when there is only one male and no pressure to compete for resources. Domesticated cats might act as a proxy for a conspecific when introduced during a sensitive period, such as a juvenile requiring maternal or sibling interactions and relationships to support their welfare. It appears that under these settings, cats from a different species can take the role of a conspecific. Is the acceptance of domestic cat companionship by lynx really an indication that they should be socially housed? Perhaps not, as there are also anecdotal reports (EAZA, 2004) that even lynx kept together for long periods of time may attack each other. This is where the absence of empirical data becomes a problem. Anecdotes do not tell us how serious this risk is, the scenarios to be avoided, or early indicators that sociability is absent or breaking down; just as we do not know how many stray cats got into enclosures holding solo lynx and were attacked or even eaten. So, while cohousing with conspecific and heterospecific cat species can occur, the extent to which this is acceptable or positive for welfare remains unclear.

Moving toward social housing poses several hazards to animals' welfare that are particularly acutely felt in relation to charismatic wildlife. So, facilities may take a conservative approach and solo house lynx, especially males, or they may go in a different direction and allow their lynx to mix with individuals of other exotic cat species such as serval.

Keyword: conspecific proxy

Example: Cheetah & dog

The idea of socializing a wild cat species with a domesticated animal was taken further with the use of dogs as social partners for cheetahs. This practice was established in 1990 by Dr Laurie Marker, founder of the Cheetah Conservation Fund. The domesticated dog provides a trusted companion that demonstrates stability and calm, especially around humans. The success of this approach is even more surprising given the differences between felids and canids, and the fact that canids would, in an evolutionary sense, have been competitors of cheetahs in the wild (e.g., the African wild dog, *Lycaon pictus*; Hayward, 2008).

Nevertheless, it was found that a juvenile cheetah would accept a dog as a companion, but only *that* dog. So, the ability to be a conspecific proxy, if this is indeed what is happening, depends on the dog and cheetah meeting during the cheetah's sensitive period. San Diego zoo introduced this technique and explored it in more flexible ways. For example, by keeping two female cheetah littermates together with two companion dogs; there would be the opportunity to keep the cheetahs together or to separate them, with the support of a dog companion each if necessary (Maple, 2017).

A key reason for this pairing has been to facilitate using cheetahs as an "ambassador animal" that engages the interest of the visiting public (Salamone et al., 2018). This raises the question of how being used in this role, even with the assistance of a companion, compromises the welfare of the cheetah, and the extent to which ambassador-cheetahs benefit cheetahs in their natural environment. It is also worth considering to what extent humans and their pets act as heterospecific friends versus conspecific proxies for each other, based on whether we or they can understand each other's ethology or expectations to adapt to each other's ways. How could we test this hypothesis?

Keyword: Laurie Marker

13.4 PLANNED DOMESTIC ANIMAL RELATIONSHIP

There are many familiar arrangements where different species of domestic animal are kept together. The nature of these relationships varies, with common types being:

> Guardian animals, such as guardian dogs with livestock or in some cases, donkeys or llamas. These animals will aggressively repel predators not only for themselves but also for the flock or herd that they live with and take responsibility for.
>
> Much like the cheetah and dog, some animals are chosen for being a calming and reassuring influence. Some highly strung thoroughbred horses have a goat companion who helps calm their nerves when they travel to races.

Co-owned pets and livestock are often kept in mixed-species groups. Aquariums, for example, combine fish that originate from radically different environments. In some cases, naturally symbiotic species can also be kept together, such as tanks with clownfish and sea anemones or pistol shrimp and sand gobies, where they retain their natural interactions in a captive setting.

A multi-species household is as complex to manage as a multi-species zoo exhibit, with the added complication that the "zookeeper" lives on site. The pet owner holds absolute responsibility for understanding the needs of the individuals and their relationships. The 2022 American Veterinary Medical Association (AVMA) pet ownership and demographics sourcebook estimated 37 million US households with an average of 1.78 cats (*Felis catus*), suggesting that multi-cat households are common. These cats are increasingly "indoors-only" because of the safety aspects of traffic, predatory wild animals, and the protection of native wildlife. In some municipalities (e.g., the city of Fort Collins in Colorado), it is illegal to have a pet "at large"—so no outdoor cats unless they are leashed. This places the welfare needs of the indoor animal squarely in the hands of the human owner. Needs for exercise, social contacts, and natural behaviors must be supplied, which makes the home a *multi-species habitat*.

In naturalistic settings, cats, like many larger felines, have social groups based on related females and are highly territorial, with relatively large territories. Cats can find it difficult to maintain separate territories inside the home, although closely related cats (like mothers, offspring, and siblings) may exist more harmoniously. By contrast, there is a strong belief that raising canine siblings together causes a problem called *littermate syndrome*, which provokes either aggression or codependence, although this has yet to be formally researched. Cats as a species can be very socially adaptable but are also susceptible to stress, which can provoke respiratory infections and *idiopathic cystitis* (cystitis is an inflammation of the bladder, and idiopathic means there is no apparent medical cause). In fact, stress in cats is tightly linked with physical illness, and you will often notice a stressed cat sneezing because of reactivation of a respiratory virus. There are pros and cons for housing cats together with indoor versus outdoor access. Indoor housing can reduce the single cat's opportunities for stimulation or force several housed cats into proximity without adequate territorial space.

The granularity of negotiating a multi-cat, multi-species household can be daunting, and so we thought we'd share some anecdotal evidence (TR) around adopting two cats, and then introducing them to two dogs, and then moving them all (Figure 13.1).

Keywords: multi-species habitat, littermate syndrome

Two cats, two dogs, and much behavior

Adopting cats M & P: In December 2019, two young cats were adopted from a shelter. They'd been rescued some months earlier, but the arrival of a human baby meant they were relinquished. They were male cat "M" (six months old) and a female cat "P" (nine months old). M broke out of his cardboard carrier on the way home, an early portent of character. The paperwork said "bonded pair", but they obviously weren't siblings. In retrospect, I suspect that the emotional glue that held them together was adversity, and with time, its effects wore off.

Toilet training went well. M was inquisitive, outgoing, and cuddly. He had obviously been socialized during the sensitive period. P was less social, lithe, and flighty. There were no obvious issues with resources, and they had a shared

Figure 13.1 Left: Clockwise from top left, cat M, puppy P, cat P, and dog E. Right: cat M enjoying his newfound freedom by practicing his climbing skills. Photographs reproduced with the permission of Janet Patterson-Kane.

litter box and individual feeding stations—they were indoor cats. Six months later, we had to move, and to complicate matters, a new face was in the frame.

Adopting dog E: "E" was a stray from Houston, Texas, rescued after having several litters and a relocation to Colorado. She looked like a dingo; her paperwork hopefully described her as a "shepherd mix". To mitigate any territorial claims, everyone moved to the new house together, and it worked well. The cats arrived, and "lo and behold", the house came with a dog! The cats slotted into new, bigger territories, and E settled down. Cat P made friends with easygoing E.

Adopting puppy P: Four months later, puppy "P", a 10-week-old purebred border collie with American Kennel Club (AKC) papers, rounded out the household. P's main drama was playing out in her colon, with uncontrolled diarrhea. Fastidious cat P "cleaned" up the pads left for accidents, folding them and pushing them into a corner before the pup could use them. Eventually, we figured out that our pup's microbiome was unhealthy, which was confirmed and fixed with a fecal transplant (literally the poop from dogs with a healthy microbiome in a pill; Tuniyazi et al., 2022). Pup P has managed three-dimensional poops ever since and soon formed an alliance with cat M—they play together.

As the cats matured, they became more territorial. M wanted out, crying incessantly. No amount of vertical space would do for him. So, the bird feeding station was wire fenced and the grass kept short so he was visible when outside. Using AirTags to track him was futile, as he kept pulling them off. Luckily, cat M was not a skilled hunter and showed none of the athleticism of cat P. Going outside did wonders for his body, and his tummy "pouch" began to go. He became stronger, and his expanded territory reduced conflict with cat P. He seemed content, and every evening would come in on demand; his time outdoors was timed to avoid the dusk/dawn predation hours. Cat P, still more elusive and disinterested in going outside, was provided with hiding spaces that she seemed to rotate through (boxes under the bed, cupboard doors left open). Cat M would encounter conspecifics outdoors and made friends with one—they frequently hung out together. His vaccinations were changed to reflect his outdoors lifestyle. In the end, it wasn't tragedy that curtailed his freedom; it was foxtail grasses getting caught on his fur and the dangers they presented. As we had no control over those outside our yard, his experiment with the great outdoors ended.

The second move was halfway across the country. Feline bathroom behavior in the new house deteriorated. Two litter boxes and cat feeding stations were provided in the basement, but someone was standing on top of a litter box and urinating. Litter boxes were rearranged, getting the biggest ones with privacy flaps, and placing them out of sight of each other. Add two cat trees and a window cradle, and things seem to have settled down. The streets are too busy for cat M to go outside, so he's back indoors. With one entrance to the basement, there was scope for resource guarding, but that hasn't materialized. The cats seem to have agreed-upon territories. There are mixed positive and negative valence interactions between them, but the nose bumps are less noisy and therefore less obvious than the negative hisses.

Our indoor/outdoor arrangement seemed to work well while it lasted. Healthwise, cat M's modest weight issues were resolved by going outdoors and have not returned now that he's back inside. Cat P developed painful resorptive lesions on her teeth, and removing those teeth markedly improved her playfulness. Given the remarkable prevalence of this barely studied disease, many cats are likely to be withdrawn because they have a painful mouth, irrespective of anything else going on.

Dog E has remained her easygoing delightful self, and puppy P, now adult, still plays very happily with cat M. (Note that the American Association of Feline Practitioners [AAFP] and the International Society for Feline Medicine have taken cat welfare very seriously and have incredibly informative websites about how best to care for your cat.)

As an update, a recent paper from the AAFP contains an in-depth study and set of recommendations for a multi-cat household (Rodan et al., 2024). As a result, we realized that cat P's reclusive behavior was partly due to her not being able to defend her territory from cat M. So, we rearranged the furniture, which improved her ability to monitor her territory. She got a cat tree, and she is going to get a collar-activated cat flap in the door. She is already noticeably happier and sits in her travel cat box surveilling her room.

13.5 MANAGING HARMFUL INTER-SPECIES INTERACTIONS

Interactions between species have the potential to cause harm, for example through the transmission of disease. Cattle are susceptible to tuberculosis, which can be transmitted by local wildlife, such as the common brushtail possum in New Zealand and (supposedly) the badger in the United Kingdom, although this is disputed. One popular method of reducing this risk was to kill the implicated wildlife species. However, an interaction-based approach to transmission begs the question: why and how are cattle and these wildlife species meeting? Possums are inclined to stay in wooded areas, and badgers stay close to their sett (also called a den or burrow).

Research determined that killing badgers increases population mobility, as survivors will migrate into vacated areas. This extra travel increases their chances of encountering and infecting cattle. That said, even the notion of who infects whom, and how, remains extremely contentious as cattle testing is notoriously inaccurate (see Part II, Figure 6.2). Recent work has now confirmed that badger culling has been pointless (Torgerson et al., 2024), which comes as no surprise to scientists who, long ago, accused British government bureaucrats of cherry picking the data to support its case for the cull. The United Kingdom's Badger Trust has called for cattle-based TB controls and improved hygiene and husbandry rather than a badger "slaughter". Meanwhile, keeping resident badgers alive and vaccinated reduces how infectious they are to cattle, to each other, and to their own offspring, while possums only contact cattle when they become ill and disoriented and stagger into a pasture: an event that can be prevented by installing possum-proof fencing.

Good fencing makes good neighbors for farmers and coyotes as well. And perhaps, *not* doing the thing that comes naturally to many livestock owners, which is to shoot a coyote on sight, *is* the thing to do. It may sound strange, but sometimes "good" coyote neighbors that don't predate stock should be left alone, as their presence will deter other, perhaps more adventurous and ravenous coyotes from moving in (Shivik et al., 2014). Coyotes live in a fission/fusion society—meaning they can live individually or in a group (Flores, 2016). Persecution can provoke fission, dispersal, and the settlement of new territories.

Sheep-killing coyotes are generally older and less timid coyotes that need to feed larger litters (Jaeger et al., 2001). And if a rancher kills those coyotes, the "fusion" pack is fractured. Fewer coyotes means more livestock available for any new coyote arrivals, which provokes the females to birth more pups to fill the niche. The result is an even bigger demand for food, and in situations of insufficient wild prey species, more livestock are taken. So, the strategy of killing this highly successful predator can backfire. As noted by the New Hampshire Wildlife Coalition, "the harder coyotes are hunted, the faster their population grows".

Keywords: fission/fusion society

READINGS

Predator–Prey

Edwards, M. C., Ford, C., Hoy, J. M., FitzGibbon, S., & Murray, P. J. (2021). How to train your wildlife: A review of predator avoidance training. *Applied Animal Behaviour Science*, *234*, 105170.

Example: Spontaneous predation in zoos and aquaria

Anderson, R., Waayers, R., & Knight, A. (2016). Orca behavior and subsequent aggression associated with oceanarium confinement. *Animals, 6*(8), 49. https://doi.org/10.3390/ani6080049

Capella, J. J., Abramson, J. Z., Vilina, Y. A., & Gibbons, J. (2014). Observations of killer whales (Orcinus orca) in the fjords of Chilean Patagonia. *Polar Biology*, *37*, 1533–1539.

Giljov, A., Karenina, K., & Kochnev, A. (2017). Prey or play: Interactions between walruses and seabirds. *Acta Ethologica*, *20*(1), 47–57.

Kirby, D. (2012). A killer show. *Serial-All Animals 2009–15, 14*(4), 7.

Kuczaj, S., Lacinak, T., Garver, A., & Scarpuzzi, M. (1998). Can animals enrich their own environment. In *Proceedings of the third international conference on environmental enrichment* (pp. 168–170). San Diego, CA: Shape of Enrichment Inc.

Kuczaj II, S. A., Gory, J. D., & Xitco Jr., M. J. (2009). How intelligent are dolphins? A partial answer based on their ability to plan their behavior when confronted with novel problems. *Japanese Journal of Animal Psychology*, *59*(1), 99–115.

Nordquist, C., & Hutchins, M. (1985). Killer whale shares food with gulls at the Vancouver Public Aquarium. *Zoo Biology*, *4*(4), 367–374.

Veasey, J. S. (2022). Differing animal welfare conceptions and what they mean for the future of zoos and aquariums, insights from an animal welfare audit. *Zoo Biology*, *41*(4), 292–307. https://doi.org/10.1002/zoo.21677

Veasey, J. S., & Hammer, G. (2012). Managing captive mammals in mixed species communities. In D. G. Kleiman, K. V. Thompson, & C. Kirk Baer (Eds.), *Wild mammals in captivity; Principles & techniques* (2nd ed., pp. 151–161). Chicago University Press.

Zimmermann, T. (2014). *Killer in the pool*. New Word City.

Conspecific proxy

Krelekamp, C. J. (2004). *Husbandry guidelines Eurasian lynx (Lynx lynx sspp.)* (pp. 9–34). European Association of Zoos and Aquaria (EAZA).

Example: Cheetah & dog

Hayward, M. (2008). Prey preferences and dietary overlap amongst Africa's large predators. *South African Journal of Wildlife Research*, *38*, 93–108. https://doi.org/10.3957/0379-4369-38.2.93

Maple, K. (2017). *Connecting the 'spots' between visitors and cheetahs with companion dogs*. ABMA Board of Directors.

Salamone, K., Meurer, J., & Legoll, K. (2018). San Diego Zoo Global's Ambassador Cheetahs (Acinonyx juba-tus): Training, welfare and their role in conservation. In the proceedings of the annual ABMA conference. *Angie Llanas Conference Committee Co-Chair Christina Burgess Conference Committee Co-Chair* (p. 141).

Two cats, two dogs, and much behavior

Rodan, I., Ramos, D., Carney, H., DePorter, T., Horwitz, D.F., Mills, D., & Vitale, K. (2024). AAFP intercat tension guidelines: Recognition, prevention and management. *Journal of Feline Medicine and Surgery*, *26*(7). https://doi.org/10.1177/1098612X241263465

Tuniyazi, M., Hu, X., Fu, Y., & Zhang, N. (2022). Canine fecal microbiota transplantation: Current application and possible mechanisms. *Veterinary Sciences*, *9*(8), 396. https://doi.org/10.3390/vetsci9080396

Managing harmful inter-species interactions

Flores, D. (2016). *Coyote America: A natural and super-natural history*. Basic Books.

Jaeger, M. M., Blejwas, K. M., Sacks, B. N., Neale, J. C. C., Conner, M. M., & McCullough, D. R. (2001). Targeting alphas can make coyote control more effective and socially acceptable. *California Agriculture*, *55*, 32–36.

Shivik, J. A. (2014). *The predator paradox – ending the war with wolves, bears, cougars, and coyotes*. Beacon Press.

Torgerson, P.R., Hartnack, S., Rasmussen, P., Lewis, F., & Langton, T.E.S. (2024). Absence of effects of widespread badger culling on tuberculosis in cattle. *Sci Rep, 14(1),* 16326. https://doi.org/10.1038/s41598-024-67160-0

Chapter 14

Interactions—Animal and Human

14.1 HUMANS AS INTERACTING ANIMALS

There is a good case to be made for referring to human–animal interactions as simply part of that realm of heterospecific interactions where one of the species just happens to be human. However, the scientific study of this type of heterospecific interaction vastly exceeds the study of any others. So, for the purposes of clarity, it makes sense to use the term most used in the field, that of *human–animal interactions*.

These interactions are divided over two chapters: the first focuses on how animals affect humans, and the second, vice versa. This is also obviously a false dichotomy. When people in a community are bothered by urban coyotes, are these coyotes really causing anxiety to humans, or is it the human encroachment on coyote habitat that's bothersome? It goes both ways, but in most cases, one part of the exchange is the subject of more study and holds the power. This first section focuses on some important areas where animals are thought to affect people, specifically in healing, animal-assisted interventions, humane education, tourism and conservation, and the use of animals in general as understood through the lens of the "ancient contract" (the last section of this chapter).

Keyword: human animal interactions

14.2 ANIMALS AS HEALERS

Science has long recognized the benefits for humans of interacting with animals. These benefits go beyond our ability to exploit animals for tangible gain as commodities, and enter the realm of psychology and spirituality. That is, interaction with animals is typically part of what makes the human experience complete or enriched. Interacting with animals can produce therapeutic benefits for people and a uniquely effective type of support for those who fall between

DOI: 10.1201/9781003375456-18

the gaps of human society (Kerr-Little et al., 2024). For beneficial interactions, some of the mechanistic aspects of the underlying biology have been deduced. For example, some interactions with animals stimulate the release of *oxytocin*, a hormone produced in the hypothalamus, which is stored and released on demand by the posterior pituitary. Oxytocin is associated with social bonding and positive mood (Beetz et al., 2012).

That said, some people are not very interested in animals or fear being harmed by them. There are many common memes that say or imply that we should be suspicious of people who don't like pets. These sayings include: "Dogs are not our whole life, but they make our lives whole" (Roger Caras); "I'm suspicious of people who don't like dogs, but I trust a dog when it doesn't like a person" (misattributed to Bill Murray); and "I think having an animal in your life makes you a better human" (Rachael Ray). Psychological studies have shown no inferiority in people who are disinterested in pets, or in animals in general. Any differences in personality between owners of different pet species and non-pet owners tend to be subtle and nuanced, and people may appreciate animals in different ways, such as seeing them in the wild (Johnson & Rule, 1991). Of course, seeing the spiritual value of animals is uncontested in many Indigenous societies where they are associated with clan and kinship systems.

A recently expanding area of research investigates how being an animal guardian is not purely beneficial (Kogan et al., 2023a). For example, research on the challenges people encounter during illness and pet-related guilt (modeled on parental and caretaking guilt) shows that pet owners experience variable amounts of guilt about potential deficits in the care of their animals, which is associated with feelings of anxiety and depression (Kogan et al., 2023b). Of course, significant distress may also be experienced due to pet loss, pet-incurred debt, and the worry that the elderly have over who might look after their pet when they pass. These are all very real concerns for many people.

Keyword: oxytocin

Example: Nature cures

The healing power of nature is strongly expressed in most cultures and traditions. Drawing on animals to improve human physical, mental, or spiritual wellbeing is a very broad set of traditions. This includes using animal products as medicine (zootherapy) and employing animals as a means for spiritual growth, such as in shamanism (Stutley, 2003), animism, and humane education. Human *peak experiences* often occur in the shared natural environment (peak experience being a transcendent moment of joy, understanding, and self-development, a term associated with Abraham Maslow and Humanist psychology). Performatively, this might be called being in the zone or flow.

However, just quiet moments of wonder also apply, in which the self takes a back seat, the "monkey brain" shuts up and time stands still. This author (TR) remembers a vacation on Westray, one of the Orkney Islands off the Northern tip of Scotland. During a walk on the beach in quite blustery weather, rain then sun, I was alone yet felt watched. Turning seaward, I locked eyes with my stalker; a seal, vertical inside a wave, looking straight at me. So, we literally "stood", looking at each other, one wave bound, one on land. Land to sea, human to animal. Perfect.

Across the history of medicine, remedies have been developed that return a person to a natural environment to cure a specific disease or general malaise. Hot springs and other sources of water were often reputed to be healing. In Greco-Roman philosophy, this was represented by retreats to the countryside, which was often idealized as a structured but gentle and worry-free place. (The "Arcadian Myth" was the earthly paradise, named after a sparsely populated Greek mountain province, that is now more commonly used in reference to an idyllic life in a farm and nature setting.) In the United Kingdom, healing spaces could also be found in verdant park-like surroundings, with farm animals or pets, where one could stroll and recuperate. Some used large and imposing central buildings in a farm setting, popularized as the Kirkbride plan (Pérez-Fernández & López-Muñoz, 2019), while others preferred housing people in scattered cottage-like buildings. Famous asylums that attempted to provide this environment include Overbrook in New Jersey (1896–2007) and the Rockwood Asylum in Kingston, Ontario (1859–2000).

In America, the idea of a healing space shared with animals reached its zenith in the wild. Henry David Thoreau (1817–1862) is one writer who contributed to this lasting connection between American nature, the robust responsibility of American citizens, and the importance of civil disobedience. A Harvard graduate, Thoreau threw himself into building a small house and living off the land, writing a series of essays that became the literary classic *Walden* (1854). In this work, animals are vibrantly present as competitors, food, and neighbors, and notably, ones he found more amiable than any human. Thoreau wrote:

> we need the tonic of wildness … At the same time that we are earnest to explore and learn all things, we require that all things be mysterious and unexplorable, that land and sea be infinitely wild, unsurveyed and unfathomed by us because unfathomable. We can never have enough of nature.

A modern version of this is the *new nature writing* genre of literature, exemplified by Richard Mabey's (2007) *Nature cure: A story of depression and healing*. Mabey, recovering from a bout of depression, revels in the mesmerizing ecosystem of the watery fenlands of eastern England.

Where the British tended to dream of an idyllic farm life, by the late 19th century, the post-colonial American vision was more of a "wilderness cure". In America, a more muscular ideal was pursued, and the ideal man became the outdoorsman or cowboy, in tune with his horse, dog, and stock, and conqueror of the perils presented by wildlife. The mythical archetype, developed during the golden years of the Adirondacks (1830–1865), was the backwoodsman. This man (they were generally men) would be spiritually cleansed and returned to a state of innocence by his struggles in the wilderness; a life that was morally superior to a decadent city lifestyle (Recarte, 2010). One outgrowth of this tradition was wilderness exposure camps for youth, first developed in Germany and the United Kingdom as "Outward Bound", which were adapted in the United States as "Youth Leadership Through Outdoor Survival" at Brigham Young University (Utah) in 1966. These seemed to combine the inspiring aspects of nature with the challenge of undertaking demanding physical tasks far from the support of civilization.

More extreme versions arose in American wilderness camps for troubled youth. Here, a certain demanding harshness was considered part of the treatment, often of adolescents with what were considered serious antisocial or violent behaviors. On reflection, some of the participants in these "boot camps" consider them to have been more like child abuse by proxy, where the experiences included being underfed, exhausted, and experiencing extremes of temperature and poor hygiene. Exposés of preventable deaths in these camps led to the development of a third-party accreditation system via the Outdoor Behavioral Healthcare Council. The complex connections in American culture between health, hardship, abuse, being in nature, and exploiting nature provide a complex background for attempts to advocate for wildlife and invasive species: for example, to try to limit trapping for fur, manage wild horse populations, or reintroduce locally extinct species. Within one community, beliefs will range widely from spiritually exalting an animal such as a wolf to thinking the only good wolf is a dead one.

Keywords: Arcadian myth, wilderness cure, Henry David Thoreau

14.3 CONFLICT—PREDATORS IN HUMAN CULTURE WARS

The reintroduction of wolves into sparsely populated rural areas can cause passions to run high. But this is more of a human issue, where the animal is emotionally pivotal to the argument and yet is largely, if not completely, incidental to human lives. Large predator reintroduction presents an unusual mix of interactions between wolves and their prey (especially elk) and between wolves and humans. Some people (e.g., tourists) see wolves as an attraction and enjoy their presence, but the thought of living with them provokes anxieties in others, even when largely unfounded.

Colorado Proposition 114 was to develop a plan to reintroduce the gray wolf and to compensate for any livestock losses, and it narrowly passed in November 2020. Supporters pointed to the positive trophic cascades for the ecosystem after reintroducing wolves to the Yellowstone National Park, effects that seem less set in stone than this author had imagined (see Hobbs et al., 2024). In December 2023, Colorado Parks and Wildlife (CPW) relocated 10 gray wolves from Oregon onto state-owned land in the Summit and Grand counties of Colorado's Western Slope. Ranchers, hunters, and outfitters (commercial hunting guides) vehemently opposed Prop. 114; they felt that sheep, cattle, and elk would be killed and that their outdoor recreation businesses would suffer. Some Coloradans voiced concerns for the safety of their children.

Analyses of the worldwide data for wolf attacks between 2002 and 2020 (Linnell et al., 2021) showed that the risk *to humans* posed by deadly wolf predation is not zero but is too low to calculate. There were two fatalities in 18 years in North America (from rabies) with no deadly (wild) wolf attacks against humans (from predation) in the lower 48 in the last 100 years. Moose are far more dangerous to people and much less predictable than wolves. But a moose isn't a wolf and doesn't have its reputational baggage.

Coloradan wolves are covered by rule10(j) of the Endangered Species Act (ESA). That rule allows agencies to manage wolves as if they were threatened instead of endangered. In effect, rule 10(j) green-lights non-lethal deterrence (e.g., using noise makers or a fladry—flaps of fabric on the fence line) or if absolutely necessary, lethal control. Human opposition to the wolves was a mix of risk perception and social ideology. The risk you have some control over is more tolerated than the risk that is imposed. If the risk is imposed by a politician, then you can expect some blow-back in our current political climate. The cognitive shortcuts (heuristics) that we take when assessing risk are nuanced and will probably differ for rural versus urban Coloradans.

Many opponents of Proposition 114 felt aggrieved because the vote was pushed through by urban voters on the Front Range (a north–south-running mountain range tracking Denver's metropolitan corridor), whereas the wolves would be reintroduced to their "backyards" on the other side of the Continental Divide. Personality types are also of relevance in *group processes* (individual behaviors within small working groups). Some people affirm their social identity in groups that practice *positive ingroup bias*, a process that asserts superiority over other (out)groups. Humans already resort to exceptionalism to assert their superiority over animals (Plous, 1993). Ingroups go further and use *infrahumanization* to claim authority over other groups of people. In contrast, human social groupings that embrace their similarities to others and animals have positive benefits for animal welfare through increased empathy and a desire to protect animals (Amiot & Bastian, 2019).

The *Social Dominance Human-Animal Relations Model* (SD-HARM) shows that speciesism and generalized prejudice against other people share common ideological roots (Dhont et al., 2016). Understanding these group dynamics can inform messaging and productive interactions, for example, by referencing shared beliefs nested within larger, more contested issues (e.g., discussing a love of the outdoors before turning the conversation to more controversial topics). As far as the Colorado wolves are concerned, one has already been killed by a mountain lion, and radio collar activity would suggest that some of them risk unwittingly visiting Wyoming, where they might be shot on sight. One pack (the pair and four pups) has been captured after a rancher applied for a permit to kill after the wolves predated his livestock. The permit was denied, in part because the rancher had failed to accept help from Colorado Parks and Wildlife, and because of the discovery of an open "dead pit" of carcasses on the ranch that could attract the wolves. CPW held off recapturing the wolves while the female was denning. All six were then recaptured, but the male, who was underweight and with leg injuries, died a few days later. This project remains a work in progress, with CPW affirming their willingness to work with ranchers (https://cpw.state.co.us/living-wolves#230548828-902623122).

Keywords: group processes, infrahumanization

Janet Kessler and the coyotes of San Francisco

We decided to include this special contribution from a self-taught coyote expert, Janet Kessler. Janet, known as the "Coyote Lady" by locals in San Francisco, has 17+ years of coexistence with urban coyotes and is a walking treasure trove of "coyote-lore". Janet has been interviewed many times by media outlets and featured in a short film that you can watch for free on the WaterBear media platform (https://www.waterbear.com/watch/dont-feed-the-coyotes). Indeed, one of us (TR) bumped into Janet when on the trail of coyotes in a canyon in San Francisco a few years ago, and we became friends. Janet's data is the mother lode of coyote behaviors, life stories, and most precious of all, her stunning photo-reportage. This is where animal watching turns into narrative ethology; storytelling across the human–animal divide. One of the key differences between Janet and others with self-taught expertise is her outreach. Janet has her website and an Instagram platform, and is often around to chat to people about coyotes and give talks. We thought we'd let Janet speak for herself, and she wrote this piece for us. The brief was an introduction to her work with some biographies of the coyotes that she knows so well. The aim was to demonstrate their individuality (Figure 14.1). We have not edited the text, and some will no

Figure 14.1 Identity parade—putting faces and names together for nine coyotes of San Francisco that Janet will describe. You can't help but notice the differences, and with time, their personalities reveal themselves as well. Some humans, however, are not very good at "reading" or recognizing individual animal faces. Imagine how you might use the morphological features of the coyote's face to create a facial recognition tool. Look at this website to see how incredibly detailed facial landmarks are being created for use with cats: https://www.tech4animals.org/catflw. You can follow the trials and tribulations of these coyotes at coyoteyipps.com and the Instagram platform, including the latest on Scout and her territorial headaches. Space to live in San Francisco is hard to come by, for coyotes and for humans, and Janet has meticulously mapped the territories of these coyotes, revealing a city divided by more than twenty family units. For details, see https://coyoteyipps.com/2024/09/. These photographs are reproduced with the permission of janetkessler/coyoteyipps.com.

doubt "police it" for perceived anthropomorphism. We have left it to Janet to describe and interpret what she is seeing—she's the expert.

By Janet Kessler

Coyotes reappeared in San Francisco in 2002 after many years of absence, and people continue to be in the dark about them. I created a website and an Instagram platform to inform people about WHO these creatures really were: I wanted to help bring light to their behavior and offer simple guidelines for easy coexistence. https://www.instagram.com/coyoteyipps/ and https://coyoteyipps.com/.

My passion, and the work that stemmed from this, began on an extended pre-dawn walk with my dog in 2007. I had just sliced through a tendon in my finger which put an end to my pedal-harp practice for a while. On that early morning, I "met" my first coyote and was blown away by her acute awareness, personality, curiosity, but especially by her ability to cross the species-divide in her curiosity about us: it was as though I had met another person — personhood. I began to see this coyote regularly on my morning walks. The more I saw, the more I learned and the more I wanted to know, so this became my prime mover for the next couple of decades.

I've now been documenting coyote behavior, family life, and their population in San Francisco, including interactions with pets and people, over the last 17 years, and have been called "a pioneer in the photo-documentation of coyote behavior and family life" by the journalist who first interviewed me. My information is neither generic nor second-hand. At the same time, it's neither impersonal nor personal: I know the coyotes without forming mutual relationships with them, without interfering or interacting.

Early on I asked Professor Ben Sacks at UC Davis if he would perform DNA analysis from the scat I was collecting from each individual coyote (observing many defecations as they occurred allowed me to know which coyote belonged to many of the scats). All my findings had been visual up to that date, but I wanted more "scientific" data to support what I had seen. Ben found students to do the analysis, so my work became the lynchpin for those studies. My compensation was that I would be added as an author to those studies. The first paper (Caspi et al., 2025) has just been published.

I basically came to know and identify each individual coyote, their personalities, their families and their territories. It's from identifying each coyote that I was able to study their behaviors and family interactions, and I was able to map their territories here in the city. On some days I saw nothing; but on most days, covering several territories, I came across a handful of coyotes. "Knowing" doesn't mean I had a personal relationship with any of them — I didn't. Mine is a hands-off study and there was always a wall of separation. Most people don't see this behavior because coyotes are highly secretive: they don't want to be seen. But also, their activity is mainly in the evening when we aren't around to see it, and when it's too dark to see it, even though they are diurnal animals, the same as we are — they are not nocturnal.

What I'd like to emphasize here is how individual coyotes are. Of course, that's how I got to know them, so that's how I see them: as individuals, and not simply as one big lump: "coyotes". I've put together nine photos to show how different each one is. Most people will want to identify a coyote by its markings. This doesn't work because, of course, the coats are shed in the summer, but also, different lighting conditions and how the environment reflects that light can cause coyotes to look quite different under different weather and environmental conditions. So, I identify them by their faces — their facial features — no different than from the way we have facial recognition for other humans: that's where their character lies, these are the "windows into their souls".

These individuals have their own stories and I can never know the WHOLE of any of their stories, but each seems to have revealed something special to me about themselves, and I thought I would briefly summarize a little of that here.

SCOUT's story is a famous one. Based on what I've written on my blog, she's been incorporated into a documentary film and immortalized in a section of a book. Over her nine year life, so far, this female has morphed from a fun-loving, carefree and spirited loner for 4 years, more interested in watching the human and dog world than anything else, into that of a dedicated mate and mom, deeply absorbed in her own family life and that of her offspring, and in defending them. Her story includes successfully fending off an attempted territorial takeover in her loner days, and repeatedly moving her pups — carrying them one by one — to safer locations.

CHERT was Scout's mother. Here she is 11 years old with brand new pups. Chert was also free-spirited, but circumstances — the animosity and disappearance of her mother — caused her to move into the alpha position before she was ready. I honestly don't think she knew what hit her when she had her first pup. Chert never dispersed and has always lived on the same territory she was born on. She was best buddies with her brother, Ivan. I recently found out through DNA that it wasn't Silver, but Ivan who fathered her first offspring. Silver kicked Ivan out shortly thereafter, and Chert then stuck with one mate (who was her father) until he died, and since then has had 3 other mates. Her first pup was a singleton: Scout. Chert used to comfort Scout when her de-facto father was too hard on her. Did Silver know Scout wasn't his -- might this be why Scout was picked on and eventually dispersed early at only 9 months of age?

MOUSE. I met Mouse right after she lost her first litter: someone didn't want coyotes under their porch and feared they would den there, so he boarded it up, only the pups had already been born there. Frantic commotion by the coyote parents was not understood until it was too late. Her mate disappeared: either the trauma of losing the pups caused him to leave her, or he was killed by the city for becoming particularly defensive against all dogs. She soon attracted another mate, Hunter. She became a fierce protector of her next litters. Sadly, she was killed by a car. Cars are chief coyote killers in urban areas.

HUNTER became a suitor to both Scout and Mouse at the same time. He would spend time with one and then the next day would be at the other's territory. I had never seen so much joy from Scout when he was with her. He's the one who turned her attention from the human world to the coyote world: their friendship was physical and psychological: the interactions were intense and heartwarming for observers to see. BUT he decided to stick with Mouse, probably because of the chaos at Scout's territory caused by her dealing with the attempted territorial takeover. Hunter suffered a broken ankle early on in life which healed.

Years later, after his mate Mouse was killed by a car, that wound came back to haunt him. He was driven from that territory by a new pair of coyotes.

IVAN dispersed when he was kicked out, and he was forced to forget about his first love (his sister). He was the epitome of an alpha. It's from him that I learned what it means to be an alpha: an alpha isn't a "bully" lording it over his family, but rather the one who stands back and accepts everybody and makes sure all goes well. He had to deal with a situation of antagonism between his mate and his daughter who grew to dislike each other. He would regularly approach the shunned daughter to spend time with her and groom her. He and his mate eventually left this territory, in effect ceding it to this daughter, but she was eventually driven out by the mate she had tried to attract. I also watched him play with the runt of one of his litters. The runt wasn't going to make it, and Ivan probably knew this, but he encouraged the pup to play and play-hunt alone as his more robust brothers roughhoused nearby.

CYRANO was two years old in this photo. I doubt if he survived long: you can see the wound on his snout. In addition, his two back ankles were bulbous from early injuries — they never healed properly, and one foreleg ankle was stiff from yet another wound. This fellow was shy and fleeting; at the same time he was curious in a very subtle way. He avoided his brothers — of course they always dominated him, the weaker one. And then one day I never saw him again.

STUMPF is in here for his unusually lean looks, even in the dead of winter; during the summer he looks absolutely skeletal. He has always been more secretive and evasive than the other coyotes I've observed, but I think this is an important aspect to point out. Be that as it may, I regularly saw his silhouette in the distance as he accompanied his two yearling sons, teaching them the ins and outs of city life and hunting together with them in the evenings.

MA'AM's mate was a ne'er do well who she herself tried "saving" from his involvement with humans. His name was Monte. She was another very secretive coyote, but I saw her at various times trying to get her mate to leave the lure and enticement of human contact. She did this through fierce eye contact which her mate well understood — he averted his eyes to avoid it. She used her body and forearms to push him to leave areas where he was hanging around. She ended up relocating them five miles away to get away from the human situation — it's the females who tend to choose territories and densities — but this didn't stop her mate from begging and spending time away from her. He wasn't around for her. So she found another mate: this is the first "divorce" I had ever come across.

MONTE began his life normally but was changed by humans handfeeding him. I hope folks remember that we humans often shape WHO the coyotes become by our own interference. I've seen other coyotes fed, but this one succumbed to the lure: I've always thought that his innate personality — very easy going in nature — contributed to how he reacted to humans. He began to hang around listlessly, allowing dogs and people to mingle only a few feet from himself. His mate attempted extracting him from the situation by moving five miles away: he dutifully followed her but found other feeders. After she "divorced" him, he returned to his old homestead where a son had been holding down the fort. He found a new mate and began raising a new litter when his life was brought to an end by our Park Department for baring his teeth at a child in his denning area. Coyotes, of course, communicate by showing their teeth — it makes us humans pay attention. But, denning area or not, the city disposed of him. The feeding had caused him to feel much too much at ease around humans.

14.4 HARMONY—BIOPHILIA AND ZOOBIQUITY

In 1984, Edward O. Wilson proposed his *Biophilia Hypothesis*, arguing that humans are hardwired by evolution to have an innate love of nature (Wilson, 1984). A recent meta-analysis of the emotional evidence for his theory (used to inform *biophilic building design*—human–nature connections indoors) would agree (Gaekwad et al., 2022). The "nature" concept doesn't need to involve specific animals as such but is a more general regard of the natural environment.

The 2013 book *Zoobiquity: The astonishing connection between human and animal health* (Natterson-Horowitz & Bowers, 2013) argues for another connection, that between human and animal health. This work showed the commonality between many of the health problems experienced by human and non-human animals and pointed to the need to break down some of the artificial barriers between veterinary and human medicine (the One Health approach). This has been proposed many times before, and the relationship remains heavily unidirectional.

A critique of this approach is that it might underestimate the evolutionary complexity of disease (Greek, 2012). However, Natterson-Horowitz seems to have fully embraced the idea of investigating the intersections of evolution and ecology in her co-authored review "The future of evolutionary medicine: Sparking innovation in biomedicine and public health" (Natterson-Horowitz et al., 2012). This new medicine (sometimes referred to as Darwinian medicine) broadens the usual discussions of "evolutionary war" between host and pathogen by, for example, considering the disease burden provoked by an unbalanced ecology.

Changes are also afoot in *veterinary clinical research*. Veterinary clinical studies look at spontaneous illness in animals to learn about effective treatments, with the animal treated as the patient and not a research subject. The investigation of animal health is undertaken to benefit the animal in a fashion that does not reduce their standard of care and may, or may not, produce data relevant to human patients. This is a deliberate attempt to put a finger on the scale for animals. When a disorder is relevant to both animals and humans, medical trials can provide data for both parties. Currently, this is playing out with a bone cancer that is much more common in dogs than in humans (Mannheimer et al., 2023; LeBlanc & Mazcko, 2020).

Biophilia and zoobiquity are examples of particularly popular attempts to shift our paradigms to embrace how humans and other living things are related. These works are only partially novel, as they restate an ancient thread of understanding about our interconnectedness with the natural world that is still practiced in many communities outside the Global North. Their effect and impacts on how we manage and understand domestic and wild animal populations are arguably rather limited, but there is much scope for development.

Keywords: biophilia, veterinary clinical trial/research, zoobiquity

14.5 ANIMAL-ASSISTED THERAPY AND ENCOUNTERS

Currently, there are many active fields of healing people via the use of direct interaction with specific animals (Serpell, 2010). This includes the use of a variety of species to improve a person's ability to function (e.g., socially, emotionally, and cognitively). In the United States, these are generally divided into "activities" without a formal therapeutic goal but with some attempt to provide health or wellness benefits. *Animal-assisted therapies*, with a treatment goal, are often provided by a licensed professional and emotional support animals that reside with the client. These therapies include therapeutic riding/hippotherapy, animal-assisted play therapy, and animal-assisted psychotherapy, which often uses dogs but may include any species. (Service animals are a separate category, as they are trained primarily to assist with a disability rather than as a treatment modality.) There are many excellent resources to learn more about animal-assisted therapies and how they can be employed. Good places to start would include the International Association for Human-Animal Interaction Organizations (https://iahaio.org/) and Pet Partners (https://petpartners.org/).

Formal consideration of the welfare of the other side of the equation, the animals used as therapy partners, was somewhat belated. However, it is now understood that animal participation in animal-assisted programs should be voluntary and at least benign for their welfare and preferably beneficial. This is sometimes referred to as the mutual benefit model, or the *doctrine of mutual benefit*. Across all modalities of animal-based healing, the welfare of the animal has been a secondary and rather overlooked issue. However, across settings, this aspect is being more carefully considered, with attention paid to whether the use is intrinsically unsuitable for the animal as a species or as an individual. It is expected that any domesticated animal engaged in activities designed to be therapeutic to people should have their welfare actively monitored and protected. There should be an appropriate selection of animal/species, with the requisite skills, temperament, and opportunities to retreat, to rest, and to be retired. As the animal's owner or the client may have a vested interest in continuing the activity, a third-party welfare assessment or compliance with standards (such as The Standards of Practice in Animal-Assisted Interventions; therapyanimalstandards.org) may be needed to ensure that the animal's interests are properly represented.

Experiences with wildlife are often the most difficult to manage, as wild animals are less able to tolerate human contact without experiencing some kind of negative impact. Activities based on captive wildlife, such as dolphin encounters, are becoming less acceptable. Encounters with wildlife in nature are also being scrutinized for when they might amount to wildlife harassment (defined as an act that disregards the animal's welfare and causes a disruption of their normal behavior or risk of injury). Guidelines for responsible tourism are available at World Animal Protection US (https://www.worldanimalprotection.us/our-work/wildlife-friendly-travel/).

Keywords: animal-assisted therapy, wildlife harassment, the doctrine of mutual benefit

14.6 HUMANE EDUCATION

In America after World War I, there was a resurgence of interest in providing moral education to children. One approach to this was the use of pet keeping and animal stories in the context of humane education, with the aim of teaching kindness to living things. In the early 20th century, many states made humane education a compulsory part of the educational curricula, and these laws remain on the books in California, Florida, Illinois, Maine, New Jersey, New York, Oregon, Pennsylvania, and Washington (but were repealed in Louisiana in 2013). Their influence subsequently waned, and many of the laws are modest in scope and not actively enforced. California's law asks that "[e]ach teacher shall endeavor to impress upon the minds of the pupils ... kindness toward domestic pets and the humane treatment of living creatures".

Humane education is currently defined as "an approach to teaching that draws connections between human rights, animal protection, and environmental sustainability" (https://humaneeducation.org/what-is-humane-education/). Humane education is made available by many animal welfare and community organizations as supplementary education, through community events, and to visitors to locations such as animal sanctuaries.

Keyword: humane education

14.7 INJURIES FROM ANIMALS (BREED-SPECIFIC LEGISLATION)

Animals also represent a significant source of injuries and death for people. These include workplace injuries such as from cattle, which are becoming more serious as the average age of those in the farming profession increases. Equestrian activities include risks of serious injury from falls. Hitting wildlife on roads or in the air can be particularly dangerous, as are attacks or other incidental damage from wildlife. Interactions with pets, such as trips and falls, bites, and destructive behaviors such as chewing on cables and tubing (such as tubing for the continuous positive airway pressure [CPAP] machines that help people breathe) are all hazards that must be managed. These risks are generally accepted as part and parcel of sharing our lives with pets, and our planet with animals. Reasonable precautions to mitigate these risks include the use of fencing to move livestock (a single file chute), riding helmets, seatbelts, bird deterrents, rules against allowing dogs to wander the streets in packs, and flexible corrugated covers for electronic cables and tubing. However, there are also methods for reducing these risks that place an unreasonable burden on the animal and/or are not based on sound evidence: for example, the removal of claws or teeth, or illegal poisoning. One highly contested approach is the use of breed-specific legislation to try to reduce the rate or severity of dog bite injuries. Breed-specific legislation refers to any law that limits or prohibits ownership of certain animals, typically particular breeds of dogs. These types of laws are currently commonly used to reduce ownership of pit bull–type dogs.

The dogs associated with serious bite attacks have varied and historically included Great Danes, Doberman pinschers, and German shepherds. In Australia, there was legislation for a time relating to greyhounds that became a hazard when being trained in public parks. In the north of America, sled-type dogs were more often implicated. It seems logical that if a certain breed or type of dog is implicated in serious bite attacks, then suppressing ownership of that type would increase public safety. However, the role of breed is generally not found to be empirically important relative to the dog being neutered, trained, and responsibly owned. For this reason, breed-specific legislation is quite widely opposed, including by the American Veterinary Medical Association, the American Kennel Club, the National Animal Control Association, and the National Association of Insurance Commissioners. These groups favor dangerous animal legislation that is not breed specific but instead focuses on the responsibility of the individual animal owner or guardian.

Keyword: breed-specific legislation

14.8 THE ANCIENT CONTRACT

How do we justify our treatment of animals, and how did humans and animals find themselves in the position of exploiter and the exploited? Is this really how the relationship was envisioned, and did it break down somehow? The polemic text that describes the groundwork for the human–animal contract is referred to as *the Ancient Contract* after an article (1989) and then a book by Stephen Budiansky called *The covenant of the wild* (1992). The book describes a hypothetical contract that animals made with humans, which suited both parties (think of a fairytale or fable, imagining how this might have come about if both parties could speak and negotiate in advance).

The idea of the contract is meant to remind people of the "deal" we made with animals. We provide for them in return for the proportional benefits that we extract from animals. But now, we have technologies that can compel animal cooperation. We use engineered walls to confine them, motorized transport to move them large distances, and tooth or horn removal to reduce their ability to resist. And many domesticated animals would struggle to adapt to life without human support. Others have suggested that the metaphor of a contract might have the reverse effect of making us feel complacent about exploiting animals. They argue that domestication has made animals dependent on us, and therefore vulnerable to us, and that this is not a moral justification for its continuation.

There is also discussion and a certain amount of controversy about the notion of human self-domestication, whereby pro-social companions were selected (who suppressed aggression and reactivity) in early human societies to preserve the cohesion of small bands of hunter gatherers. Genome analyses of modern humans and domesticated animals versus their ancestral or wild forms (Neanderthal, wolf, wisent) show a statistically significant sharing of genes associated with domestication traits, such as docility and a more gracile physiognomy (Figure 14.2; Theofanopoulou et al., 2017) .

Others have argued that selection for self-control (not necessarily the loss of aggression) was a driver of human cognitive evolution (see Shilton et al., 2020 for a thorough discussion). What is clear is that some evolutionary tweak in human development 30,000–70,000 years ago resulted in the acquisition of exceptional language skills, abstract thinking and symbolism, and other cooperative skills that represented a major cognitive upgrade; this is the start of behavioral modernity. Reputation management could drive social cohesiveness in small bands of hunter gatherers, but myths, beliefs, and storytelling were

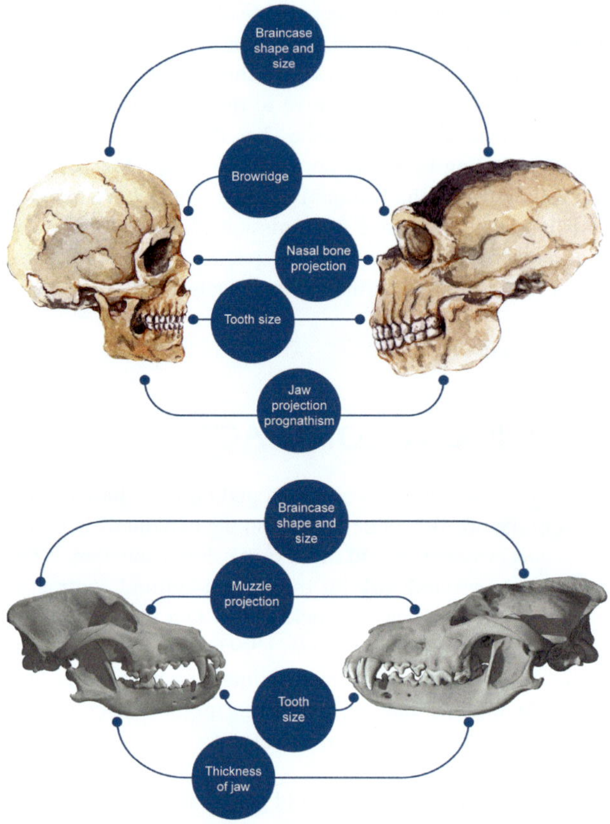

Figure 14.2 Salient craniofacial differences between anatomically modern humans and Neanderthals (top) and between dogs and wolves (bottom). Published (unchanged from original) under CC BY 4.0 https://creativecommons.org/licenses/by/4.0/, in Theofanopoulou et al. (2017). Self-domestication in Homo sapiens: Insights from comparative genomics. *PLoS One, 12*(10):e0185306. With permission to publish from Dr C. Theofanopoulou.

needed to provide the social "glue" for larger groups. While animal "superpowers" may have been to fly, run fast, or breathe underwater; our superpower was language. The associated cognitive wiring to tell each other stories, create hierarchies and tribes, and ultimately devolve power to a very few is contributing to the next cognitive revolution. This appears to be to swap the real world for a virtual one, a uniquely human "species-specific" behavior.

Keyword: ancient contract

READINGS

Animals as healers

Beetz, A., Uvnäs-Moberg, K., Julius, H., & Kotrschal, K. (2012). Psychosocial and psychophysiological effects of human-animal interactions: The possible role of oxytocin. *Frontiers in Psychology*, *3*, 26183.

Johnson, S. B., & Rule, W. R. (1991). Personality characteristics and self-esteem in pet owners and non-owners. *International Journal of Psychology*, *26*(2), 241–252.

Kerr-Little, A., Bramness, J. G., Newberry, R. C., & Biong, S. (2024). Dog ownership for people with substance use disorder: Self-reported influence on mental health and substance use. Preprint. https://www.researchsquare.com/article/rs-4247627/v1

Kogan, L. R., Currin-McCulloch, J., Bussolari, C., & Packman, W. (2023a). Cat owners' disenfranchised guilt and its predictive value on owners' depression and anxiety. *Human-Animal Interactions*. https://doi.org/10.1079/hai.2023.0044

Kogan, L. R., Currin-McCulloch, J., & Cook, L. S. (2023b). Breast cancer treatment and recovery: Pets' roles as emotional buffers and stressors. *BMC Women's Health*, *23*, 540. https://doi.org/10.1186/s12905-023-02662-z

Nature cures

Mabey, R. (2007). *Nature cure: A story of depression and healing*. University of Virginia Press.

Pérez-Fernández, F., & López-Muñoz, F. (2019). The Kirkbride buildings in contemporary culture (1850–2015): From 'moral management' to horror films. *Hist Psychiatry, 30*(3), 336–351. https://doi.org/10.1177/0957154X19839912. Epub 2019 Apr 17. PMID: 30995127.

Recarte, C. A. (2010). The myth of the Adirondack Backwoodsman: From the golden years to consumer society. *Miscelánea: A Journal of English and American Studies, 42*. https://doi.org/10.26754/ojs_misc/mj.20109413

Stutley, M. (2003). *Shamanism: An introduction*. Routledge.

Thoreau, H. D. (1854). *Walden or life in the woods*. Left of Brain Books: Boston, Ticknor and Fields.

Conflict—predators in human culture wars

Amiot, C. E., & Bastian, B. (2019). Psychological mechanisms involved in human-animal interactions. How do humans cognize about animals. In B. Fischer (Ed.), *The Routledge handbook of animal ethics* (p. 21). Routledge.

Dhont, K., Hodson, G., & Leite, A. C. (2016). Common ideological roots of speciesism and generalized ethnic prejudice: The Social Dominance Human–Animal Relations Model (SD–HARM). *European Journal of Personality, 30*(6), 507–522.

Hobbs, N. T., Johnston, D. B., Marshall, K. N., Wolf, E. C., & Cooper, D. J. (2024). Does restoring apex predators to food webs restore ecosystems? Large carnivores in yellowstone as a model system. *Ecological Monographs, 94*(2), e1598. https://doi.org/10.1002/ecm.1598

Linnell, J. D. C., Kovtun, E., & Rouart, I. (2021). *Wolf attacks on humans: An update for 2002–2020*. NINA Report 1944 Norwegian Institute for Nature Research.

Plous, S. (1993). Psychological mechanisms in the human use of animals. *Journal of Social Issues, 49*(1), 11–52.

Janet Kessler and the Coyotes of San Francisco

Caspi, T., Serrano, M.G., Vanderzwan, S.L., Kessler, J., Schell, C.J., and Sacks, B.N. (2025). " Impervious Surface Cover and Number of Restaurants Shape Diet Variation in an Urban Carnivore." Ecosphere 16(1): e70152. https://doi.org/10.1002/ecs2.70152

Harmony—biophilia and zoobiquity

Gaekwad, J. S., Sal Moslehian, A., Roös, P. B., & Walker, A. (2022). A meta-analysis of emotional evidence for the biophilia hypothesis and implications for biophilic design. *Frontiers in Psychology, 13*, 750245. https://doi.org/10.3389/fpsyg.2022.750245. PMID: 35693493; PMCID: PMC9186521.

Greek, R. (2012). Book review, Zoobiquity: What animals can teach us about health and the science of healing. *Animals (Basel), 2*(4), 559–563. https://doi.org/10.3390/ani2040559. PMCID: PMC4494279.

LeBlanc, A. K., & Mazcko, C. N. (2020). Improving human cancer therapy through the evaluation of pet dogs. *Nature Reviews Cancer, 20*(12), 727–742.

Mannheimer, J. D., Tawa, G., Gerhold, D., Braisted, J., Sayers, C. M., McEachron, T. A., Meltzer, P., Mazcko, C., Beck, J. A., & LeBlanc, A. K. (2023). Transcriptional profiling of canine osteosarcoma identifies prognostic gene expression signatures with translational value for humans. *Communications Biology, 6*(1), 856. https://doi.org/10.1038/s42003-023-05208-z

Natterson-Horowitz, B., Aktipis, A., Fox, M., Gluckman, P. D., Low, F. M., Mace, R., Read, A., Turner, P. E., & Blumstein, D. T. (2012). The future of evolutionary medicine: Sparking innovation in biomedicine and public health. *Frontier Scientific, 1*, 997136. https://doi.org/10.3389/fsci.2023.997136. Epub 2023 Feb 28. PMID: 37869257; PMCID: PMC10590274.

Natterson-Horowitz, B., & Bowers, K. (2013). *Zoobiquity: What animals can teach us about health and the science of healing*. AA Knopf.

Wilson, E. O. (1984). *Biophilia*. Harvard University Press.

Animal-assisted therapy and encounters

Serpell, J. A. (2010). Animal-assisted interventions in historical perspective. In A. H. Fine (Ed.), *Handbook on animal-assisted therapy* (pp. 17–32). Academic Press.

The Ancient Contract

Budiansky, S. (1989). The Ancient Contract. *US News & World Report, 106*(11), 74–79.

Budiansky, S. (1992). *The covenant of the wild: Why animals chose domestication: With a new preface*. Yale University Press.

Shilton, D., Breski, M., Dor, D., & Jablonka, E. (2020). Human social evolution: Self-domestication or self-control? *Frontiers in Psychology, 11*, Article 134.

Theofanopoulou, C., Gastaldon, S., O'Rourke, T., Samuels, B. D., Martins, P. T., Delogu, F., Alamri, S., & Boeckx, C. (2017). Self-domestication in Homo sapiens: Insights from comparative genomics. *PLoS One, 12*(10), e0185306. https://doi.org/10.1371/journal.pone.0185306. Erratum in: *PLoS One*. 2018 May 11;*13*(5), e0196700. https://doi.org/10.1371/journal.pone.0196700. PMID: 29045412; PMCID: PMC5646786.

Chapter 15

Interactions—Human and Animal

15.1 HUMAN IMPACTS ON ANIMALS

Humans impact animals in many ways, including indirectly through effects like climate change, construction, pollution, and controlling zoonotic disease. Human-caused activities (with knock-on effects for animals) are called *anthropogenic* factors. This chapter will cover just a few examples. These have been chosen to represent some interesting areas of research and application that illustrate positive, negative, and complex/neutral effects when viewed through the lens of protecting or promoting animal welfare. We have chosen to focus on more direct effects of individual humans on individual animals. Practices based on systems such as economics, culture, and religion will be covered in more depth in Part V.

Keyword: anthropogenic

15.2 GREEN CRIMINOLOGY AND GREEN SCIENCE

Green criminology is defined as the study of harm and injustice to the environment, including non-human animals, and is informed by what has been termed *green science* (Lynch, 1990; Beirne et al., 2018). This term focuses on the sciences able to track environmental degradation, such as ecotoxicology, environmental toxicology, and chemistry. Animal welfare science is not generally explicitly considered in this context, as it relates only to one facet of the overall "green" concern. However, scientific evidence relating to animal welfare is used when relevant to a particular action. Green criminology may address deceptive greenwashing, the protection of natural entities (e.g., environmental personhood), the criminalization of Indigenous environmental activists, or the rights of marginalized groups. Fresh challenges will come from a recent decision from the Supreme Court of the United States (June, 2024) that overturned 40 years of precedent wherein courts deferred to expert federal opinion (termed

DOI: 10.1201/9781003375456-19

Figure 15.1 Scientific expertise weighs in at a solid zero now that judges can decide not to defer to federal expertise when it comes to deciding how to apply the law in the face of uncertainty. As a result, stockholders are already challenging the federal mandate to use electronic ID tags on certain cattle/bison that cross state boundaries. This is a straightforward biosecurity measure that could be under threat, even as bird flu continues to move through the dairy herd. Image by Tina Rich.

Chevron deference; Figure 15.1) for legal gray areas. This decision could enable incalculable harm to the American people and animals through challenging current laws and provisions that protect us all.

Green criminology is also a field within the *critical studies* paradigm. *Critical theory* is used to reveal and challenge harmful power structures—you are being critical of the status quo. Examples include critical animal studies and critical race theory. The overarching goal is to change and improve society in dramatic ways, such as the pursuit of animal liberation. This contrasts with the use of "green" in terms like *green consumerism* (Lynch & Stretesky, 2011). The latter approach aims to improve a system without seriously disrupting it: for example, by encouraging consumers to support products made in more environmentally friendly ways.

A green criminology approach considers animals as individuals and deploys a broader notion of victimhood and harm (e.g., harm to the environment). For example, the pain felt by a rhinoceros shot for its horn will be considered in the context of *that* rhinoceros as a victim rather than as an asset for the continuation of an endangered species. Green criminology is not focused on charismatic or rare animals but any animal that is being exploited (Wyatt et al., 2022). Imagine if we treated every goby fish as if they were as important as a rhinoceros!

Keywords: green criminology, critical animal studies

The marine aquarium fish trade

In the aquarium trade, marine tropical fish are overwhelmingly collected from the wild rather than bred in captivity. Eighty percent of the trade in marine ornaments is sourced in the Philippines and Indonesia and transported to the United States (Shuman et al., 2004; Townsend, 2011). Collection methods such as cyanide fishing (Vaz et al., 2017) damage the environment and threaten many species of fish, with rare fish attracting higher prices and incentivizing unsustainable collection.

Currently, there is extremely little research on the welfare implications for the pet trade on ornamental fish, even though it handles over an estimated billion animals annually. This knowledge gap probably reflects a lack of motivation on the part of funders in tackling this issue. A visit to many pet stores will show fish such as comet goldfish, intended to be "feeders", languishing in deceased states. Data used for industry standards are mainly drawn from research into aquaculture fish and zebrafish (Lee et al., 2022). The zebrafish (*Danio rerio*), a small tropical freshwater fish, is a popular animal model used in laboratory research and is a relatively robust species with simpler care needs than marine ornamentals. There are very obvious gaps between the approved standards and the reality experienced by fish. For example, the MAC (Marine Aquarium Council) standard of a 1% capture and mortality loss (1% dead on arrival [DOA] and 1% dead after arrival [DAA]) is tiny compared with the much higher estimate (although not precisely known) of 10–40%. Can anyone look at these numbers and think that the welfare of aquarium fish is being taken seriously?

15.3 ANIMAL CRUELTY

The typical definition used in academia of cruelty to animals is "socially unacceptable behavior that intentionally causes unnecessary pain, suffering, or distress to and/or death of an animal" (Ascione, 1993). The legal definitions are determined by state law and vary considerably in terms of the actions, penalties, and animals covered (Table 15.1). A few exceptions are covered by federal law, including animal crushing for the purpose of making fetish videos for interstate or foreign commerce. Other federally regulated aspects of animal handling relate to commercial activities that cross state lines, are international, or relate to animals on federal lands. These include The Animal Fighting Venture Prohibition Act, 18 U.S.C. §49; The Horse Protection Act, 15 U.S.C. §1821–1831; The Humane Methods of Slaughter Act, 7 U.S.C. §1901–1907; and The Transportation of Animals, also known as the Twenty-Eight Hour Law, 49 U.S.C. §80502.

Table 15.1 Categories of animal cruelty handled criminally at the state level, or more rarely, by smaller jurisdictions

Active cruelty (animal abuse or maltreatment) Poisoning	Causing physical harm Organized illegal activities	Non-accidental injury ("battered pet")
		Fighting (dog, cock etc.)
		Animal crushing
Cruelty by omission	Withholding routine care	Lack of shelter
		Starvation
		Lack of necessary grooming
		Failure to vaccinate or provide other preventative care
	Failing to respond to need	Lack of therapeutic veterinary care
		Small- or large-scale neglect associated with mental illness or incapacity ("hoarding")
	Institutional/ Commercial setting	Improper husbandry by a breeder ("puppy mill", "backyard breeder")
		Improper husbandry by a shelter, rescue, or sanctuary
		Farm animal neglect

Table 15.1 outlines some general categories of animal cruelty that are handled criminally at the state level, or occasionally by smaller jurisdictions, with some examples of the terminologies used to describe them. One important area of animal welfare science is the development of evidence that can be used in criminal cases, such as forensic veterinary medicine and forensic behavior evaluation.

One area where there has been increasing interest is *animal hoarding*. First studied in the 1980s, animal hoarding was specifically called out as a special manifestation of hoarding under the Obsessive Compulsive Disorder in the *DSM* (Diagnostic and Statistical Manual of Mental Disorders 2014), specifically: "the accumulation of a large number of animals and a failure to provide minimal standards of nutrition, sanitation, and veterinary care and to act on the deteriorating condition of the animals (including disease, starvation, death) and the environment (e.g., severe overcrowding, extremely unsanitary conditions)".

Hoarding is a situation that illustrates that a caretaker having strong attachment to an animal is not always enough to keep them safe. Animals in hoarding situations often suffer severe neglect, and their human caregivers have little insight into the harm that they are causing. Some of these cases relate to diagnosed or suspected hoarding disorder, but a larger category encompasses all cases where a care deficit would not have occurred under the care of a reasonable person. The *reasonable person test* is commonly used in the law to help a jury determine whether actions are negligent by comparing them with the actions of a hypothetical individual acting with an appropriate or normative degree of care and attention.

Animal hoarding is a condition that is remarkably difficult to deal with, or intervene in, and individuals can move between jurisdictions to evade detection. Large-scale cases require an integrated response from experts in human and animal services as well as specialist attorneys who can pursue treatment plans, diversion, or prosecution, depending on the circumstances. Most animal hoarders are widowed, divorced, or single; 70–80% are women; half are aged over 55 or live alone; men will tend to hoard dogs, and women are more likely to hoard cats; 80% of hoarders will have diseased, dying, or dead animals on their property, which is usually a fire hazard and unsanitary; 16% of those homes will ultimately be condemned. Many hoarders will have a relative who also hoards, suggesting a strong genetic component (Dozier and Ayers, 2017).

However, the more sensational aspects of hoarding may cause it to attract disproportionate interest relative to the more mundane forms of illegal and legal neglect that constitute the vast majority of animals suffering from human cruelty (Caldwell et al., 2025).

Keywords: animal cruelty, hoarding (animal), reasonable person test

15.4 HUMANS AS POSITIVE INTERACTORS—RAT TICKLING

Humans should give more attention to whether they can bring positive experiences into the lives of animals. This is, however, quite rarely studied. One exception is an interesting line of research into *rat tickling*. Certain reproducible behavior patterns in animals are strongly associated with evidence of positive subjective states. One of these is the phenomenon of *play*. While it is most common in juveniles, many animals will play throughout their life when offered an appropriate opportunity.

As discussed in Part I, one of the main theories of why play occurs is that it is practice for the unexpected. Regardless of its function, play is often associated with a lot of indicators of positive affect (Reinhold et al., 2019). Rats are one of many species that show "rough and tumble play" or *roughhousing*. This is when they will chase, grapple, and pin each other down, as shown in this YouTube video: https://www.youtube.com/watch?v=dNHP6qRoa2w. This behavior is associated with distinct vocalizations that have been characterized as "rat laughter". It has been shown that humans can engage in rough and tumble play with rats using their hand as a proxy for another rat. Playing with a rat in this way can be used as enrichment, as a reward, and to regulate mild aversive events like an injection by providing play interactions before and after (i.e., it alters how they relate to their handler; Cloutier et al., 2015; LaFollette et al., 2018). By understanding the situations that make animals feel good, we can choose to deliberately create these situations for them. This not only affects how they feel at that time but can change and improve how the animal experiences interactions with people that might otherwise be stressful.

Keywords: play, rat tickling

15.5 WILDLIFE INTERACTIONS

One area where there has been strong pressure to modify interactions for the benefit of animals is in zoos and aquariums or in the wild. Historically, menageries kept animals in small environments that gave maximum exposure to the human gaze. However, as an understanding of the need for privacy grew, especially for more sensitive or shy wildlife species (i.e., vigilant prey species), zoos have worked to mitigate stress by improving enclosure design, creating off-display areas, and changing the proximity to visitors by using screens, etc. (Sherwin & Hemsworth., 2019). Increasingly, some animals are kept only in "off exhibit" areas if they have a low tolerance for visitor activity.

In a zoo or aquarium, animals directly interact with two types of people: the zoo staff, which may include regular skilled handlers, and the public, which is represented as an ever-changing group of unfamiliar people. This "visitor effect" can range from negative to positive for animals. For example, orangutans were reported to prefer an unscreened platform area where visitors can see them and vice versa (Bloomfield et al., 2015). Perhaps they were interested in their audience? During the COVID restrictions, many zoos were able to see how animals behaved in the extended absence of visitors, with different results depending on the species and setting (Williams et al., 2022; Brereton et al., 2023; Williams et al., 2021).

Keyword: visitor effect

15.6 SOCIAL COHESION WITH ANIMALS

There are multiple indications that many domesticated species are much more tolerant of, and more easily influenced by, humans. This indicates that they have come to rely on humans as a trusted source of information, resources, and assistance. Three specific behavioral indicators of this are animals copying human yawning, following a human pointing gesture, and soliciting aid from humans via gaze alternation.

When humans are together and one individual yawns, there is an increased likelihood that other people who see this will also yawn. This is a form of *social contagion*, where a feeling, belief, or behavior spreads through a group of individuals. Yawn contagion (Palagi et al., 2020) occurs not only in groups of humans but also in groups of other social animals of the same species, such as lions (Casetta et al., 2021), sheep, chimpanzees, and dogs (Joly-Mascheroni, et al., 2008; O'Hara & Reeve, 2011). While the exact function of both yawns and the social contagion of yawns remains unclear, it is thought to indicate social cohesion. Contagion is also stronger between individuals who are more closely attached and may be associated with a degree of empathy. Researchers have been curious as to whether non-human animals are susceptible to human yawn contagion, but studies attempting to show yawn contagion from people to dogs have recorded mixed results.

Another way of studying our social links with dogs is whether they understand what we mean when we point at something. Repeated studies have shown that dogs will depend on human pointing as a source of information about where to find resources such as food (Virányi et al., 2008; Takaoka et al., 2015). They will go to a position indicated by a pointing human even when they know the food is actually somewhere else (i.e., they follow deceptive pointing and override their olfactory stimuli). Of course dogs may also respond to the gesture as a command. By comparison, wolves have great difficulty in interpreting human pointing, although they can achieve some basic understanding of the gesture if they are hand raised by humans during their sensitive period, or they receive special clicker training. Our close relatives, chimpanzees (captive), also have problems in understanding the human social cues around pointing (Kirchhofer et al., 2012), although captive chimpanzees will understand pointing cues in a competitive task but not a cooperative one (Bräuer et al., 2020).

These yawning and pointing tests focus on the role of humans as the leader and initiator. Conversely, many wild species have shown the ability to initiate interactions where they request help from or provide aid to humans. Take, for example, the many stories of dolphins saving drowning sailors, or modern examples of dolphins trapped in nets approaching human divers for assistance. Individuals across species seem able to identify each other's species-specific specialties. One specific strategy used by animals (and preverbal toddlers) to solicit help is *gaze alternation*. This is when an animal approaches a human and alternates between looking at something that is a problem for them and looking at the human (Marshall-Pescini et al., 2013). This serves to solicit help from humans. Gaze alternation seems to be a less species-specific strategy, as it is found in livestock, chimpanzees, and other wildlife like kangaroos (Nawroth et al., 2016; McElligott et al., 2020).

These behaviors may seem merely like amusing games or tricks. But researchers argue that they have hidden depths as the grammar of basic communication between species giving and receiving the assistance that they are biologically equipped to provide. The closeness of the associations is revealed by how specific and reliable the signals are. For example, we see reproducible patterns from the evolutionary generation of joint activity (like hunting), with pointing (dogs), equivalence of function (such as the language of Alex the parrot), and

Figure 15.2 UK pony culture on full display. These are sparring Konik ponies. We know very little about animal cultures, and yet studies like John Hartigan Jr's ethnography of a wild horse community in Spain reveals how traumatic their interactions with humans are. Photograph reproduced with the permission of Steve Beardall.

direct genetic relationships (patterns of social attention between humans and great apes).

Perhaps if we learned more about how animals naturally communicate with each other, it would build our understanding of when and how to become part of the conversation. A richer approach to this problem would involve not only ethology and disciplines invented to consider animals, but also techniques developed to study human societies, such as anthropology, ethnography, and sociology (see Guillo & Hamilton, 2015; Kopnina, 2017).

The comparison of ethology and the human-focused social sciences reveals a false divide whereby animals are presumed to act according to their nature and humans according to their culture, and different techniques have been mandated for the study of either. When John Hartigan Jr. published his ethnography of a wild horse community in Spain (*Shaving the beasts: Wild horses and ritual in Spain,* 2020), it stood as a prominent example of studying animals as members of their own culture, which is poorly understood and frequently disrupted by their human neighbors (Figure 15.2). When humans attempt to study culture in animals, we have a habit of projecting our cultural ideas or fetishes onto animals, for example, the stallion defending his harem. This over-simplification negates the authority of feral mares, often found to be the most dominant band members. Stallions fight each other, and the mares look after themselves and take the lead role in maintaining the social order. This is where anthropomorphizing animal culture really becomes the problem.

Keywords: gaze alternation, yawn contagion, multi-species ethnography

READINGS

Green criminology and green science

Beirne, P., Brisman, A., Sollund, R., & South, N. (2018). Editors' introduction to the special issue:'For a green criminology'—20 years and onwards. *Theoretical Criminology, 22*(3), 295–297.

Lynch, M. (1990). The greening of criminology. The Critical Criminologist 2, 3, 4, 11, 12; reprinted in N. South., and P. Beirne, (Eds.) (2006) *Green criminology* (pp. 165–170). Ashgate.

Lynch, M. J., & Stretesky, P. B. (2011). Similarities between green criminology and green science: Toward a typology of green criminology. *International Journal of Comparative and Applied Criminal Justice, 35*(4), 293–306.

Wyatt, T., Maher, J., Allen, D., Clarke, N., & Rook, D. (2022). The welfare of wildlife: An interdisciplinary analysis of harm in the legal and illegal wildlife trades and possible ways forward. *Crime, Law and Social Change, 77*(1), 69–89.

The marine aquarium fish trade

Lee, C. J., Paull, G. C., & Tyler, C. R. (2022). Improving zebrafish laboratory welfare and scientific research through understanding their natural history. *Biological Reviews*, *97*(3), 1038–1056.

Shuman, C. S., Hodgson, G., & Ambrose, R. F. (2004). Managing the marine aquarium trade: Is eco-certification the answer? *Environmental Conservation*, *31*(4), 339–348.

Townsend, D. (2011). Sustainability, equity and welfare: A review of the tropical marine ornamental fish trade. *SPC Live Reef Fish Information Bulletin*, *20*, 2–12.

Vaz, M. C. M., Esteves, V. I., & Calado, R. (2017). Live reef fish displaying physiological evidence of cyanide poisoning are still traded in the EU marine aquarium industry. *Scientific Reports*, *7*, 6566. https://doi.org/10.1038/s41598-017-04940-x

Animal cruelty

Ascione, F. R. (1993). Children who are cruel to animals: A review of research and implications for developmental psychopathology. *Anthrozoos*, *6*(4), 226–247.

Dozier, M. E., & Ayers, C. R. (2017). The etiology of hoarding disorder: A review. *Psychopathology*, *50*(5), 291–296.

Caldwell, S., Patterson-Kane, E., Brandler, E., Gupta, M., & Lockwood, R. (2025). animal cruelty in New York city: cruelty cases presented to the ASPCA in partnership with the NYPD 2013–2022. *Animals*, *15*, 662.

Humans as positive interactors—rat tickling

Cloutier, S., Wahl, K. L., Panksepp, J., & Newberry, R. C. (2015). Playful handling of laboratory rats is more beneficial when applied before than after routine injections. *Applied Animal Behaviour Science*, *164*, 81–90.

LaFollette, M. R., O'Haire, M. E., Cloutier, S., & Gaskill, B. N. (2018). Practical rat tickling: Determining an efficient and effective dosage of heterospecific play. *Applied Animal Behaviour Science*, *208*, 82–91.

Reinhold, A. S., Sanguinetti-Scheck, J. I., Hartmann, K., & Brecht, M. (2019). Behavioral and neural correlates of hide-and-seek in rats. *Science (New York, N.Y.)*, *365*(6458), 1180–1183. https://doi.org/10.1126/science.aax4705

Wildlife interactions

Bloomfield, R. C., Gillespie, G. R., Kerswell, K. J., Butler, K. L., & Hemsworth, P. H. (2015). Effect of partial covering of the visitor viewing area window on positioning and orientation of zoo orangutans: A preference test. *Zoo Biology*, *34*(3), 223–229. https://doi.org/10.1002/zoo.21207. Epub 2015 Feb 25. PMID: 25716803.

Brereton, J. E., Jones, E. M. L., McMillan, C., & Perkins, K. (2023). Visitors and observers otter-ly influence the behavior and enclosure use of zoo-housed giant otters. *Zoo Biology*, *42*(4), 509–521. https://doi.org/10.1002/zoo.21755. Epub 2023 Jan 27. PMID: 36705293.

Sherwen, S. L., & Hemsworth, P. H. (2019). The visitor effect on zoo animals: Implications and opportunities for zoo animal welfare. *Animals (Basel)*, *9*(6), 366. https://doi.org/10.3390/ani9060366. PMID: 31212968; PMCID: PMC6617010.

Williams, E., Carter, A., Rendle, J., Fontani, S., Walsh, N. D., Armstrong, S., Hickman, S., Vaglio, S., & Ward, S. J. (2022). The impact of COVID-19 zoo closures on behavioural and physiological parameters of welfare in primates. *Animals (Basel)*, *12*(13), 1622. https://doi.org/10.3390/ani12131622. PMID: 35804521; PMCID: PMC9265073.

Williams, E., Carter, A., Rendle, J., & Ward, S. J. (2021). Understanding impacts of zoo visitors: Quantifying behavioural changes of two popular zoo species during COVID-19 closures. *Applied Animal Behaviour Science*, *236*. https://doi.org/10.1016/j.applanim.2021.105253

Social cohesion with animals

Bräuer, J., Hanus, D., Pika, S., Gray, R., & Uomini, N. (2020). Old and new approaches to animal cognition: There is not "one cognition". *Journal of Intelligence*, *8*(3), 28. https://doi.org/10.3390/jintelligence8030028

Casetta, G., Nolfo, A. P., & Palagi, E. (2021). Yawn contagion promotes motor synchrony in wild lions, Panthera leo. *Animal Behaviour*, *174*, 149–159.

Guillo, D., & Hamilton, P. (2015). What is the place of animals in the social sciences? *Revue française de sociologie*, *56*(1), 135–163.

Hartigan Jr, J. (2020). *Shaving the beasts: Wild horses and ritual in Spain*. University of Minnesota Press.

Joly-Mascheroni, R. M., Senju, A., & Shepherd, A. J. (2008). Dogs catch human yawns. *Biology Letters*, *4*(5), 446–448.

Kirchhofer, K. C., Zimmermann, F., Kaminski, J., & Tomasello, M. (2012). Dogs (Canis familiaris), but not chimpanzees (Pan troglodytes), understand imperative pointing. *PLoS One*, *7*(2), e30913.

Kopnina, H. (2017). Beyond multispecies ethnography: Engaging with violence and animal rights in anthropology. *Critique of Anthropology*, *37*(3), 333–357.

Marshall-Pescini, S., Colombo, E., Passalacqua, C., Merola, I., & Prato-Previde, E. (2013). Gaze alternation in dogs and toddlers in an unsolvable task: Evidence of an audience effect. *Animal Cognition*, *16*, 933–943.

McElligott, A. G., O'Keeffe, K. H., & Green, A. C. (2020). Kangaroos display gazing and gaze alternations during an unsolvable problem task. *Biology Letters*, *16*(12), 20200607.

Nawroth, C., Brett, J. M., & McElligott, A. G. (2016). Goats display audience-dependent human-directed gazing behaviour in a problem-solving task. *Biology Letters*, *12*(7), 20160283.

O'Hara, S. J., & Reeve, A. V. (2011). A test of the yawning contagion and emotional connectedness hypothesis in dogs, Canis familiaris. *Animal Behaviour*, *81*(1), 335–340.

Palagi, E., Celeghin, A., Tamietto, M., Winkielman, P., & Norscia, I. (2020). The neuroethology of spontaneous mimicry and emotional contagion in human and non-human animals. *Neuroscience & Biobehavioral Reviews*, *111*, 149–165.

Takaoka, A., Maeda, T., Hori, Y., & Fujita, K. (2015). Do dogs follow behavioral cues from an unreliable human? *Animal Cognition*, *18*, 475–483.

Virányi, Z., Gácsi, M., Kubinyi, E., Topál, J., Belényi, B., Ujfalussy, D., & Miklósi, Á. (2008). Comprehension of human pointing gestures in young human-reared wolves (Canis lupus) and dogs (Canis familiaris). *Animal Cognition*, *11*, 373–387.

Part V

Human Factors

In this color photograph of a cave painting, we see prehistoric aurochs, horses, and at the bottom of the painting, a rising margin of human handprints. This image illustrates how modern humans, and their ancestors, have been a pervasive and inescapable influence in animal lives. This cultural interplay, past and present, is addressed in Part V. Photograph of cave painting (#322028831) by Enrique Alaez Perez/Shutterstock.com.

DOI: 10.1201/9781003375456-20

Chapter 16

Individual Factors

16.1 INTRODUCTION TO HUMAN FACTORS

In previous sections, the concern was to use language *prescriptively*, i.e., what the words "animal welfare" should mean to scientists and how they should be defined in a dictionary of scientific terms. However, language is also descriptive (*descriptive linguistics*), and the semantics (the study of meaning) is relevant. When the wider community uses these words and similar terms, what do they mean? That meaning might vary in time and place, but ultimately, it will determine how people treat animals and how they generate wider anthropogenic drivers of welfare. These drivers include societal norms, morality, laws, spiritual beliefs, and the generation of pervasive influences such as pollution, encroachment, and the exploitation of natural environments. The following chapters will examine these under four general categories and then cross-culturally. We will cover:

1. Individual attitudes and beliefs
2. Prevailing norms and social movements
3. Codified conduct, ethics, and economics
4. Professional settings
5. Working across cultures

Keyword: descriptive linguistics (vs. prescriptive)

16.2 HUMANS AS INDIVIDUALS

Animal welfare science rests on the assumption that animals have welfare and that our actions should be guided by our understanding of it. So, for animal welfare science to meet its goals, we must understand the human societies in which we are embedded. One starting point is to appreciate *what it is* that convinces people that animals have welfare and how they are persuaded to support that welfare. Advancing animal welfare requires a deeper dive into understanding human behavior, so we must acknowledge from the outset that humans do

DOI: 10.1201/9781003375456-21

not enter the picture as blank slates. For this reason, a deeply held conviction cemented by faith, tradition, and other beliefs can be harder to influence than a more ephemeral attachment to an idea.

One example of a more fleeting human susceptibility to an idea is the "animal fad"—the use of a pet as a stylish accessory. While any particular trend in pet ownership might be transient, the tendency to leap into them is enduring. Even seemingly innocuous fads can have their issues, such as the current fashion to own a designer crossbreed or the purchase of a pandemic puppy (Figure 16.1; Bryson et al., 2024; Packer et al., 2023). Back in 1975, a veterinary consultant to the American Humane Society (Hummer, 1975) suggested that the demand in exotic pets exposed them to incompetent ownership and other collateral issues for the environment; even the spread of monkeypox and Seoul virus was attributed to the exotic pet trade (Shivambu et al., 2023; Atasoy et al., 2024). This powerfully illustrates how human behaviors can have mixed and sometimes unexpected outcomes.

The film industry, media, and increasingly, social media and pet influencers also play a part in promoting animal fads, with pet owners sometimes engaging in cruel or inappropriate scenarios to get "likes" or attention. There are the teacup-sized pigs that weren't, or the clown fish that needed more looking after than expected. We see upticks in acquisitions of Dalmatians, setters, and Danes after featuring in movies, only to see a similar "wave" in rescue shelters a little later.

In this context, being loved by people can be more dangerous than being hated or ignored. This is also demonstrated by some social media that exploits our love of animals. If cat videos are anything to go by, animal content is some of the most popular material online, which has also driven the creation of fake animal rescue videos. This is when people deliberately place animals at risk (from a predator, of drowning, or other hazards) in order to film themselves as the rescuing hero of the story. Host platforms like YouTube are effectively immune to any consequences from hosting these videos as service providers are not treated as the publisher of their online content. This rule is colloquially known as Section 230 (47 U.S.C. § 230) and was part of the Communications Decency Act of 1996. As an idea of the magnitude of the problem, a recent court filing (Freethinker v. Google LLC) noted more than 100 YouTube channels hosting 2000 videos of animal cruelty with millions of subscribers and 1.2 billion views. Any particular animal fad needs to be dealt with immediately, but we must always consider what it is about human nature that creates

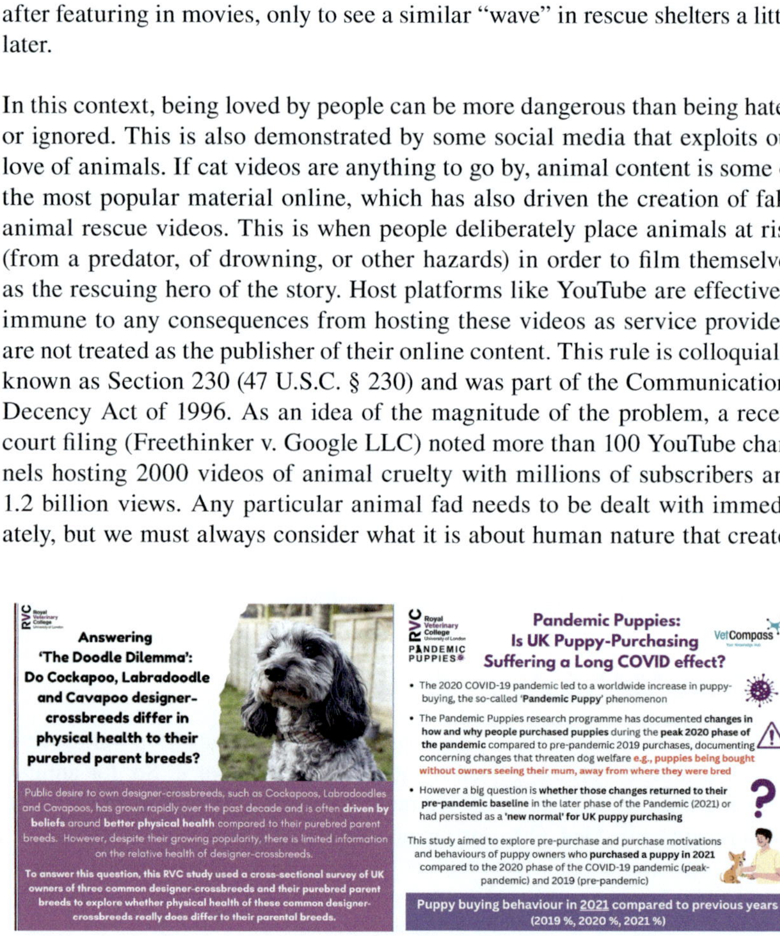

Figure 16.1 Sections of two infographics produced by VetCompass™. The first investigates the "doodle" craze and the presumed health benefits of these designer crossbreeds. The next researches purchasing habits around the pandemic puppy phenomenon, and was conducted with the Pandemic Puppies Project. VetCompass™ is a not-for-profit research project owned by the United Kingdom's Royal Vet College. The name is shortened from *Veterinary Companion Animal Surveillance System*, which gives you a sense of their mission in converting data to outcomes that support the health and wellbeing of companion animals. VetCompass™ offers infographics as an open access resource with the CC-BY license, available online at https://www.rvc.ac.uk/vetcompass/audio-visual-resources/research-infographics.

this pattern of inventing new ways to harm animals. More recently (February 2025), an algorithm used by Instagram malfunctioned and directed images of animal abuse and other violence to user feeds. The parent company, Meta, issued an apology; an underwhelming response for those traumatized by the content that they may have seen. Clearly, we need legislation that is fit for purpose in protecting users from social media platforms that are, even now, dialing back on content moderation while enabling the compilation of vast libraries of animal abuse.

Keyword: animal fads

16.3 WHY DO SOME PEOPLE CARE MORE ABOUT ANIMALS THAN OTHERS?

There are many reasons why it might be helpful to know which people in a community are attracted to animal welfare issues, as they might be mobilized in direct actions, as donors, as adopters, or as consumers of valued-added welfare products. A person's current belief systems can be assessed using surveys that show where their beliefs fit relative to others in the community. For example, important drivers of attitudes to animals are a person's emotional responses and whether they find the animal useful or inconvenient (Serpell, 2004).

It is common to be judgmental about people who view animals through the lens of whether they are useful to them. In previous sections, we have discussed the welfare of the animal. But here, we pivot to the welfare of people, because it is often these people who will determine the quality of life of their animal neighbors. If animals are threatening your life, home, or family, then it is perfectly reasonable to have negative feelings about them. If the degree to which you exploit an animal will create wealth and security for you, then it is very easy to arrive at a belief system that legitimizes that exploitation. This often causes conflict between animal advocates and the communities that live in direct contact with the animals in question.

Overarching models of human personality are a first step in conducting research into causes of welfare concern. However, they will not be generalizable to every community, nor will they explain every individual's actions, even in the communities in which they were developed. For a closer examination of what is meaningful to individuals or small groups, qualitative approaches are needed. The most productive research often moves back and forth between objective and generalizable data collection and deeper dives into personal identities and motivations. While qualitative research is still quite rare in animal welfare science, this will hopefully change, as the results can be powerful.

Qualitative research explores the granularity of human behaviors and why they are performed. There can be some skepticism about qualitative methods, but this might reflect an anxiety that the data retrieved are the result of creative guesswork and are not really scientific. Qualitative methods are actually a range of approaches that bring transparency and a systematic approach to describing an individual's lived experience: that is, the meaning people attach to their own beliefs, motivations, and actions. This work must be approached with a minimum of preconceived ideas and an awareness of personal bias and beliefs that might distort our understanding of the participants.

It is difficult to summarize the whole domain of qualitative research. But the essence of it is to identify the right people and find a way for them to share what life is like for them and how it is to live in their world. This may include almost any kind of expression—speech text, art, imagery, or how they present themselves, but the focus is always on understanding their lived experience. Once this information is collected, sophisticated data analysis methods can be used that may involve dedicated software (ATLAS.ti, QDA Miner, or NVivo) and some degree of user expertise to reveal important themes.

For example, Garland-Lewis et al. (2024) provided people who were experiencing homelessness with disposable cameras. This allowed them to document moments when their pets were of importance to them and the barriers that they faced relating to having pets (Figure 16.2). This enhanced and shaped the stories they also shared verbally. An exhibition of text and imagery was held in the community in which the people lived before a peer-reviewed paper was published in the open access journal PLoS ONE.

One productive area for combined qualitative and quantitative research has been to investigate beliefs and feelings people have about farm animal welfare. For example, Canadian research shows that consumers agree that they know very little about farming, which matches farmers' perceptions. Moving to more quantitative studies, the researchers found that a more informed group of citizens became somewhat reassured that farm animals' biological needs were being met, but concerns lingered about naturalness issues, such as access to pasture. So, contrary to the farmers' expectations, a more educated public might become a more disapproving public! This research is beginning to unpick the factors that prevent consensus as to how best to cater to farm animals' needs (Spooner et al., 2014; Ventura et al., 2016; Bassi et al., 2019): for example, the role of tradition in maintaining painful procedures and the gaps between consumer support for humane approaches and our apparent unwillingness to pay for them.

Keyword: qualitative research

Figure 16.2 Unstable living circumstances and pet ownership can be a strain, and this study set out to explore the human–animal bond in those circumstances. The cat sunning themselves on the dashboard belongs to Dee Powers. Dee commented: "We're all people, you know? And people have pets. And pets are a huge comfort to these people's lives. I will eat ramen for the rest of my life before I betray my promise to give them a forever home". Dee & Walter and Chanel. The full quotation can be read at Garland-Lewis G, Scott E, Ramirez V, Rabinowitz P (2024) Shifting the Focus: A Photovoice exploration of the benefits and barriers of having a pet while experiencing homelessness. *PLoS ONE 19*(3): https://doi.org/10.1371/journal.pone.0295588. Published under CC BY 4.0, https://creativecommons.org/licenses/by/4.0/. Photo (with no changes) reproduced with the permission of Dee Powers.

Psychological scales

There are many standardized surveys attempting to quantify something relevant to animal welfare–related behaviors. A *psychological scale* is a survey that seeks to quantify a target psychological quality with a rating from zero to some maximum number: for example, a ten-question scale with yes/no questions generating a score from zero to ten. Most scales use self-report, but some can be filled out by a third party. A few examples include:

- Empathy Toward Animals Scale, an assessment of empathetic concern adapted to apply to animals (trait—empathy; https://evaluation-archive.naaee.org/tools/empathy-toward-animals-scale)
- Animal Empathy Scale (AES; Paul, 2000), feelings about animals and their treatment (trait—empathy)
- Lexington Attachment of Pets Scale (Johnson et al., 1992), to quantify attachment to pet dogs and cats (trait—attachment)
- Pet-Related Stress Scale (PRSS; Matijczak et al., 2024), a measure of general, multi-species pet-related stress among US pet owners (trait—stress)

Scale scores can be correlated with other factors, such as attitudes toward other humans and animals (e.g., speciesism, xenophobia) and opinions about issues like pet neutering or opposition to intensive farming. Scales can reveal systematic differences between people related to gender, age, and nationality. However, conclusions, including any apparent links between survey instruments and real-world behavior, should be treated with caution.

Scales are developed by and for people from specific cultural backgrounds. Used outside this setting, they are prone to bias and misinterpretation. For example, it has been found that attachment styles vary greatly according to factors such as religion and national origin. A scale based on one particular view of attachment, and the assumption that it is purely biological, may rate some styles as low attachment when they are simply different. The role of cultural differences in human relationships has been extensively researched, but less so in terms of attachment to animals. These biases play out when minority groups are labeled as "not caring much about animals" and can lead to people being treated inappropriately when seeking to adopt pets.

Keyword: psychological scale

16.4 COMMUNITY BEHAVIOR CHANGE

When animals might benefit from a change in attitudes and behaviors across an entire community, it is important to recognize that this change will be made up by many *individuals* changing their minds and/or practices. At the population level, these changes might play out as small incremental gains for animals: for example, evacuating from disasters with pets rather than abandoning them, taking pets to regular veterinary preventive care visits, or abandoning painful practices such as tail docking. However, for the individual person, the change is likely to be quite sudden.

Effective public information campaigns seek to connect with what is important and convincing to individuals. For example, the "Trueno" Preparedness Campaign (Morales, 2019) for Pet Owners in Costa Rica (2012–2014) was a television public service announcement produced and disseminated by World Animal Protection featuring Trueno the dachshund. Trueno was prepared for disaster and potential evacuation with 1) an ID tag, 2) a kennel and leash, 3) a secure emergency kit complete with any prescription medications, and 4) someone who could take care of Trueno in case of emergency. The viewer is put in the position of Trueno's family, doing what they need to keep him safe; something they can easily apply to their own life.

The traction achieved by these public action campaigns is extremely variable. In the United States, animal welfare organizations have consistently advocated that people take their pets with them when they evacuate from natural disasters. However, a dramatic shift toward supporting co-evacuation with pets, including admitting pets to transportation and *cosheltering*, only occurred after Hurricane Katrina struck New Orleans in 2005. (Cosheltering refers to accommodating animals in shelters, not only in the context of disaster evacuation but also in shelters for the homeless, migrants, or people leaving a domestic violence situation.) This change was driven by media coverage about people dying because they had refused to be separated from their pets.

After Katrina, a tipping point in public opinion had been reached. Now, the entire family, including pets, would be helped during an emergency. This change came from the realization that the welfare of pets and their human family is intertwined, and a danger to one was a danger to all. In 2006, the PETS Act mandated that state and local emergency preparedness plans address the needs of individuals with household pets and service animals. In recent media reports of hurricanes hitting Florida, you might have noticed that pet-friendly shelters can be few and far between. FloridaDisaster.org advises people to evacuate with pets and then tells them that most shelters are not pet friendly, although they accept service animals. They do, however, provide information for advance planning. Going further, some researchers feel that pet ownership can go from being a liability to being *the* reason why people might recognize the vulnerability of their family and plan more effectively for disaster scenarios (Thompson, 2015). Grass roots change in a community *can* be achieved with sufficient community cohesion and the right approach. A striking example was the town of Palisade as it fought the invasive Japanese beetle that could have devastated its local agribusinesses (Figure 16.3). Achieving change needs planning, and the next section provides a useful toolkit for that.

Keyword: cosheltering

Figure 16.3 From 2002, the small Coloradan town of Palisade fought the Japanese beetle. Left: Palisade has an extraordinary microclimate protected by its Book Cliffs, warm air funneling from the interstate I-70, and irrigation from the Colorado River. This makes Palisade Colorado's "peach capital", and the town is famed for its fruit and wine. Right: The attractive but devastating beetle. This beetle could have destroyed the local orchards and vineyards. According to local media, all but 1 of the 712 households willingly joined in the pesticide and trap campaign; the single hold out was concerned that the pesticides might harm his miniature chihuahuas. Palisade won this round against the beetle, but it's back, this time in nearby Mesa County. Image by Tina Rich.

Theory of change and how to make things happen

Many frameworks are available to help us understand how individual behaviors make up population-level shifts. One of these is referred to as *Theory of Change*, which became prominent in the 1990s as a form of program evaluation (Chen, 1990). The Theory of Change (ToC) approach often starts by identifying a goal for significant behavior change (i.e., what would success look like?) and then back-maps a causal pathway from "success" to your start point. This method fleshes out the "missing middle", which is the constellation of activities, indicators, and steps that measure, map, or frustrate progress along that causal pathway. To be successful, a ToC must identify all assumptions and include all the relevant stakeholders so that their obstacles, costs, and benefits can be considered. This is a living plan, and it can be updated as fresh obstacles or other factors arise.

ToC has been used to advance a range of animal welfare goals, such as reducing bovine tuberculosis transmission (Maye et al., 2020), improving the coexistence of people and dingoes in Australia (van Eeden et al., 2021), and helping non-profits become more effective in promoting the welfare of domestic dogs and cats (Reed & Upjohn, 2018; Terry & Stephens, 2023). However, in academia, the practice is often to focus down on a single problem and to tune out everything else, which is the opposite of what is needed to complete a ToC. Academics in STEM disciplines (science, technology, engineering, and mathematics) might be particularly hostile to ToC, as it involves participatory processes that are unusual in the "hard sciences". That said, when trying to explain a project, it can quickly become apparent that it is unfeasible—a realization that is best achieved sooner rather than later.

Keyword: theory of change

16.5 PET OWNER COMPETENCY

One area where the role of individual human behavior has been closely studied is pet ownership. What makes people responsible pet owners versus those who pose a risk to their pets or make their animals a danger to themselves and others? There is a huge diversity of data points and theories that are used to understand and influence pet ownership styles. A few are discussed below.

Demographics: These are measurable aspects of a person that can be used to understand the structure of a population, e.g., age, sex, education, nationality, and ethnicity. For example, women and younger people are often more empathetic toward animals, while those engaging in animal cruelty are more likely to be male, with motivations that differ according to age, geography (an urban versus rural setting), and mental illness. That said, the feelings of groups like older rural men are often poorly understood. Some men tend not to speak lovingly about their pets, or will even insult them, but then care for them conscientiously and show affection when alone with their animals.

One study using a theoretical example to see when people would seek vet care found that demographic variables had no significant effect, suggesting that differences in accessing care may relate more to a lack of resources and other barriers (Park et al., 2021). Whatever group you happen to fall into, not having money or transport will be an obvious problem in seeking vet care, and the shortfall between aspiration and what can be managed will be an obvious stressor. This shows that great care needs to be taken when interpreting the association of demographic variables with pet care behaviors, as bias can easily cloud our interpretations (Smith, 2010; Agishtein & Brumbaugh, 2013).

Personality: Many efforts have been made to determine how elements of individual personality associate with pet ownership behavior. One study found that owners of dog breeds considered "aggressive" tended to have high scores for psychoticism on the Eysenck Personality Questionnaire Short Scale (Wells & Hepper, 2012). If this is broadly the case, then it suggests that the potential of these breeds to injure people might be driven not so much by the dogs themselves but by the behaviors of their owners.

Attachment: Attachment theory measures whether relationships are secure and developed, and was initially formulated to study the needs of children and then adapted to pets (Beck & Madresh, 2008; Mikulincer & Shaver, 2007). For example, attachment to pets is associated with evacuating *with them* during disasters. Many tools exist to assess the degree and style of attachment between a person and an animal. Examples include the Dog Attachment Questionnaire (DAQ), the Short Attachment to Pets Scale (SAPS), and the Pet Attachment Questionnaire (PAQ). Pets demonstrate attachment by their desire for proximity to the owner, who offers a "safe haven and security to explore", and from whom separation causes distress (for details see Mikulincer & Shaver, 2007).

Keywords: attachment, personality

Ownership of potentially hazardous animals and endangered species

There are debates in communities about the appropriateness of owning animals that are considered more hazardous (Nyhus et al., 2003; Lucca, 2013). The ownership and trade patterns of dangerous animals can be problematic in multiple ways. For example, tigers are endangered and are one of the species most implicated in human deaths, both in their native habitats as well as in captivity. In the United States, there are typically one or two fatal attacks annually, which is high given an estimated population of 5000–10,000 animals. Even individuals who are extremely familiar with working with tigers, and with specific tigers, are at risk of serious injury or death from them.

Why do people keep tigers and want to be close to them? Some generalizations can be made: the zoo worker who accidentally shares exhibit space with a tiger, or people developing a false sense of security around their pet tigers. There are also more unusual situations which are challenging to understand, such as the Zanesville, Ohio "massacre".

On October 18, 2011, Terry Thompson, a legal and permitted private owner of a collection of 56 exotic animals, apparently deliberately cut open many of their enclosures before committing suicide by gunshot. As the animals dispersed, a bear charged a deputy. This animal and 47 others, including 18 endangered Bengal tigers, were killed by gunshot by law enforcement due to the danger they presented to the public. Terry Thompson was the subject of many sensationalist investigative journalism articles. They include "Animals: The horrific true story of the Zanesville zoo massacre" by Chris Jones of Esquire and "18 Tigers, 17 lions, 8 bears, 3 cougars, 2 wolves, 1 baboon, 1 macaque, and 1 man dead in Ohio" by Chris Heath of GQ.

These stories present a litany of reasons as to why Mr Thompson was vulnerable and under stress, allowing the reader to play armchair psychologist. We have a tendency to think that because an event can be tidily explained, it should also have been predicted in advance. Fewer column inches were given to Thompson's love of his animals and his standing in the exotic animal community, to the extent that this community would come to question the official account of October 18th.

It is common for law enforcement, psychologists, and other authorities to have to try and predict whether a person might become a danger to themselves or others (their *dangerousness*). This is commonly attempted by using some combination of actuarial data and clinical judgment. Given the extreme difficulty of this task, it is not surprising that predictions often tend to be "off" (Ogloff & Davis, 2020; Haden, 2016). Many theories have attempted to use past behavior to confidently predict future actions. For example, there have been a range of theories asserting that "dark traits" will inevitably lead to violent crimes. One of these is the *Dark Triad* of bed wetting, fire setting, and animal cruelty, suggested as reliably linked in male adolescents (Wax & Haddox, 1974).

This approach developed in the 1970s, when the true extent of crimes such as child abuse and murder (including serial murder) were finally being studied and discussed. But it was also a time of anxiety and excessive speculation about the dangers of largely fictitious conspiracies relating to communists, satanists, and even alien invaders. This led to some systematic misinterpretations of animal cruelty evidence and a wave of improper imprisonments, especially around ritualistic child sexual abuse. In a final report about the latter (Goodman et al., 1995) to the National Center on Child Abuse and Neglect, the authors concluded that "many more children [were] being abused in the name of God than in the name of Satan".

More extensive research shows vast webs of associations between acts of violence and property damage, a lack of resources, being subject to social stigma, health and mental health challenges, and a lack of resilience. This new wider picture came to be referred to as either general deviance theory (referring to the perpetrator under pressure exhibiting more antisocial behavior) or general strain theory (referring to their experiences; Agnew & Brezina, 2019). While some factors may be better predictors of future violent behavior than others, the overall picture remains muddy, and most clinicians do not make strong claims about being able to see into the future of individual patients or defendants.

The main problem with trying to manage future risks is that safety for some is bought by curtailing the freedoms of others. And the United States is home to a culture that is particularly focused on freedoms. In the case of the Zanesville massacre, the freedom to own large exotic animals was curtailed by the Ohio Dangerous Wild Animal Act, which commenced in 2014 (Lucca, 2013). This act banned the possession of a long list of species designated as dangerous wild animals, although existing owners could be grandfathered in with the right permits, care standards, and insurances.

Keywords: dangerousness, Dark Triad (a.k.a. MacDonald Triad)

Thought experiment: Domesticating tigers

Consider the idea of the ancient contract, as described in Part IV. If you think about the current situation of tigers (*Panthera tigris*), what arguments could be made for or against tigers becoming a domesticated species? (For the purposes of a thought experiment, consider the consequences of both options fully without letting fundamental objections as to the plausibility or morality of either option end the discussion. Brown & Fehige, 2023). Consider the interests of both tigers and humans in this change of relationship, and reflect on as many aspects of the situation as possible, but with a focus on the welfare of individuals and how they would change. (And remember, you can use one of the theoretical frameworks from Part I.)

READINGS

Humans as individuals

Atasoy, M. O., Naggar, R. F. E., Rohaim, M. A., & Munir, M. (2024). Zoonotic and zooanthroponotic potential of monkeypox. *Advances in Experimental Medicine and Biology*, *1451*, 75–90. https://doi.org/10.1007/978-3-031-57165-7_5. PMID: 38801572.

Bryson, G. T., O'Neill, D. G., Brand, C. L., Belshaw, Z., & Packer, R. M. A. (2024). The doodle dilemma: How the physical health of 'Designer-crossbreed' Cockapoo, Labradoodle and Cavapoo dogs' compares to their purebred progenitor breeds. *PLoS One*, *19*(8), e0306350. https://doi.org/10.1371/journal.pone.0306350. PMID: 39196904; PMCID: PMC11355567.

Hummer, R. L. (1975). Pets in today's society. *American Journal of Public Health*, *65*(10), 1095–1098. https://doi.org/10.2105/ajph.65.10.1095. PMID: 1163708; PMCID: PMC1776006.

Packer, R. M. A., Brand, C. L., Belshaw, Z., Pegram, C. L., Dale, F., Stevens, K., & O'Neill, D. G. (2023). Is UK puppy purchasing suffering a long COVID effect? Ongoing negative impacts of the COVID-19 pandemic upon puppy purchase motivations and behaviours in 2021. *Animals*, *13*(13), 2186. https://doi.org/10.3390/ani13132186

Shivambu, N., Shivambu, T. C., & Chimimba, C. T. (2023). Zoonotic pathogens associated with pet and feeder murid rodent species: A global systematic review. *Vector-Borne and Zoonotic Diseases*, *23*(11), 551–560. https://doi.org/10.1089/vbz.2023.0024. Epub 2023 Aug 29. PMID: 37643287.

Why do some people care more about animals than others?

Bassi, E. M., Goddard, E., & Parkins, J. R. (2019). "That's the way we've always done it": A social practice analysis of farm animal welfare in Alberta. *Journal of Agricultural and Environmental Ethics*, *32*(2), 335–354.

Garland-Lewis, G., Scott, E., Ramirez, V., & Rabinowitz, P. (2024). Shifting the focus: A photovoice exploration of the benefits and barriers of having a pet while experiencing homelessness. *PLoS One*, *19*(3), e0295588.

Serpell, J. (2004). Factors influencing human attitudes to animals and their welfare. *Animal Welfare*, *13*(S1), S145–S151. https://doi.org/10.1017/S0962728600014500

Spooner, J. M., Schuppli, C. A., & Fraser, D. (2014). Attitudes of Canadian citizens toward farm animal welfare: A qualitative study. *Livestock Science*, *163*, 150–158.

Ventura, B. A., Von Keyserlingk, M. A., Wittman, H., & Weary, D. M. (2016). What difference does a visit make? Changes in animal welfare perceptions after interested citizens tour a dairy farm. *PLoS One*, *11*(5), e0154733.

Psychological scales

Johnson, T. P., Garrity, T. F., & Stallones, L. (1992). *Lexington Attachment to Pets Scale (LAPS)* [Database record]. APA PsycTests.

Matijczak, A., Tomlinson, C. A., Applebaum, J. W., Kogan, L. R., & McDonald, S. E. (2024, July). Development and validation of the pet-related stress scale. *Pets*, *1*(2), 70–87.

Paul, E. (2000). Empathy with animals and with humans: Are they linked? *Anthrozoos: A Multidisciplinary Journal of The Interactions of People & Animals*, *13*, 194–202. https://doi.org/10.2752/089279300786999699

Community behavior change

Morales, E. (2019). *"Trueno" preparedness campaign for pet owners in Costa Rica 2012–2014*. Contributing Paper to GAR 2019.

Thompson, K. (2015). For pets' sake, save yourself! Motivating emergency and disaster preparedness through relations of animal guardianship. *Australian Journal of Emergency Management*, *30*(2), 43–46.

Theory of change and how to make things happen

Chen, H. T. (1990). *Theory-driven evaluations*. Sage.

Maye, D., Enticott, G., & Naylor, R. (2020). Theories of change in rural policy evaluation. *Sociologia Ruralis*, *60*(1), 198–221.

Reed, K., & Upjohn, M. M. (2018). Better lives for dogs: incorporating human behaviour change into a theory of change to improve canine Welfare Worldwide. *Frontiers in Veterinary Science*, *5*, 93.

Terry, M., & Stephens, R. (2023). Simplifying the cat's cradle: Using theory of change with a behaviour change model to develop new strategic thinking for a UK animal welfare charity. In L. Simeone, D. Drabble, N. Morelli, & A. Götzen (Eds.), *Strategic thinking, design and the theory of change: A framework for designing impactful and transformational social interventions* (pp. 218–231). Edward Elgar Publishing.

van Eeden, L., Dickman, C., Crowther, M., & Newsome, T. (2021). A theory of change for promoting coexistence between dingoes and livestock production. *Conservation Science and Practice*, *3*(3), e304.

Pet owner competency

Agishtein, P., & Brumbaugh, C. (2013). Cultural variation in adult attachment: The impact of ethnicity, collectivism, and country of origin. *Journal of Social, Evolutionary, and Cultural Psychology*, *7*(4), 384.

Beck, L., & Madresh, E. A. (2008). Romantic partners and four-legged friends: An extension of attachment theory to relationships with pets. *Anthrozoös*, *21*(1), 43–56.

Mikulincer, M., & Shaver, P. R. (2007). The attachment behavioral system: Basic concepts and principles. In *Attachment in adulthood: Structure, dynamics, and change* (pp. 3–28). The Guilford Press.

Park, R. M., Gruen, M. E., & Royal, K. (2021). Association between dog owner demographics and decision to seek veterinary care. *Veterinary Sciences*, *8*(1), 7.

Smith, C. (2010). Owner demographics and medical care for pets. *Journal of the American Veterinary Medical Association*, *236*(12), 1289.

Wells, D. L., & Hepper, P. G. (2012). The personality of "aggressive" and "non-aggressive" dog owners. *Personality and Individual Differences*, *53*(6), 770–773.

Ownership of potentially hazardous animals and endangered species

Agnew, R., & Brezina, T. (2019). General strain theory. In M. Krohn, N. Hendrix, G. Penly Hall, & A. Lizotte (Eds.), *Handbook on crime and deviance* (pp. 145–160). Springer.

Goodman, G. S., Qin, J., Bottoms, B. L., & Shaver, P. R. (1995). *Characteristics and sources of allegations of ritualistic child abuse. Final report to the national center on child abuse and neglect.*

Haden, S. C. (2016). The challenge of predicting dangerousness. In L. Wilson (Ed.), *The Wiley handbook of the psychology of mass shootings* (pp. 96–114). John Wiley and Sons.

Lucca, C. M. (2013). Keeping lions, tigers, and bears (oh my) in check: The state of exotic pet regulation in the wake of the Zanesville, Ohio Massacre. *Villanova Environmental Law Journal*, *24*, 125.

Nyhus, P. J., Tilson, R. L., & Tomlinson, J. L. (2003). Dangerous animals in captivity: Ex situ tiger conflict and implications for private ownership of exotic animals. *Zoo Biology: Published in affiliation with the American Zoo and Aquarium Association*, *22*(6), 573–586.

Ogloff, J. R., & Davis, M. R. (2020). From predicting dangerousness to assessing and managing risk for violence: A journey across four generations. In J. S. Wormith, L. A. Craig, T. E. Hogue (Eds.), *The Wiley handbook of what works in violence risk management: theory, research and practice* (pp. 79–98). John Wiley and Sons.

Wax, D., & Haddox, V. (1974). Enuresis, fire setting, and animal cruelty in male adolescent delinquents: A triad predictive of violent behavior. *The Journal of Psychiatry & Law*, *2*(1), 45–71.

Thought experiment: Domesticating tigers

Brown, J. R., & Fehige, Y. F. (2023). Thought experiments. In E. N. Zalta, & U. Nodelman (Eds.), The Stanford encyclopedia of philosophy. Online at: https://plato.stanford.edu/archives/win2023/entries/thought-experiment

Chapter 17

Norms

17.1 NORMS

There are many ways to identify cultural behavior, but probably the best start point is to consider the cultural norms. *Norms* are the informal assumptions and rules that govern behavior in a community and take into account culture (a term that encompasses societal beliefs and practices), morality, and tradition. Healthy social norms foster conformity but are not so rigid that they cannot accommodate diversity, subcultures, or gradual change. The deliberate modification of norms by a central authority is referred to as *social engineering*. For example, the Iranian government tried to suppress pet keeping as a product of "Westernization". Organized attempts to counter this interference would be referred to as *reverse social engineering* (Delon, 2018). Social engineering can be problematic if attempts to control people are made without considering their best interests (i.e., human hacking). But what if callousness and violence are the norm? Post-communist Romania struggled with its past of cruelty to both people and animals. As noted by Plant (2016), unwell or non-working elders and unwanted children were classed as "*varsa*" (weeds) and were often abandoned along with surplus companion animals. Joining the EU in 2007 should have improved animal welfare in Romania, although the World Animal Protection's Animal Protection Index (which compares animal welfare policy and legislation, online at https://api.worldanimalprotection.org/) still gives Romania a "D" rating—bad for Europe but the same as the United States (Iran gets a "G" rating, the lowest).

As amply demonstrated in Romania, one constant is that isolation and disability increases the vulnerability of *both* people and animals by removing the umbrella of societal protection. We also see this play out for institutionalized individuals who, once separated from their families, become vulnerable to abuse. Likewise, animals separated from owners are more likely to fare badly in multiple settings, and if a dog, may simply be shot. Economic stress also has implications for populations in the United States. It was notable that after the 2008 recession, many jurisdictions stopped funding animal control. As a result, problems with free-roaming animals have led to undesirable outcomes, like law

DOI: 10.1201/9781003375456-22

enforcement destroying animals by shooting them or legislating against animal breeds. This shows how progressive pro-social societies can quickly break down when placed under stress.

Most social norms arise informally from customary practice, which is hopefully wise and based on generations of experience. At the same time, blind adherence to old habits can lead to ethical blind alleys. For example, one of the authors (EPK) was once attending a meeting of people interested in responding to the unexpected side effects of beta agonists on production animals (we covered this in Part II). Temple Grandin had posed a question along the lines of "What is the acceptable rate of hoof sloughing in cattle?" After a pause where the audience seemed to ponder the question, she responded emphatically that it was zero. ZERO. That hoof walls falling off a steer or cow was never ever acceptable, and we needed to act urgently to stop it. It was a wake-up call to resist normalizing ethically bad outcomes, no matter their frequency or cause (known or not). Or as Grandin put it, "do not let bad become normal" (Grandin, 2022).

Keywords: norms, social engineering, culture

Lab Lore

In the 1980s, a thread of research developed that looked at the laboratory culture of scientific researchers studying animals. One particularly important paper by the sociologist/anthropologist Arnold Arluke was called "sacrificial symbolism in animal experimentation" (Arluke, 1988). Arluke showed how the precise and impartial laboratory environment was distinguished by strongly symbolic language. For example, the popular term at the time for killing an animal during or after use in research was to "sacrifice" it. Use of this term persists in some labs today, although it's doubtful that those sacrificing the animals know that the translation of the Latin *sacrificium* means "to make holy".

This work shone a light on the strange and often unscientific cultures of laboratories and their implications for animal welfare. Obsolete techniques for animal handling, euthanasia, surgery, and other activities are often taught to generations of graduate students by respected senior researchers or lab heads. These cultural behaviors are known as *lab lore*—the informal beliefs and practices passed down by members of a laboratory as "the way" things are done. The effects of lab lore can be very persistent, as the recipients of this "wisdom" are often early career scientists who are unlikely to challenge their supervisors or mentors.

For example, when I was first instructing undergraduates as a PhD student, the students asked this author (EPK) "Are the rats happy?" Initially, I gave the same answer that had been given to me, that rats in captivity are healthy and live much longer than those in the wild. I belatedly realized that this did not really answer the question, and I genuinely didn't know whether the rats were happy or not. I could not even say with certainty what would make a rat happy and how we would know. This realization led to a change in direction in my PhD research and ultimately, a change of track from being predominantly an operant psychologist to being focused on a career in animal welfare science.

Keywords: lab lore, Arnold Arluke

17.2 SOCIAL MOVEMENTS

A *social movement* is a widespread effort to change how society operates that is not led by any specific person or organization. The original social movements in America included the *Labor Movement* (to protect the interests of workers), the *Stamp Act Resistance Movement* (opposing taxation without representation), and the *Third-Party Vote Movement* (to develop an alternative to the two-party system—note: not every social movement is successful!). Many social movements were spawned by the changing realities of the industrial revolution and new problems such as child labor and dramatic increases in pollution. Indeed, the aligned interests of powerless groups is a feature of these movements, and breakthroughs for one group can translate to others (Anderson, 2011). Animals were involved in some older and transitional social movements, including *Women's Suffrage* (women's right to vote), the *Temperance Movement* (promoting moderation or abstinence in the use of liquor), and the *Anti-Vivisection Movement* that opposed animal use in research or teaching (Figure 17.1), for which some animals were seized from animal shelters. (Pound seizure laws allow shelter animals to be taken for research. Check out the Oklahoma statute here https://www.animallaw.info/statute/ok-impound-%C2%A7-394-delivery-animals-demand-municipal-ordinances-relating-impoundment-and.)

The original social movements were about creating livable conditions for all people, especially those from lower social classes. In the 1960s–1970s, the *New Social Movements* aimed to prevent discrimination against groups and promote environmentalism. Animal rights and the early animal welfare movement were a small part of this era. We are currently in a third wave—*the Post Millennial Social Movements*. These are characterized by being heavily intersectional and decentralized. Many are represented strongly in online spaces such as #MeToo and Black Lives Matter. Science is seen as an increasingly contested source of influence, and how animal welfare science engages with these movements remains unclear.

Science policy is enmeshed in social movements, while scientific practices are traditionally thought of as culturally neutral and "fact-based". However, even

Figure 17.1 In 1919, the Anti-Vivisection Organization founded by Frances Power Cobbe, now called Cruelty Free International, joined with the Dogs Trust to campaign for the Dogs Protection Bill. If passed, the law would have protected dogs from being seized for medical research. A cross section of the arguments of the day can be read in Tansey, 1994; you will also get an idea of how issues of disease influenced animal welfare. Image reproduced with the permission of Cruelty Free International.

the most basic of knowledge is created and shaped by where that research occurs. During the periods of new social movements, science became somewhat more accessible, with college degrees and science professions opening to a wider range of people (although representation remains a challenge). Animal welfare science clearly involves advocating for improved animal welfare within the greater social movement, but it also tries to use science as a way to present facts that we can agree upon. The animal welfare scientist will appear on the spectrum between the impassioned advocate and the truth seeker. This will automatically cause conflict, as advocacy involves gathering only the evidence that strengthens your position, while science demands changing your position based on the entirety of the available evidence. On occasion, these stresses play out when a governmental administration attempts to subvert scientific integrity, for example, by banning words associated with climate change. These stresses, and others, have resulted in the emergence of voices such as the Alt National Park Service, founded January 2017.

In the United States, scientific bodies of knowledge have become extremely contested and can be at odds with other codes (e.g., evolution versus creationism) or become politicized (e.g., vaccination). Added to this, the scientific establishment has a problematic record and continues to provide shills for industry, is guilty of widespread discrimination (Graves et al., 2022), and must deal with the worldwide pervasiveness of scientific fraud (Cogan 2022). In this skeptical age, you cannot lean on the authority of science to drive home your point. To navigate social policy and politics, the scientist might like to refer to Pielke's publication (2007) *The honest broker: making sense of science in policy and politics*. In essence, the scientist must feel able to change their mind if the weight of the evidence suggests that they should. Attacks on scientific integrity that you should expect to encounter may include some of the following:

- Being encouraged to "cherry pick" the evidence that supports a predetermined position and ignore or under-represent unclear or contradictory findings
- Having people assume that scientific evidence you produce, or present, means you hold a particular point of view or position
- If your data fail to support a position, it is because you are unintelligent, corrupt, or even part of an unsavory conspiracy. This may also involve assumptions based on your personal identity, such as race, gender, age, etc.

Being assumed to be acting in poor faith is not necessarily common but is probably unavoidable. To protect your scientific reputation, it is good practice to support your actions using transparency, by referencing appropriate standards, and by maintaining good documentation; also, reflect on your own confirmatory bias.

Keywords: social movement, pound seizure law

Fish as friends

One example of how science has engaged directly with animal welfare as a social movement concerns fish. The very idea of industrially caught fish having welfare only really became a reality in the 21st century. People associated with advocating for fish welfare tended to be marginalized or considered extreme.

However, considerable progress has been made in building a consensus around the welfare of at least the largest class of fish—the teleosts.

Research has been conducted into pain and fish sentience (Medway 1980; Lambert et al., 2022), tracing the anatomical structures that detect harm, and studying behaviors indicating distress and learning based on aversion. While the acceptance of fish sentience has grown over the decades, some researchers resort to comparative anatomy and behavior to undermine the idea (as discussed in Sneddon et al., 2018; Striedter, 2016). Meanwhile, brain neuroanatomy in other animals continues to surprise, with human-like neocortical "wiring" now described in brains without obvious cortical-like areas (Stacho et al., 2020; Nieder et al., 2020). Deciding that an animal lacks a particular functionality based on how *we think* their brains are organized is dangerous guesswork given what is at stake.

What made the acceptance of fish sentience a reality was not just an accumulation of evidence but a shift in norms and ultimately, the economics and laws governing the fish trade. The more humane slaughter of fish (e.g., by stunning) is becoming the norm, when previously they were left to die by decompression and asphyxiation after being removed from the water (aka *dewatering*).

The accumulation of scientific evidence was just part of a cross-cultural embrace of the fish as a sensitive being. This "micro-social movement" can be seen building throughout the 1990s and 2000s, as shown by these different developments. A children's book, *The rainbow fish*, was published in 1992, with *Fish welfare* (Ed. Branson) following in 2008. By 2016, the popular science audience was engaged by *What a fish knows: The inner lives of our underwater cousins*. The movies *Finding Nemo* (2003), *Shark tale* (2004), and *Finding Dory* (2016), all increased the profile of fish. PeTA launched their *Sea Kittens* campaign in 2009, which rebranded fish as "sea kittens" and challenged people to rethink their opinions of fish. Increased legal protections (European Council Directive 98/58/EC) and various non-profits and non-governmental organizations (NGOs) have all helped to improve fish welfare. These include the United Kingdom–based RSPCA's (Royal Society for the Prevention of Cruelty to Animals) Humane Standards for Farmed Atlantic Salmon (2018); creation of the Fish Welfare Initiative and the Aquatic Life Institute (2019); and Compassion in World Farming's "Rethinking Fish" campaign. In 2021, the American Humane Certified™ program for aquaculture (ASC-Aqua, a seafood accreditation scheme) launched their own fish welfare project to develop standards, and more recently, the UK Humane League launched their "Forgotten Fish" campaign (2021).

The forefront of humane conservation shows evidence of continued growth, for example, in advancing invertebrate welfare in the Shrimp Welfare Project https://www.shrimpwelfareproject.org/, which offers free humane stunning devices to shrimp farmers.

Keywords: dewatering (fish-killing), stunning

17.3 VIOLENCE AND "THE LINK"

Many people believe that those who are cruel to animals are also dangerous to other people. This is represented in a widely distributed set of etchings called

The Four Stages of Cruelty, by the English painter and satirist William Hogarth 1697–1764. In these, the most important "victim" of animal cruelty was the culprit, who morally degenerates because of their actions. This theme of moral degeneracy, especially in youth, was a preoccupation of the times and one Hogarth returned to in other famous series such as *A Rake's Progress*; the main concern being that moral degeneracy was putting your immortal soul at risk.

In modern times, more emphasis is placed on how animal abusers harm and cause suffering to others. The connection between animal cruelty and violence to humans (domestic violence, child and elder abuse, etc.) is referred to as *The Link* (Flynn, 2000). The notion of links is used to increase the resources put into detecting and responding to crimes against animals, and it is convincing even to those not very interested in animals as victims. While The Link has been very productive in improving coordinated responses to animal welfare issues, it runs the risk of overstating the dangerousness of especially young perpetrators of animal abuse and so cutting them off from avenues of rehabilitation.

Concepts such as The Link tend to be treated as universal truths that will present the same way in every community. However, different social groups are actually deeply affected by factors such as the role animals have functionally and symbolically played in their societies. For example, Indigenous communities in the continental USA are very diverse from each other but tend to share the following characteristics regarding animals:

- First nation creation societies often have animals as "actors" rather than victims
- Animals are seen as family and teachers for adults
- Traditionally, animals are seen as having spirits that must be considered during acts such as hunting and criminality

Traditional Indigenous American norms and laws prohibited the physical and spiritual mistreatment of animals, but this was taken as a sign of weakness and inferiority by colonizers, who saw the absence of written law to mean "no law" (Deer and Murphy, 2017). Instead, a patriarchal culture of violence toward animals and women was imposed, which has had serious repercussions for women ever since. Deprivation and inequality still haunt these communities, with federal reports indicating that the highest rates of interpersonal violence in the United States involve Native peoples (Rosay, 2016). Further, the highest female incarceration rates in the United States are of American Indians and Alaskan Natives (2020 census data, US Department of Justice).

External actors have also sought to exploit legal loopholes on Indigenous territory to avoid state-mandated environmental protections, one example being the Sun Prairie swine farms and the Rosebud reservation (1998–2003), in which discrimination and animal cruelty were also reported. Indigenous law has always protected animals; for example, the Creek had laws to protect dogs as of 1883. Of course, issues of poverty and violence play out in Indigenous cultures worldwide. For the Norwegian Sámi, the feeling was that if they asked for help to deal with domestic violence they would be stigmatized, which is probably a common feeling.

If you intend to work with an Indigenous community that you are not a part of, think about the Two-Eyed Seeing paradigm and the ten calls to action that we discussed earlier. A few additional and accessible options are listed below:

- Trauma-informed and culturally safe training—free training programs for the animal services sector (Vancouver Humane Society) https://vancouverhumanesociety.bc.ca/trauma-informed-training/
- The Association for Animal Welfare Advancement DEI training opportunities https://theaawa.org/dei/
- AVMAaxon webinars on DEI https://axon.avma.org/page/diversity-equity-inclusion-courses

Keyword: cultural competencies

17.4 ARCHEOLOGY AS A SOURCE OF ATTITUDES

Archeology has repeatedly shown how important animals were for ancient humans. We hunted them, used their bodies for tools, domesticated them, revered them in artwork, and ritualistically killed them, sometimes in a rite of passage (Edrey, 2008; Brown et al., 2019). Bioarchaeology, the study of human/biological remains in the context of their surroundings, was popularized in 1972 by the English archeologist Grahame Clark and independently, by the American anthropologist Jane Buikstra. The United Kingdom and the United States differed in their interpretations of bioarchaeology, but a unifying theme was that animals were conceptualized as natural resources rather than as populations of direct interest. Reading this picture in reverse, we see a long history of people influencing animals, and so animal welfare inevitably rears its head. For example, a *zooarchaeologist* might interpret the marks left on a horse's teeth and skull as evidence of poor welfare from behaviors such as crib-biting (Bartosiewicz and Gál, 2018). This suggests that these behaviors have ancient origins, perhaps associated with the advent of stabling.

Ideas about the ancient past strongly influence current behavior, such as choosing to use a dominance-based dog training method or feed a diet based on the needs of wolves. As our knowledge of these origins evolved, it became clear that our stories of the past were overly simplistic and must now accommodate new understandings. For example, modern wolves demonstrate that they are fundamentally sociable and not ruled primarily by dominance and fear, and that the path to becoming a dog includes genetic changes to allow the digestion of new starchy foods.

Keywords: zooarchaeologist, Grahame Clark, Jane Buikstra

Example: Archeology and the domestication of dogs

Dogs are considered the first domesticated species, with the dates for domestication from Pleistocene wolves having been relentlessly pushed back to between 40,000 and 15,000 BC, the Last Glacial Maximum (Serpell, 2021). The traditional approach of archeology (as well as biology and genetics) was to represent domestication as a deliberate process to "bend the wolf to our will", albeit after tolerating their scavenging around human encampments (as depicted in Konrad Lorenz's *Man meets dog* [1953]). The current variant of this notion (*Commensal Scavenger hypothesis*) reframes domestication as more self-inflicted (Coppinger and Coppinger, 2002), with wolves adapting to a scavenger niche (Zeder, 2012).

A competing idea (*the Pet Keeping hypothesis*) was that wolf puppies were acquired to be pets, and the tamer and more biddable ones were kept. Modern conceptions place a much greater emphasis on humans and wolves associating due to mutual benefit—a hunting coalition, if you will, with domestication as a much more nuanced and ongoing process (Losey, 2022).

Many Native peoples revere the wolf, and it was forbidden to hunt them, or dogs, as they had taught humans how to hunt and were considered family (Deer & Murphy, 2017: referenced in Violence and "the link"). This perspective of mutualism is partly why archeology is influencing developments in many areas, including agriculture. Images of humans and dogs hunting together can be found on rock art (petroglyphs), with some exceptional examples from Saudi Arabia (viewed in this video https://www.science.org/content/article/these-may-be-world-s-first-images-dogs-and-they-re-wearing-leashes).

However conceived, the "wolf–human" alliance provided the space for divergent evolution into dogs that could more fully exploit the relationship. Given how widely and for how long dogs and humans have been associated, this connection is symbiotic and transformative for both species. Any specific line of animal with heritable traits developed through artificial selection is considered a *breed*. Archeological findings suggest multiple times when domestication occurred, with some early examples that remained geographically isolated to specific *ancient breeds* (informal terminology for dog breeds existing prior to kennel club "pedigrees"). Equivalent terms would include primitive dogs as described by the Fédération Cynologique Internationale (FCI) or pre-modern breeds. These landraces are dogs that sprang from different and earlier styles of human–animal interaction, where the "rules of engagement" might have been quite different.

Ancient breeds are considered more aloof and harder to train, and are often supplanted by modern breeds where available. Ancient breeds may also experience welfare problems associated with the expectation that dogs respond to training methods that allow them very limited autonomy. Conversely, modern breeds of dog may experience poor welfare by virtue of their tameness, making them easier to neglect, abuse, and exploit. Dogs also seem to have "paid" for their alliance with humans in that they are less socially cohesive with each other than wolves (Cafazzo et al., 2018). The wolf pack stayed firm, the "dog pack" less so. It all comes down to expectations and the niches that dogs are expected to inhabit, noting that many dogs, irrespective of expectations or lineage, do very well in urban and other settings.

Keywords: breed, ancient breed (dog)

17.5 IMPORTANT CULTURAL PRACTICES

Animal welfare advocates often find themselves opposed to a cultural practice that seems to harm animals. This is partly due to being unfamiliar with the history, cultural meaning, benefits, and constraints that led to this activity occurring, and may also be associated with explicit or implicit bias against that cultural group. Most of you are probably aware of opposition to whale or dolphin hunting, seal clubbing, or bull fighting. Other conflicts arise in the United States, where practices have limited acceptance and the dividing lines between law enforcement and the majority are explicit (e.g., dog/cock fighting) or fuzzier (e.g., rodeo, dog ear/tail cropping).

Moral philosophers suggest that there are distinct reasons why we pass judgment on others. Specifically, these may relate to whether harm is done (e.g., stealing food) or whether the act is considered impure (e.g., eating dog meat). The use of dog meat eating as a common example of moral impurity (Gray & Keeney, 2015) shows how strong this taboo is in Western and specifically American contexts. However, according to Farzad Abedi (2021), even in 491 BC, the Persian King Darius asked the Carthaginians to refrain from eating dog meat, or sacrificing their children, as neither practice was acceptable to his troops. So, sticking purely to the question of dog meat, this taboo of eating dog meat has circled the globe at various times in history.

In America, we tend to designate some animals as food and others as pets, with no exceptions. For some species, the transition between food source and pet may have occurred quite recently. For example, in the United States, the rabbit only moved into the pet category after World War II. In 2014, Wholefoods began selling rabbit meat, and this was met with considerable backlash from pet rabbit owners and organizations. The immediate corporate response was to reference the welfare standards for farms supplying the rabbit meat, which missed the point. The public response was not about the humane treatment of rabbits; it was a rejection of the very concept that rabbits are meat animals. Rabbit meat stopped being offered a few years later due to lack of demand.

In other regions, the lines between "pet" and "food" are drawn in different places. For example, horse meat is consumed in many areas of Europe. Elsewhere, the popularity of dog meat fluctuates according to cultural and economic trends (Dugnoille, 2018). Dog meat eating is quite common in areas of Vietnam, and dog meat, "Thit Cho", restaurants started to spring up in the early 2000s (Avieli, 2011). Their clientele of politically engaged middle-class men, saw their consumption of traditional *Confucius foods* as part of a show of masculinity, power, and allegiance to the current regime and its elites. (Confucius foods follow the recommendations of the highly influential Chinese philosopher Confucius [551–479 BCE]).

In each culture where it occurs, the consumption of dog meat has particular associations, just as Americans feel that eating beef or steak is associated with being manly and patriotic. Understanding the motivations underlying an activity are important when attempting to advocate for change. Researchers in disciplines such as ethnology and anthropology can provide this information using well-established quantitative and qualitative methods that help to account for any bias on the part of the researcher.

Keywords: tradition, Confucius foods

17.6 SATIRE AS PROTEST

The strong aversion of most of the US public to eating dog meat is skillfully "repackaged" by pro-vegan satirists for Elwood's Organic Dog Meat (https://www.elwooddogmeat.com/). (Satire is a literary form that ridicules the norms or conduct of a group of people.) On their website (Figure 17.2), forms of messaging used to promote other meat products are repurposed to describe dog meat. A second satirical website, Seymour Loudermilk and Associates (https://hatevegans.com), advertises a fictitious law firm set up to protect innocent omnivores who are provoked to existential crisis by vegans. Potential clients are extolled to "be more, eat more with Seymour".

Figure 17.2 Logo of the satirical website advertising Elwood's Organic Dog Meat (https://www.elwood-dogmeat.com/). The reach of satirical websites like this can be extraordinary. An internet search of "dog meat eating countries" now erroneously includes Switzerland, in part because of another satirical exercise around a fake restaurant called La Table Suisse (https://www.youtube.com/watch?v=q9lPyTOHUIE). The Elwood's Organic Dog Meat logo is reproduced with the permission of Molly Elwood.

These websites are potent reminders of our norms relating to acceptable standards of care of animals in different settings and the broader *meat paradox*—liking to eat meat but not wanting to kill animals (Tian et al., 2016). In American society, it is impossible to ignore the sentience of our pets, but this is more commonly denied or overlooked for agricultural animals or wildlife.

The use of animals in farming has a rich history in satire. Perhaps the most famous is *A modest proposal* (1729) by Jonathan Swift, which touted the idea of the poverty-stricken Irish monetizing their children as a food source. *Animal farm* (1945) by George Orwell used farm animals to represent the oppression of the proletariat, specifically under Stalinism. And *The British miracle meat* (2023) was a satirical documentary by Matt Edmonds that reprised the Modest proposal with lab-grown human meat. Cash-strapped pensioners could donate flesh at £250/donation, enough, according to "the reporter", to cover the power bill for two weeks. The purpose was to compare human exploitation to that of farm animals.

Keyword: satire

17.7 CITIZEN SCIENCE

Scientific research that includes the active participation of members of the public is called *citizen science*. Typically, the citizen scientist will collect data, observations, or samples. Digital platforms such as Zooniverse.org exist to recruit citizen scientists and show considerable activity around wildlife and conservation topics. An example of citizen science in action is a project concerning "bird strike" (Loss, 2023). Every year, up to a billion birds are killed in the United States by striking windows, but many people are unaware of the problem because the injured bird will fly away before collapsing and dying (Figure 17.3). Systematic real-world observations as part of a *natural experiment* can assess the severity of bird strike at different locations so that we can plan less hazardous built environments.

Not all citizen scientists are organized or directed by academic scholars. Some projects are led by independent scholars or self-educated researchers. Often, the best science is instigated and carried out by members of the community where the need exists. Local experts can also prove to be crucial for their in-depth knowledge and access.

Figure 17.3 This Eurasian siskin clattered into a window at a zoological station in Finland that this author (TR) was teaching at. A little dazed, it seemed to recover after a while and flew away. Hopefully it wasn't one of the ones that died later on. Reproduced with the permission of Janet Patterson-Kane.

When a project is led by someone in a position of authority, such as university faculty, the academic point of view is ordinarily privileged compared with that of the citizen scientist. Care should be taken to make sure that the lay collaborators are not being exploited (Hall et al., 2024) and that animal welfare is respected (Palmer et al., 2021). Ethical guidelines such as those produced by the European Citizen Science Association are aimed at addressing the power imbalances in these relationships and can be used in the design of these projects (https://ecsa.citizen-science.net/documents/).

Keywords: citizen science, natural experiment

READINGS

Norms

Delon, N. (2018). Social norms and farm animal protection. *Palgrave Communications*, *4*(1), 1–6.

Grandin, T. (2022). Bad becoming normal is detrimental to beef cattle welfare. *Journal of Applied Animal Ethics Research*, *4*(2), 151–157.

Plant, M., Schaik, P., Gullone, E., & Flynn, C. (2016). "It's a dogs life": Culture, empathy, gender, and domestic violence predict animal abuse in adolescents–implications for societal health. *Journal of Interpersonal Violence, 34*. https://doi.org/10.1177/0886260516659655

Lab lore

Arluke, A. (1988). Sacrificial symbolism in animal experimentation: Object or pet? *Anthrozoos: A Multidisciplinary Journal of The Interactions of People & Animals*, *2*, 98–117. https://doi.org/10.2752/089279389787058091

Social movements

Anderson, J. (2011). Protection for the powerless: Political economy history for the animal welfare movement. *Stanford Journal of Animal Law & Policy*, *4*, 1.

Cogan, E. (2022). Preventing fraud in biomedical research. *Frontiers in Cardiovascular Medicine*, *9*, 932138. https://doi.org/10.3389/fcvm.2022.932138

Graves Jr, J. L., Kearney, M., Barabino, G., & Malcom, S. (2022). Inequality in science and the case for a new agenda. *Proceedings of the National Academy of Sciences of the United States of America*, *119*(10), e2117831119. https://doi.org/10.1073/pnas.2117831119

Pielke Jr, R. A. (2007). *The honest broker: Making sense of science in policy and politics*. Cambridge University Press.

Tansey, E. M. (1994). Protection against dog distemper and dogs protection bills: The medical research council and anti-vivisectionist protest, 1911–1933. *Medical History*, *38*(1), 1–26. https://doi.org/10.1017/s0025727300056027

Fish as friends

Branson, E. J. (Ed.). (2008). *Fish welfare*. John Wiley & Sons. https://doi.org/10.1002/9780470697610

Lambert, H., Cornish, A., Elwin, A., & D'Cruze, N. (2022). A kettle of fish: A review of the scientific literature for evidence of fish sentience. *Animals*, *12*(9), 1182.

Medway, L. (1980). *Report of the panel of enquiry into shooting and angling* (1976–79). University of Bradford.

Nieder, A., Wagener, L., & Rinnert, P. A. (2020). A neural correlate of sensory consciousness in a corvid bird. *Science, 369* (6511), 1626–1629. https://doi.org/10.1126/science.abb1447. PMID: 32973028.

Sneddon, L. U., Lopez-Luna, J., Wolfenden, D. C. C., Leach, M. C., Valentim, A. M., Steenbergen, P. J., Bardine, N., Currie, A. D., Broom, D. M., & Brown, B. (2018). Fish sentience denial: Muddying the waters. *Animal Sentience, 21*. https://doi.org/10.51291/2377-7478.1317

Stacho, M., Herold, C., Rook, N., Wagner, H., Axer, M., Amunts, K., & Güntürkün, O. A. (2020). A cortex-like canonical circuit in the avian forebrain. *Science, 369*(6511), eabc5534. https://doi.org/10.1126/science.abc5534. PMID: 32973004.

Striedter, G. (2016). Lack of neocortex does not imply fish cannot feel pain. *Animal Sentience, 3*(15). https://doi.org/10.51291/2377-7478.1037

Violence and "the link"

Deer, S., & Murphy, L. (2017). "Animals may take pity on us": Using traditional tribal beliefs to address animal abuse and family violence within tribal nations. *Mitchell Hamline Law Review, 43*(4), Article 1. https://open.mitchellhamline.edu/mhlr/vol43/iss4/1

Flynn, C. (2000). Why family professionals can no longer ignore violence toward animals. *Family Relations, 49*, 87–95. https://doi.org/10.1111/j.1741-3729.2000.00087.x

Rosay, A. B. (2016). *Violence against American Indian and Alaska Native Women and Men: 2010 Findings from the National Intimate Partner and Sexual Violence Survey.* National Institute of Justice Research Report. NCJ 249736. U.S. Department of Justice, National Institute of Justice.

Archeology as a source of attitudes

Bartosiewicz, L., & Gál, E. (Eds.). (2018). *Care or neglect? Evidence of animal disease in archaeology.* Oxbow Books.

Brown, D., & Anthony, D. (2019). Late Bronze Age midwinter dog sacrifices and warrior initiations at Krasnosamarskoe, Russia. In B. A. Olsen, T. Olander, & K. Kristiansen (Eds.), *Tracing the Indo-Europeans: New evidence from archaeology and historical linguistics* (pp. 97–122). Oxbow Books.

Edrey, M. (2008). The dog burials at achaemenid ashkelon revisited. *Tel Aviv: Journal of the Institute of Archaeology of Tel Aviv University, 35*, 267–282. https://doi.org/10.1179/033443508788921276

Example: Archeology and the domestication of dogs

Cafazzo, S., Marshall-Pescini, S., Lazzaroni, M., Virányi, Z., & Range, F. (2018). The effect of domestication on post-conflict management: Wolves reconcile while dogs avoid each other. *Royal Society Open Science, 5*(7), 171553. https://doi.org/10.1098/rsos.171553. PMID: 30109041; PMCID: PMC6083655.

Coppinger, R., & Coppinger, L. (2002). *Dogs: A new understanding of canine origin, behavior and evolution.* University of Chicago Press.

Losey, R. J. (2022). Domestication is not an ancient moment of selection for prosociality: Insights from dogs and modern humans. *Journal of Social Archaeology, 22*, 131–148. https://doi.org/10.1177/14696053211055475

Serpell, J. A. (2021). Commensalism or cross-species adoption? A critical review of theories of wolf domestication. *Frontiers in Veterinary Science, 8*, 662370. https://doi.org/10.3389/fvets.2021.662370

Zeder, M. A. (2012). The domestication of animals. *Journal of Anthropological Research, 68*, 161–190.

Important cultural practices

Abedi, F. (2021). Study of the political, social and cultural dimensions of the decree of Darius I to the people of Carthage. *Social History Studies, Institute for Humanities and Cultural Studies (IHCS), 11*(1), 125–161. https://doi.org/10.30465/shc.2021.37886.2263. A translation is available online, but is not free to access. The abstract is in English.

Avieli, N. (2011). Dog meat politics in a Vietnamese town. *Ethnology, 50*(1), 59–78.

Dugnoille, J. (2018). To eat or not to eat companion dogs: Symbolic value of dog meat and human–dog companionship in contemporary South Korea. *Food, Culture & Society, 21*(2), 214–232.

Gray, K., & Keeney, J. E. (2015). Impure or just weird? Scenario sampling bias raises questions about the foundation of morality. *Social Psychological and Personality Science, 6*(8), 859–868.

Satire as protest

Tian, Q., Hilton, D., & Becker, M. (2016). Confronting the meat paradox in different cultural contexts: Reactions among Chinese and French participants. *Appetite, 96*, 187–194. https://doi.org/10.1016/j.appet.2015.09.009

Citizen science

Hall, D. M., Avellaneda-Lopez, P. M., Ficklin, D. L., Knouft, J. H., & Lowry, C. (2024). Citizen silence: Missed opportunities in citizen science. *Bioscience, 74*(5), 319–321.

Loss, S. R., Li, B. V., Horn, L. C., Mesure, M. R., Zhu, L., Brys, T. G., & Farnsworth, A. (2023). Citizen science to address the global issue of bird–window collisions. *Frontiers in Ecology and the Environment, 21*(9), 418–427.

Palmer, A., Reynolds, S. J., Lane, J., Dickey, R., & Greenhough, B. (2021). Getting to grips with wildlife research by citizen scientists: What role for regulation? *People and Nature, 3*(1), 4–16.

Chapter 18

Codified, Ethical, and Economic Conduct

18.1 CODIFIED CONDUCT AND STRUCTURED BELIEF SYSTEMS

The informal norms that drive human behavior exist in a dynamic relationship with written rules such as laws and regulations, ethics and moral codes, religion, and other codified spiritual doctrines. A *code* is different from a norm in that it is recorded with some degree of enforcement to ensure compliance. Behavior that does not conform to codes is often considered *deviant* (the term is quite commonly used in behavioral sciences but carries connotations of immorality or perversion). Codified conduct has been discussed throughout this book, including laws like anti-cruelty laws, regulations like kennel club registry, and spiritual beliefs like Confucius food consumption. However, there are two areas that require more comprehensive coverage. These are the forms of formal ethics most often applied to questions of animal welfare and the economic systems that support commercial animal exploitation.

Keywords: deviance, codified

18.2 ETHICAL PARADIGMS

Historically, a lot of academic philosophy shied away from practical questions like "Should you dock a dog's tail?", instead favoring illustrative but abstract scenarios like where to direct a trolley that will run over different numbers of people. (The trolley example is a typical *thought experiment* that involves various ethical dilemmas [Kamm, 2015]). Philosophy is an extremely diverse field, but as a discipline, it tends to coalesce around how people are different from animals, and how ethics are rational rather than emotional. Within this framework, there are a few major schools of thought that are important for animal

DOI: 10.1201/9781003375456-23

welfare work, which are briefly outlined below: utilitarian, deontological, and duty of care. We will also discuss how ethical approaches are used in applied situations.

Scientific research tends to appear in peer-reviewed journals on a piecemeal basis, while for ethics and philosophy, book-length works play a more prominent role. Three works, *The case for animal rights* by Tom Reagan (1983), *Animal liberation* by Peter Singer (1975), and *Beyond animal rights* edited by Josephine Donovan and Carol Adams (1996), are hugely influential in animal welfare science and are recommended reading for anyone working in this field. It might seem that these relate more to animal rights than to animal welfare, but these concerns are more closely related than they might seem.

Keyword: thought experiment

18.3 UTILITARIAN

Most professional and government settings use a *utilitarian* approach to ethics. A decision is utilitarian when the risks and benefits are weighed to choose the option that benefits the greatest number of people or individuals. The exploitation of others is permitted if it serves the greater good (be that for a larger number of individuals, such as the community or the country). Utilitarian approaches will arrive at quite different outcomes depending on how values are assigned and how the cost/benefit calculation is carried out.

For example, the use of animals in research is often governed by a utilitarian assessment carried out by a scientific and ethical review body (e.g., an Institutional Animal Care and Use Committee/IACUC). The potential benefits of doing the research are weighed against the imposition or harm done to the animals used in the research. A decision is then reached, with reference to the beliefs of the surrounding community, as to whether this use is justified. In this context, a very large number of animals, especially rodents and zebrafish, are used in research with a range of outcomes. It is debatable as to whether the case for the individual animal can be effectively made.

Peter Singer, a prominent utilitarian, asserts that animals should be given the same moral consideration as humans. Singer takes a non-speciesist approach in which the capacity to suffer is, again, key. Taking this thesis to its obvious conclusion, Singer also somewhat controversially suggests that there would be some situations in which human subjects would be satisfactory research subjects (although rarely and with informed consent). More recently, Singer has been influential in the movement of *Effective Altruism*, i.e., prioritizing causes that yield the greatest good, as described in his 2015 book *The most good you can do.*

Singer's book *Animal liberation* has had a powerful and ongoing impact on the public debate about the use of animals for human purposes. Famous activists who were inspired by this book include Ingrid Newkirk, cofounder of *PeTA*, and Matt Ball of *One Step for Animals*, as well as generations of animal advocates. *Animal liberation* is sometimes referred to as the "bible" of the animal rights movement. Since publishing *Animal liberation* in 1975, Singer has had ample time to assess the utility of the research that has been sanctioned by countless

IACUCs. The result is not great, and millions of animals have suffered and had their lives cut short for poor-quality research with little real-world benefit. The suggestion is that we should better align research quality with research ethics, so that improved outcomes can be achieved (Brill et al., 2021; Ioannidis, 2016, 2018). Singer took such a hardline approach to utilitarianism that he is often considered an animal rights philosopher. However, the idea of rights properly belongs to deontological approaches, as described below.

Keyword: utilitarianism

18.4 DEONTOLOGICAL

A deontological approach is concerned with the rights and duties of individuals. Where rights exist, they cannot be ethically disregarded, regardless of the outcome. Therefore, *deontology* is an ethical approach of understanding our duties and obligations based on the rights and dignity of others. These rights include the right to bodily integrity and the right to life.

The most famous deontologist addressing animal rights was Tom Regan. In *The case for animal rights* (1983), Regan asserted that because animals are "subjects of a life" (i.e., with lives of subjective experience), they cannot be exploited by others. That is, an animal is like a human, because they exist for themselves and are not to be a resource or tool in the hands of others. Rather than closely following a particular school of ethics, most people will tend to shift their reasoning based on the situation: for example, taking a moderate utilitarian approach to a mainstream activity, like whether to neuter a dog, but condemning outright an activity that they find disgusting, like animal crushing.

As outlined in Hal Herzog's 2010 book titled *Some we love, some we hate, some we eat*, our attitudes are influenced by many factors and are rarely logical or consistent. This takes into account not only rights and consequences but feelings, obligations, and relationships. While this haphazard approach is condemned by most philosophers and ethicists, it is, in fact, a natural part of the human condition.

Keyword: deontology

18.5 DUTY OF CARE

The exclusion of feelings from these ethical approaches was directly challenged by a series of works from feminist ethics, and particularly by Carol Adams and Josephine Donovan. This new approach arose in the early 1980s with the feminist ethic of care theory. *Duty of Care* ethics assert that we have particular obligations to people, especially those that we have agreed to care for (e.g., children or the elderly). These ideas were first expressed in an anthology of essays, *Beyond animal rights* (1996), and were updated in *The feminist care tradition in animal ethics* (2007).

It is also assumed that even those toward whom you have no duty of care should not be exploited. In *The sexual politics of meat* (1990), Adams pointed out the converging themes in oppressing women and animals in terms of objectification and use. Taken together, the feminist ethic of care validates emotion as a motivation for ethical action. Responding with sympathy and care to those in need is a legitimate response and is in no way irrational. And it extends to many groups of exploited and suffering people and animals. This idea developed into the concept of *intersectionality*—specifically, how structures of oppression apply across categories such as gender and species.

Intersectional approaches can be controversial. Initiatives that try to help humans *and* animals struggle to get the balance right. Human subjects often take the focus away from animals (e.g., evacuating animals from disaster zones might be seen as taking resources away from humans in peril). Conversely, when any equivalence is drawn between human minorities and animals, this triggers memories of racist and sexist abuse: for example, the use of misogynist slurs like "bitch" or "pig".

Keyword: duty of care

18.6 WHAT KIND OF LIFE?

Animal welfare is generally seen as a state that can range in valence from very poor to very good. However, there are a number of milestones along this scale that we might aim for. "A Life Worth Living" is a very basic requirement for animal welfare. It is often the standard used when completing a quality-of-life assessment for an animal approaching the end of their life. Basically, any animal whose life is below this level might be better off dead.

Another common goal in animal welfare is "the good life". This is better than simply a life worth living, but often comes with two very different interpretations of what a natural life is. One sees nature as a threat that we need to protect animals from, and the other embraces natural life as the only true setting for a happy life. In the first instance, a good life can be demonstrated by metrics that show that the animal is healthy and safe. This approach aligns well with a rational/industrial understanding that ostensibly protects the animal by providing a predictable and safe environment without the worry of malnutrition, disease, fighting, predation, parasites, and extreme weather. In these scenarios, animals are shown to have good core biological function using completely objective measures (e.g., body condition, growth, reproduction).

The contrasting idea comes from a romantic or agrarian position. This typically equates animals with natural settings, with freedom of movement and association, and free from the physical and behavioral signs of close confinement (lesions, lameness, stereotypy). In this context, nature is benevolent rather than a source of hazard. Groups that align with an agrarian ideal tend to include consumers without direct experience with farming, smaller farms, behavioral researchers, and movements including organic farming. There are also *aspirational* reference points, which can be referred to as *thriving* or *eudaimonia*. Like many other terms that we use (e.g., *telos, euthanasia)*, the idea of *eudaimonia* comes to us from Aristotle. Here, "*daimon*" indicates spiritual wellness and advancement to a higher state of being.

Keyword: eudaimonia

18.7 PRACTICAL ETHICS

Animal welfare science often involves people needing to make actual decisions rather than discuss imaginary trolley trajectories and outcomes. Practical ethics is where we structure our ethical deliberations so that we can decide what to do in a real-world situation. Ideally, scientific evidence can support a shared understanding of the situation and the likely consequences of making different choices.

Stephen Toulmin (1922–2009) was an English philosopher who emphasized a more practical approach to ethics and the construction of a sound argument. Animal welfare scientist David Fraser (2011) applied Toulmin's approach to animal ethics based on the four main areas where welfare concerns arise. These are: "(1) to provide good lives for the animals in our care, (2) to treat suffering with compassion, (3) to be mindful of unseen harm, and (4) to protect the life-sustaining processes and balances of nature". These principles direct our attention to current problems and needs that relate to interacting with an animal in an ethical way rather than making sweeping and abstract assertions about what is normative in society.

Keyword: practical ethics

Example: The ethical matrix

Any animal welfare issue will involve a unique set of stakeholders, cultural structures, and anxieties, which means that people often have trouble understanding each other and might talk at cross purposes. One ethical tool that can help groups in decision making was developed by Ben Mepham in 1994 and is referred to as "the ethical matrix" (described in Mepham, 1996; Mepham et al., 2006). This approach pulls out two dimensions of an issue: 1) the stakeholders involved and 2) the ways they might conceptualize the issue according to ethical principles—and be affected by the decisions made. The relevant principles are captured in three key elements:

- "Wellbeing is considered to be the best way of representing utilitarian concerns ('maximizing the good')
- Autonomy represents deontological concerns (treating others as 'ends in themselves')
- Fairness represents respect for justice, an interpretation advocated by the philosopher Rawls" (Mepham et al., 2006)

The goal of the matrix is to define the ethical issues, identify the basis for disagreement, and encourage a fair and transparent decision-making process. An important part of the matrix is that the stakeholders are encouraged to "stand in someone else's shoes" or even assume the role of the animal and consider whether they would make the same recommendation. They must seek to cloak their own identity in reaching their decisions.

This ethical matrix (Table 18.1) reflects the decision making around whether to amputate a leg. A dog has bone cancer in his front left leg. The X-ray looks like classic osteosarcoma. Without amputation, it might be best to put the dog

Table 18.1 Ethical matrix for a surgical intervention

	Wellbeing	Autonomy (sometimes replaced with "Dignity")	Fairness (sometimes replaced with "Justice")
Dog's perspective	Surgery will end pain from the cancer, although there will be pain during the recovery period. Potentially a longer healthy life. Could be a lower quality of life if there is a failure to adapt to being a tripawd. If the cancer has already metastasized, then lifespan would be short anyway.	Initially autonomy lessened until recovery, then better as the painful leg will no longer be a problem. Appropriate harness and care can aid safety and lessen anxiety. Other than the cancer, the dog is healthy.	Is it fair to expose the dog to a serious surgery for an uncertain outcome that they cannot conceptualize?
Owner/s' perspective	Attached owners want to continue sharing life with their dog.	Consider whether they are extending the life of the dog to avoid grief and loss rather than acting purely in the interests of the dog.	Is it reasonable to expect owners to deal with the ongoing anxiety and trauma of surgery, chemotherapy, side-effects, wound management, and the physical requirement to lift and carry the dog for an uncertain outcome?
Veterinarian's perspective	Wishes to use their expertise and save the life of the dog.	Might feel good about saving the dog's life and completing a successful treatment. Feels that the dog would do well on three legs.	Veterinarian may feel traumatized about putting the dog to sleep when they feel the dog could be saved.
Perspective of a friend of the family	Due to a conviction that the dog wishes to keep living, advocates for the surgery and will be distressed if this does not occur.	Friend needs to feel they have done everything they can to advocate for the dog.	If the friend is so sure that surgery would go well, is it fair to undergo surgery if it goes badly and the dog suffers, dies in surgery, or dies soon after?

to sleep because the cancer weakens the leg—it could fracture at any time. The veterinarian cannot take a biopsy to assess the severity of the cancer before surgery, as that would be painful—and the leg needs to be amputated anyway. So, the leg would be removed and the cancer examined afterwards by a veterinary pathologist. If the cancer is high grade, the dog may die before the fur has time to grow back because the cancer will already have metastasized. If the cancer is less aggressive, then chemotherapy post surgery could give the dog another year. What would you do? Amputate or not?

The outcome: this is the patient, Satu (Figure 18.1). His fur has grown back after surgery. Satu lived for another three years—a natural lifespan. He was a statistical outlier, and his tumor was extremely rare. This author (TR) was one of Satu's owners. The other owner, an expert veterinary pathologist, ended up having to diagnose the tumor because it was so rare that nobody quite knew what it was. Janet Kessler (our coyote expert whom you met previously) was the

Figure 18.1 Satu post operation, demonstrating his athleticism. Picture reproduced with the permission of janetkessler/coyoteyipps.com.

friend advocating for Satu. Satu, like many other three-legged dogs, took his "tripawd" status as a new-normal and got on with life. We were probably more traumatized than he was!

Keyword: the ethical matrix

18.8 ECONOMICS

The focus of the discipline of economics is often upon the use of money within a market where vendors have a profit motive. As such, economics is an important part of how animals are handled within industries. However, the full scope of behavioral economics is not just the use of currency but the study of how people and organizations assign value, use limited resources, and build social capital. As such, it provides novel approaches to many aspects of animal welfare science. For example, *Economic Demand Theory* revealed the lengths to which an animal will go to improve its environment and was instrumental in changing beliefs about what some animals need. (Note that *Demand Theory* describes the relation between price and resources.)

A single paper in the influential journal *Nature* (Mason et al., 2001) revealed the lengths to which mink will go to access water to swim in, which is not normally provided when they are farmed for fur. That mink yearn for water (logical given their natural history as semi-aquatic animals) was invisible until researchers gave them the opportunity to work for it. Given that opportunity, mink demonstrated *inelastic demand*. They would work for access to water no matter the price. (Inelastic demand is associated with essential items like food, which can be equated to "needs".)

Keyword: inelastic demand

Boycotts

An obvious intersection between animal welfare and economics are factors that affect whether people will buy a product or service, or whether they might not only refuse to buy something but will actively encourage others to do the same for ethical reasons (a "boycott").

The tactic of boycotting is named after its first high-profile "victim". Captain Charles Boycott was an English agent for Lord Erne, a landlord for large areas in Ireland, who planned to evict poor tenants impoverished by failed harvests in 1880. The Irish National Land League encouraged the Irish locals in County Mayo to withdraw their economic contributions to his household. There was to be no seasonal labor, no service in shops, and no mail delivery. Boycott was economically shunned by the Irish he lived among, and by December of that year, he had retreated to England. The use of this tactic was widely reported by newspapers, causing the term "boycott" to enter common use.

One classic example of a dramatic market shift due to boycott is the in-home consumption of veal, which began to decline in the late 1960s. The impetus for this decline was from pressure groups demonstrating the small size and isolation of a veal calf enclosure. In 1982, the Farm Animal Reform Movement (FARM, which later became the Farm Animal Rights Movement) led a multi-year boycott against veal by using newspaper advertisements, pickets, and protests. In 1987, FARM's president and co-founder, Alex Hershaft, shut himself in a veal crate behind the White House for 24 hours; some years later, another activist would repeat the stunt for 100 hours outside the US Department of Agriculture. A contemporaneous research paper gives an interesting description of this movement and the pressing need for research on dairy animal welfare at that time (Albright, 1987).

It should be noted that if you boycott a product, the underlying problem does not magically go away. Veal production is part of the wider system of cattle use. In order to produce milk, cows need to have calves. Some of the female calves will supplement the herd, but the males are surplus to requirement and are usually slaughtered. They may still be used as veal, which continues to be sold in fine dining restaurants (a market segment notably unaffected by the boycott and resulting shift in norms). The meat may also be used in processed products where the source of the meat is not immediately apparent to the consumer.

One way to avoid a welfare deficit in male calves is to find a way for them not to be born at all. This would avoid them having a life not worth living and spare producers the expense of disposing of them as a waste product once demand for veal plummeted. Options are being explored to prevent the conception of male cattle by sorting sperm based on X-bearing sperm being larger than Y-bearing sperm (Wheeler et al., 2006). Likewise, the detection of unwanted male chicks while they are still in the egg and before they become conscious would also be of value. Understandably, there is anxiety that these technologies will be misused by human populations where female offspring are more often aborted or suffer infanticide.

Keyword: boycott

18.9 ANIMAL WELFARE LABELING

Companies can benefit from advertising their animal welfare credentials if they can communicate these to potential consumers in a trustworthy way (Sullivan, 2012). One solution to this problem has been welfare labeling schemes that are run by an independent third party. These programs create trusted science-based standards that members are expected to follow, and some also provide inspections and remove non-compliant members. Labeling schemes are arguably more important in countries such as America and Australia that resist prescriptive regulation and operate according to market forces.

For example, many countries have a national mandatory livestock identification system (e.g., cattle ear tags) that can track any animal-based product back to the originating farm. This can be especially useful in identifying and resolving food safety problems. A similar scheme (the National Animal Identification System [NAIS]) in the United States is still only voluntary. Obstacles to implementation include expense, lack of trust, and the sheer size of an agricultural and political system that remains averse to mandatory processes. Similarly, other countries have placed a ban on activities such as pig tail docking, but stateside, this remains a matter of consumer choice. That said, bans without inspection and enforcement are largely meaningless. A 2022 report by Animal Equality and the Animal Law Foundation found that routine tail docking is still performed in the United Kingdom, with 71% of British pigs going to slaughter without their tails (2013-2017). Some examples of voluntary animal welfare certification programs are shown in Table 18.2. As you can see, they are diverse in what they cover and who runs them. There is even more diversity in their aims, what their standards are, and how or if they are enforced.

Labeling schemes also exist outside agriculture: for example, cruelty-free cosmetics or *by-catch* safe fisheries (by-catch is the incidental catch of non-species targets). In addition, companies can and do make many claims without having them clearly defined or independently verified. For this reason, there have been many lawsuits seeking to challenge how a company can describe its products as well as how critics can phrase their criticisms. The outcome is that consumers are often confused by different labeling schemes and are none the wiser about production methods.

Table 18.2 Voluntary animal welfare certification programs

Name	*Operator*	*Product*	*Website*
American Humane Certified	American Humane Society (non-profit)	Cattle, bison, poultry, goats, chickens, ducks, turkeys, and swine	https://www.americanhumane.org/humane-heartland/
Animal Welfare approved	A Greener World (non-profit)	Most commonly farmed species	https://agreenerworld.org/certifications/animal-welfare-approved/
Certified Humane Certification Program	Humane Farm Animal Care DBA Certified Humane (non-profit)	Meat, poultry, egg, fiber, and dairy products	https://certifiedhumane.org/
Global Animal Partnership's (GAP) 5-Step Program	Global Animal Partnership (non-profit)	Beef cattle, dairy cattle, bison, chicken, salmon, goat, sheep, pig, laying hens, and turkey	https://globalanimalpartnership.org/
United Egg Producers Certified Animal Welfare Program	United Egg Producers (Industry)	Eggs	https://uepcertified.com/

In contrast to honest labeling to support market regulation, there are also a range of more misleading or even false representations. These activities are referred to as *welfare washing*, based on the term *green washing*, which describes the disingenuous ways companies can appear to be environmentally friendly. Bjørkdahl and Lykke (2021) described one campaign that effectively urged consumers to "keep calm and carry on consuming meat" while practicing what was described as "strategic ignorance" about animal welfare.

Keywords: welfare washing a.k.a. humane-washing, market regulation approach

18.10 CONSUMER SEGMENTATION

Research surrounding the sale of goods to consumers often breaks down the potential consumers into groups based on their beliefs and demographics. This strand of marketing is informed by research into attitudes from both industry and academia. For example, Prickett's (2010) categories were *Naturalists*, concerned about animal welfare (especially related to outdoor access and natural behaviors); *Descartes' Entourage*, who felt that the provision of food, water, and veterinary care was sufficient; and *Price Seekers*, who were concerned only with affordability. These segmentations represent a caricature of a typical group member, and the way the groups are assembled and described is partly intuitive and can be subject to bias. One common finding from consumer research is that while animal welfare is a common consumer concern, it doesn't rank as highly as food safety.

READINGS

Ethical Paradigms

Adams, C. J. (1990). *The sexual politics of meat: A feminist-vegetarian critical theory.* Continuum.

Donovan, J., & Adams, C. J. (Eds.). (1996). *Beyond animal rights: A feminist caring ethic for the treatment of animals.* Continuum.

Kamm, F. (2015). *The Trolley problem mysteries (The 2013 Berkeley Tanner Lectures and Responses to Commentators).* Oxford University Press.

Reagan, T. (1983). *The case for animal rights.* University of California Press.

Singer, P. (1975). *Animal liberation: A new ethics for our treatment of animals.* Harper Collins.

Utilitarian

Brill, S. A., Guerrero-Martin, S. M., & Metcalf Pate, K. A. (2021). The symbiotic relationship between scientific quality and animal research ethics. *ILAR Journal*, *60*(3), 334–340. https://doi.org/10.1093/ilar/ilab023

Ioannidis, J. P. A. (2016). Why most clinical research is not useful. *PLoS Medicine*, *13*(6), e1002049. https://doi.org/10.1371/journal.pmed.1002049

Ioannidis, J. P. A. (2018). Meta-research: Why research on research matters. *PLoS Biology*, *16*(3), e2005468. https://doi.org/10.1371/journal.pbio.2005468. PMID: 29534060; PMCID: PMC5865753.

Singer, P. (2015). *The most good you can do.* How Effective Altruism Is Changing Ideas About Living Ethically. Yale University Press.

Deontological

Herzog, H. (2011). *Some we love, some we hate, some we eat: Why it's so hard to think straight about animals.* Harper.

Reagan, T. (1983). *The case for animal rights.* University of California Press.

Duty of care

Donovan, J., & Adams, C. J. (Eds.). (1996). *Beyond animal rights: A feminist caring ethic for the treatment of animals.* Continuum.

Donovan, J., & Adams, C. J. (Eds.). (2007). *The feminist care tradition in animal ethics.* Columbia University Press.

Practical ethics

Fraser, D. (2011). A "practical" ethic for animals. *Journal of Agricultural and Environmental Ethics*, *25*(5), 721–746.

Example: The ethical matrix

Mepham, B. (1996). Ethical analysis of food biotechnologies: An evaluative framework. In B. Mepham (Ed.), *Food ethics* (pp. 101–119). Routledge.

Mepham, B., Kaiser, M. J., Thorstensen, E., Tomkins, S., & Millar, K. (2006). *Ethical* matrix manual. LEI.

Economics

Mason, G., Cooper, J., & Clarebrough, C. (2001). Frustrations of fur-farmed mink. *Nature*, *410*, 35–36. https://doi.org/10.1038/35065157

Boycotts

Albright, J. L. (1987). Dairy animal welfare: Current and needed research. *Journal of Dairy Science*, *70*(12), 2711–2731. https://doi.org/10.3168/jds.S0022-0302(87)80345-4. PMID: 3329200.

Wheeler, M. B., Rutledge, J. J., Fischer-Brown, A., Vanetten, T., Malusky, S., & Beebe, D. J. (2006). Application of sexed semen technology to in vitro embryo production in cattle. *Theriogenology*, *65*(1), 219–227.

Animal welfare labeling

Bjørkdahl, K., & Lykke, K. (2021). Welfare washing: Disseminating disinformation in meat marketing. *Society & Animals*, 1–19. https://doi.org/10.1163/15685306-BJA10032

Sullivan, S. P. (2012). Empowering market regulation of agricultural animal welfare through product labeling. *Animal Lion*, *19*, 391.

Consumer segmentation

Prickett, R., Norwood, F. B., & Lusk, J. (2010). Consumer preferences for farm animal welfare: Results from a telephone survey of US households. *Animal Welfare*, *19*(3), 335–347. https://doi.org/10.1017/S0962728600001731

Chapter 19

Professional Settings

19.1 WORK AND VOLUNTEERING

Animal welfare science impacts many different work and volunteering opportunities, each with particular challenges to work satisfaction. Employment practices in different sectors dramatically affect what animal welfare science work is resourced and pursued, and how it is carried out. Work satisfaction has been extensively studied, but an understanding of how important it is to find meaning in your work is only a recent development (Steger, 2019). One aspect of meaningful work is to find labor that contributes to the health and fairness of the world, and animal welfare science provides many opportunities for this kind of work.

We will not be covering in any depth the options for volunteer work; examples of these would include citizen science, internships, and charity support. These opportunities do exist and can be fulfilling and beneficial, although you should pay attention to the appropriateness of uncompensated work (especially internships) where this is exploitative or exclusionary (depriving those in need of income from important learning and networking opportunities). However, immense benefit is brought about every day by people choosing to give their time and resources in animal welfare science activities.

Keyword: internship

19.2 ACADEMIA

Academia, predominantly universities, determines what is considered a field of study and the design of curricula. Animal welfare science emerged as an option for study in the early 1980s, and opportunities in this topic have grown, accompanied by a publishing base (Figure 19.1). Academia is also an area where people pursue their professional careers. Its dominance as an authority for tertiary education and scientific research makes it a home to most of the influential theorists and researchers mentioned in this text.

DOI: 10.1201/9781003375456-24

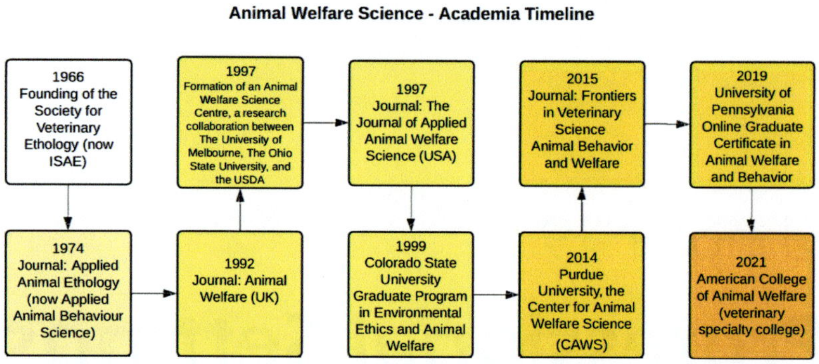

Figure 19.1 A timeline of the development of the field as an academic force. Schematic created by Emily Patterson-Kane.

That said, most university graduates will never make tenure, and an increasing reliance on non-tenured adjunct staff with inadequate compensation and benefits makes academia a less attractive workplace. While many academics are champions in outreach and application, there is a reason that one of the dictionary definitions of "academic" is "not of practical relevance; of only theoretical interest". Academia is also the home of basic, curiosity-driven, or "blue skies" research, which often has no immediate application. Other sectors may provide opportunities to create more immediate and tangible benefits for animals. Also note that minority scholars still face higher barriers to tenure and can be made to feel uncomfortable in some learning environments, especially in traditional STEM (science, technology, engineering, and mathematics) subjects.

Those interested in academic careers should take care in selecting their path through post-graduate degrees and post-doctoral positions. Following a line of research that excites you is important, but unless you're independently wealthy, you should map a path to viable employment. Investigate the average income of graduates, your supervisor's standing in the field, their reputation with current and past students, and whether former students have gone on to successful careers in the field. Idealism is an important driver for people entering animal welfare science careers; however, some enlightened self-interest can set you up for a long and effective career. Note that graduate students with prestigious supervisors will benefit from that association, which can fast-track careers and open doors to research funding. The most prestigious supervisors will attract excellent students, and these research groups can create local monopolies in a field. The research "stable" of the Nobel Prize–winning ethologist Nikolaas Tinbergen (whom we mentioned in Part I) includes Marian Dawkins, Richard Dawkins (bestselling author of *The selfish gene*), Desmond Morris (known for his book *The naked ape* and various documentary series), and the eclectic academic and broadcaster Aubrey Manning, among many others.

Also note, *academia* is generally defined as "the environment or community concerned with the pursuit of research, education, and scholarship", and I refer mainly to universities and closely associated institutions. This doesn't mean that employees in other sectors don't engage in important scholarly work, because they do, although they may be treated somewhat dismissively by academics. Just as students are encouraged to move between universities to broaden their minds, we suggest being open to moving between sectors if that will suit your professional goals (or, frankly, if that is what is available).

Keyword: academia

Peer-reviewed journals

One of the roles of the academic is to actively engage with the literature of the field. The core literature is published in reputable *peer-reviewed* journals, where qualified members of the discipline (the authors' peers) determine whether a manuscript is of sufficient quality and relevance to be published (much like a jury of "peers" decides if an accused is guilty of a crime). This may also involve scientists in other sectors, depending on their job roles and their desire to maintain a profile as a documented scientific expert, for example as an expert witness or paid consultant.

Exactly what this involves is undergoing rapid change. Traditionally, this meant publishing in high-impact journals—those with a subscriber base, highly cited articles, and the associated expense of working with a few major publishers (e.g., Elsevier, Sage, Wiley, Taylor and Francis). These publishers accept relatively few papers due to being published periodically in print form and having limited space. Many are associated with major professional organizations, with paywall-protected content and access via institutional subscriptions.

Journals of this type remain important, and it is a good idea to keep up with work published in animal welfare science as a field as well as your specialties and foundational disciplines. Traditional animal welfare science journals to be familiar with would include those listed below, but also ask your mentors and colleagues which journals they read or follow:

- *Journal of Applied Animal Welfare Science*, Animals and Society Institute, Taylor and Francis (Animals and Society Institute) https://www.animalsandsociety.org/research/academic-journals-2/jaaws/
- *Applied Animal Behavior Science*, Elsevier (ISAE) https://www.sciencedirect.com/journal/applied-animal-behaviour-science

A competing business model is to publish articles that are free to read, supported by fees paid by the published authors. These journals often publish a greater number of papers and do so in an online format that is accessible to audiences outside academia. There is less clarity about the reputational value of these journals as flagships for the field of animal welfare science, although the following might be of interest:

- *Animals*, MDPI https://www.cambridge.org/core/journals/animal-welfare/latest-issue
- *Frontiers in Animal Science*, Frontiers https://www.frontiersin.org/journals/animal-science

Other business models take a hybrid or completely different strategy. For example, the journal *Animal Welfare* (open access since 2023) has a trade publisher linked to an animal welfare association. The publisher is remunerated by institution-based charges that can be waived for low- or middle-income countries or authors without funding, and are reduced for society members.

- *Animal Welfare: the International Journal of Animal Welfare Science* (Universities Federation for Animal Welfare) Cambridge University Press https://www.cambridge.org/core/journals/animal-welfare/latest-issue

There is heated debate within and beyond academia relating to which publishing model or combination of models is the best for science and society. This includes debate about whether it is reasonable to expect experts to act as peer-reviewers for free, who should financially support peer-reviewed publishing, what procedures create the best science that is also accessible, and the ever-present issue of fraud. One academic publisher recently shuttered 19 of its journals for pushing paper mill articles (which are fake articles where authors pay for authorship).

Keyword: peer-review (refereed)

19.3 NON-PROFITS AND NON-GOVERNMENTAL ORGANIZATIONS (NGOS)

An NGO is a group that functions independently (often internationally) of any government with the objective of improving social conditions; its scope of work is typically larger than that of a non-profit that has some local legal and reporting obligations. Both types of organization are mission driven and supported by some combination of volunteer labor, donations, fees, grants, and investments. This sector includes charities, foundations, trusts, and membership organizations, including those tailored to a specific profession. Some of these also support a substantial paid staff in areas of animal welfare science and related fields (e.g., national charities and professional and disciplinary associations).

National charities offer a range of employment opportunities, especially for specialists, and many are public facing with familiar household names. They represent organizations from long-established humane societies like ASPCA (1866) and American Humane (1877) and younger, more radical organizations such as HSUS (1954; recently rebranded as Humane World for Animals), and PeTA (1980). If you look into the work of advocacy organizations, you might find that the science has been misrepresented to make a point. For a deeper dive into an organization, you can pull their IRS 990 forms (available online for tax-exempt charities and non-profits) and scan details of remuneration to the C-suite, endowment fund levels, grant funding activities, and the turnover of top executives (indicating stability). These can be as much an eye-opener as the mission statement, perhaps more so.

The emergence of multiple large national organizations was seen as signaling the maturity and professionalization of animal welfare, because you could literally have a well-compensated career working for these organizations. However, a lot of the work that they do is not connected to any particular community or their problems, and each organization is centered around particular abstract ideologies. Some groups work closely with industries to encourage gradual improvement, and others want to harass and end activities such as greyhound racing or intensive agriculture. There is a constantly changing array of non-profits in a challenging space of needs, actions, and resources, within which the large groups act as mainstays while others come and go.

Keyword: nongovernmental organization (NGO)

19.4 DUELING NON-PROFITS AND THE SILVER SPRING MONKEYS

The largest association for professional psychologists in the United States is the American Psychological Association (APA). In the early 20th century, the dominant approach in studying human processes and disorders was behavioral psychology, with many psychologists using non-human animal models even when their primary interest was human psychology. As a result, the APA supported animal use in research rather uncritically, a position not taken by Kenneth Shapiro, who formed Psychologists for the Ethical Treatment of Animals (PsyETA) in 1980. An early campaign for PsyETA was to oppose the use of monkeys at the Institute for Behavioral Research in Silver Spring, Maryland. (This research featured in one of the ethically problematic articles that we mentioned in the neuroscience section in Part II; Harry Harlow, author of the other problem study, would become president of the APA in 1958.)

This Silver Spring facility had been infiltrated by Alex Pacheco, who documented poor levels of animal welfare in research led by behavioral neuroscientist Edward Taub. (Pacheco co-founded PeTA, and this case launched PeTA's reputation.) The APA supported Taub's research in "removing" sensation from the limbs of rhesus monkeys and then studying their ability to recover limb use. In an article describing the complex history of the case, Kirk (2019) commented that "Taub firmly believed that his monkeys were a laboratory product of experimental science. So much so he failed to see how knowledge derived from the same species in a 'natural' setting could be relevant to their care and welfare".

Taub would maintain that he had a situated duty of care, informed by the needs of the experimental system versus valuing the animal in and of itself. His opinion clashed with that of PsyETA and veterinarians. Taub's funders were also in a political bind, as they were facing congressional hearings about animal use in research. PsyETA felt that the APA's position was negligent regarding an obligation to care and to compensate for the ill-effects of the research (such as a tendency to self-mutilate). Shapiro, asserting an identity as a psychologist, found such inhumane treatment to be unacceptable. Key outcomes of this case were revised legislation to improve the conditions of research animals and the requirement for a layperson on animal-use committees.

At the end of a somewhat farcical ten-year legal battle, the surgically altered monkeys who could not be rehomed in zoos were euthanized and examined. They had been kept alive so long that neuronal plasticity, a biologic feat that was hitherto unknown, was revealed. Without the intervention of animal rights advocates, this serendipitous result (subsequently used to rehabilitate human stroke victims) would never have been discovered. PsyETA would later (2005) merge into a membership group with a broader remit as the Animals and Society Institute. Taub was eventually exonerated and his funding restored.

Keywords: Layperson, PsyETA, Edward Taub

19.5 GOVERNMENTAL

As animal welfare has become codified by laws and regulations, governmental agencies have become responsible for setting, interpreting, and enforcing those standards. They also have key roles in stewarding state and federal land, animal transportation and imports, and managing many aspects of the connection between human and animal health. This has led to a range of opportunities for animal welfare scientists to work in governmental positions, including research, inspection, and enforcement. It can be incredibly difficult working within the layered constraints of government regulations, politicking, and paperwork, and without the staff needed to carry out all the inspections that are required. That said, government workers continue to do amazing work under extremely challenging circumstances.

In the United States, the federal government employs animal welfare scientists at the United States Department of Agriculture (USDA), the Animal and Plant Health Inspection Service (APHIS), and the Agricultural Research Service (ARS). Opportunities at the state and jurisdictional levels relate to wildlife biology, entomology, veterinary medicine, and paraprofessional duties (people teamed with one or more licensed or credentialed practitioner). The options for government employment are more diverse than people realize, and some job openings might not appear in the usual job posting aggregation sites, so it is important to search for them in government-specific sites such as https://www.usajobs.gov/

Since submitting this text, the Trump-Vance administration has begun a catastrophic series of sackings of federal employees, with some stranded in dire circumstances abroad. This is a moment of extreme uncertainty, and the advice would be to use caution when appraising the current employment market. Be especially vigilant if moving for a job to make sure that expenses will be honored even if you are terminated through no fault of your own.

Keyword: paraprofessionals

19.6 FOR-PROFIT

Most of the domesticated animals in the United States are part of for-profit activity at some point in their life. An important landmark in the relationship between commerce and animal welfare was the explicit realization that animals experiencing good welfare are healthier and more productive, but there is a point at which this association breaks down. There is certainly a link between an animal being well bred and cared for and how well they perform in roles such as milk production or the speed of a racehorse. However, this relationship was used as the basis to negate animal welfare, as production-oriented metrics became its surrogate, and health-focused veterinary medicine was presumed to encompass all that was necessary.

Harlow's research in the mid-20th century (Part II), raising infant monkeys without their mothers, demonstrated that the infants could be physically healthy but incredibly emotionally disturbed and subject to immense suffering. But this general principle only gradually evolved into empirical demonstrations that the association between overall welfare and productivity is moderate and sometimes even weak (Henningsen et al., 2016; Robichaud et al., 2019). Any decoupling of welfare and productivity needed to have a scientific basis, because by the mid-20th century, animal care was seen as a scientific activity, and scientific evidence would be needed to combat criticisms of emotional decision making. Now, the tradeoff is between productivity and non-use welfare values, i.e., economic gain through good animal welfare (Lagerkvist et al., 2011). This balancing act is seen as fitting the goals of animal welfare and species conservation, with Beausoleil (2018) noting that the goal should be both "Feelings and Fitness".

Dawkins (2016) believes that any conflict between animal welfare and efficient farming has the potential to be largely resolved by helping animals to live longer, healthier lives, with better products, improved farmer wellbeing, and the production of *value-added products*. Value-added products are premium items that people will pay more for than they would for the raw ingredients. This includes obvious things like being precooked, cut or seasoned, organic, locally sourced, or humanely farmed. These products may be made more convenient, more attractive, more palatable, or easier to use than the raw ingredients. Whether recoupling productivity with a multi-dimensional approach to animal welfare is possible, or even a good idea, remains to be seen, although that might reflect the limits to this author's (EPK) imagination. For consumers to drive optimal welfare, the great majority need to have the empathy, the information, and the wealth to drive the market in this direction.

It is far too easy to see industry as the enemy, although it is undeniable that commercial business has produced enormous benefits for animal welfare. These benefits included producing nutritionally balanced feed, vaccinations that ended livestock plagues, and more recently, the development of effective fish stunning devices and research to sort sperm so that we can avoid the birth of unwanted animals. A lot of animal welfare research is only possible because of direct or grant-based funding from private companies and their support of influential non-profits such as the Human-Animal-Bond Research Institute. The issues that tend to be more problematic in for-profit research relate to the open sharing of data where this might interfere with the ability to monetize the work or recruiting the expertise to even understand the most pressing issues for animals.

Traditionally, scientists have been found to be "ethically inarticulate" (Arluke 1992) and can easily get caught up in rather dubious activities. Some scientists even allow themselves to be used as professional skeptics, with the express aim of injecting unhelpful uncertainty into a debate to create a false dichotomy. In short, working in industry provides the opportunities to make massive changes to benefit animals and people if they can be aligned with the company's profit motive and, if they have one, mission.

In conclusion, for-profit activity provides many important career possibilities for animal welfare scientists. I would treat with caution anyone who vilifies a particular sector. In all sectors, there are people and groups doing good work for good reasons and those who, by intent or ignorance, are not. A key role of being part of animal welfare science is to be clear eyed, appreciate multiple perspectives, and in doing so, synthesize compelling and workable solutions for all animals, human and non-human.

Keywords: value-added products, non-use welfare

The history of animal trainer scientists

Great contributions to our understanding of animal behavior and welfare have come from animal trainers, both before and after the establishment of the scientific method. Vladimir Leonidovich Durov (1863–1934) was a Russian circus clown and satirist who worked extensively with animals. Durov created a dynasty of humane animal trainers by deploying an encyclopedic knowledge of the species and their traits rather than a one-size-fits all approach using coercion or force. Durov gave lectures on his discoveries and methods that were appreciated by academics such as Vladimir Bekhterev, considered by many to be the father of objective psychology and the great competitor of Pavlov, his arch-nemesis. Brilliance aside, Bekhterev possibly misread the room when he diagnosed Stalin (the Soviet era dictator) with paranoia; Bekhterev departed this earth shortly afterwards, and the suppression of his work is why we know so much about the work of Pavlov, but not Bekhterev. The brains of all three were ultimately reunited at the Brain Institute in Moscow.

Animal Behavior Enterprises was another example of science applied to training, this time involving early behaviorists and graduate students of B.F. Skinner (Marian Breland Bailey and Keller Breland). Via their business, the duo developed the idea of *instinctive drift* (mentioned in Part II; Breland & Breland, 1961). This is where a task that an animal is trained to do morphs over time toward something that is more biologically relevant for them. For example, raccoons that were meant to place large coins in a bank slot for treats started handling and washing the coins as if they were food. This term is typically used only in relation to non-human animals, but you can probably think of examples from human behavior.

Animal trainers have a long history of scholarship in animal behavior and welfare. Learning on the job can be incredibly valuable, but we often block these conduits of knowledge from contributing to the formal discipline of animal welfare science. The careers of these animal trainers remind us that important insights and advances often come from people with different perspectives.

Keyword: instinctive drift

19.7 DISCIPLINES AND PROFESSIONS

A *discipline* is a branch of knowledge, usually one studied in higher education, while a *profession* is an occupation that involves prolonged training and a formal qualification and ethical obligations. It is common for a student of a discipline to enter an associated profession: for example, a veterinarian or veterinary nurse/technician from veterinary medicine, or in the field of sociology, a sociologist or social worker. But it is equally common for the connection to be more indirect. Both authors have degrees that, after a comparatively short period of time, no longer matched their job titles or research areas. It is not uncommon to be employed in roles that could easily be held by someone trained in a different discipline (research, animal science, animal welfare), just as it is common for people to transition between disciplines, professions, research fields, or to occupy interdisciplinary spaces (especially if you are interested in a topic like animal welfare science). The center of professionalization in animal work is veterinary medicine. This should be a source of anxiety, because this is a profession in crisis. The profession has one of the highest rates of suicide (Stoewen, 2015), especially for women (Zimmermann et al., 2023), it is one of the least diverse, and it is economically unstable. Its ongoing existence is predicated by the continual willingness of young people to enter it, even when it would seem to be an irrational career choice. There are plans for multiple new vet schools in the United States, and yet what we really need is an understanding of what it is about some veterinarians that makes them able to stay in the profession. Why are they so resilient? Then, we could better support the veterinarians that we train rather than increasing the churn.

Many animal welfare scientists benefit from their expertise in at least one formal discipline, combinations of disciplines, or well-developed inter/trans-disciplinary spaces. A strong foundational discipline provides a wide body of knowledge and skills, including some robust strategies that can be applied to problems that are novel or resist simple solutions. This book addresses animal welfare science as a shared interdisciplinary space, with its own knowledge base, which may or may not be seen as a discipline or a profession in its own right. Despite one of us (EPK) having held a position literally called "animal welfare scientist", this is an issue I remain agnostic about.

Professions have a certain gravitas that (allegedly) separates them from other forms of employment. There are two reasons for this. One is that professions require a certain level of advanced training, and the other is that they hold a position of trust in society. From the medieval to early modern period, there were only three "learned" or "liberal" professions: law, medicine, and theology. These professions grew from associations that developed to regulate specialist training, define expectations about what that profession should or may be able to do, and remove their endorsement of unfit or unqualified members. Members of a traditional profession are expected to maintain evidence of their membership, fill the associated role in their community, and follow a code of ethics. Mature and well-formed professions will often have national and local associations, recognized publications and meetings, accredited systems for initial and ongoing education, and a system of licensure that is separate from the system of education (e.g., state veterinary license, passing the bar). In return for complying with these requirements, professionals are granted a degree of trust and prestige, exclusive access to a profitable activity, and the ability to self-regulate.

In the modern era, professions have become more attached to the idea of being rational and evidence based, with their members and leadership committed to advancing the profession through activities such as research. In some cases, the degree of professionalization is variable or incomplete—indeed, the traditional idea of a profession butts heads with modern ideas of antitrust laws (monopolies), consumer rights, and local regulations. There is no precisely agreed-upon list of modern professions, and many are specialist forms of extant "mother" professions and draw some of their standing from their antecedents' long history. (Other professions are "new", based on being developed from an industry or trade need, such as engineering, accountancy, and pharmacy.)

Perhaps the oldest animal profession is veterinary medicine. However, its support staff of veterinary nurses or technicians still lacks a uniformly agreed-upon title, which indicates a profession that is still emerging and maturing. Vet nurse/technicians have an association, and some states provide licensure and continuing education, which transitions them from being a para-profession to a profession. However, there is no uniformity between states or even between facilities as to what activities may or may not be performed by an accredited vet nurse/technician. This kind of territorial cross-over can provoke fractious debate, especially as consideration of a new role equivalent to the nurse practitioner in human medicine might increase the scope of their work.

Keywords: discipline, profession

19.8 THE CHANGING FACE OF DISCIPLINES AND PROFESSIONS

Many long-standing features of disciplines and professions have changed and will continue to change. Membership of a professional association is becoming less common, with some professionals joining associations relating to their specialty rather than their profession. Others have no association for their profession or choose not to belong to one. Nowadays, professionals have so many more options for participation, especially with online communities. Levels of outside regulation are also generally increasing, such that professionals are now only partly self-regulating. For example, when veterinarians provide services that the community opposes, like cat declawing, this may be forbidden in sections of law outside the Veterinary Practice Act. A great many new disciplines and professions are emerging and may not follow the centuries-old models established by medicine, theology, and the law. The mother professions lost most of their power because they failed to protect communities, losing our trust and their reputations: for example, the willingness of the lawyer to protect corporate interests against the public good, and the way in which the church concealed and protected sexual abuse by clergy.

The outline provided for participating in your discipline and profession may or may not be relevant or helpful for all people. However, taking part in profession-wide activities can help you to network, broaden your horizons, keep your skills up to date, and facilitate your contributions to the field's knowledge and skill base. This includes making sure that animal welfare science has a high visibility and if possible, is made part of the expected competencies of your field. However, joining your professional association is not the uniform expectation that it used to be. Many people abstain for reasons of ethics or a lack of perceived value, as associations tend to be relatively centrist, with less appeal to those whose approaches are markedly conservative, progressive, or alternative.

The professionalization of other animal careers, such as laboratory technician (a person who works in a laboratory performing analytical or experimental procedures), was associated with advancements in animal welfare, such as the wider use of environmental enrichment and the pressure to establish humane endpoints. Lab technicians in animal labs tended to be strong advocates for animals, perhaps due to spending more time with the animals and being tasked with their care as a core role (Arluke, 1992). Professionalization helped give these positions more authority and higher pay.

Many other developing fields are in the process of becoming a new discipline and/or profession. These include animal welfare data analyst/data scientist and statistical analysis specialist, all indicating the increasing role of data in science generally. Other well-established disciplines are changing and expanding their professions to incorporate animal welfare science, for example, animal care and control officers, animal trainers, and groomers. These activities might benefit from the standing and higher wages associated with a profession, and in turn, help us develop new models of professional identity fit for modern society.

Keyword: laboratory technician

READINGS

Work and volunteering

Steger, M. F., Dik, B. J., & Shim, Y. (2019). Measuring satisfaction and meaning at work. In M. W. Gallagher & S. J. Lopez (Eds.), *Positive psychological assessment: A handbook of models and measures* (2nd ed., pp. 373–388). American Psychological Association. https://doi.org/10.1037/0000138-024

Dueling non-profits and the Silver Spring monkeys

Kirk, R. G. W. (2019). The Silver Spring monkey controversy: changing cultures of care in twentieth-century laboratory animal research. *HoST - Journal of History of Science and Technology*, *13*(2), 31–59. https://doi.org/10.2478/host-2019-0012

For profit

Arluke, A. (1992). Trapped in a guilt cage. *New Scientist*, *134*(1815), 33–35.

Beausoleil, N. J., Mellor, D. J., Baker, L., Baker, S. E., Bellio, M., Clarke, A. S., Dale, A., Garlick, S., Jones, B., Harvey, A., Pitcher, B. J., Sherwen, S., Stockin, K. A., & Zito, S. (2018). "Feelings and fitness" not "feelings or fitness": The *Raison d'être* of conservation welfare, which aligns conservation and animal welfare objectives. *Frontiers in Veterinary Science*, *5*, 296. https://doi.org/10.3389/fvets.2018.00296. PMID: 30538995; PMCID: PMC6277474.

Dawkins, M. S. (2016). Animal welfare and efficient farming: Is conflict inevitable? *Animal Production Science*, *57*(2), 201–208.

Henningsen, A., Czekaj, T. G., Forkman, B., Lund, M., & Nielsen, A. S. (2016). *The relationship between animal welfare and economic performance at farm level: A quantitative study of Danish pig producers*. Department of Food and Resource Economics, University of Copenhagen. IFRO Working Paper No. 2016/05. http://econpapers.repec.org/RePEc:foi:wpaper:2016_05

Lagerkvist, C. J., Hansson, H., Hess, S., & Hoffman, R. (2011). Provision of farm animal welfare: Integrating productivity and non-use values. *Applied Economic Perspectives and Policy*, *33*(4), 484–509.

Robichaud, M. V., Rushen, J., De Passillé, A. M., Vasseur, E., Orsel, K., & Pellerin, D. (2019). Associations between on-farm animal welfare indicators and productivity and profitability on Canadian dairies: I. On freestall farms. *Journal of Dairy Science*, *102*(5), 4341–4351.

The history of animal trainer scientists

Breland, K., & Breland, M. (1961). The misbehavior of organisms. *American Psychologist*, *16*(11), 681.

Disciplines and professions

Stoewen, D. L. (2015). Suicide in veterinary medicine: Let's talk about it. *The Canadian Veterinary Journal = La Revue Veterinaire Canadienne*, *56*(1), 89–92.

Zimmermann, C., Strohmaier, S., Niederkrotenthaler, T., Thau, K., & Schernhammer, E. (2023). Suicide mortality among physicians, dentists, veterinarians, and pharmacists as well as other high-skilled occupations in Austria from 1986 through 2020. *Psychiatry Research*, *323*, 115170. https://doi.org/10.1016/j.psychres.2023.115170

The changing face of disciplines and professions

Arluke, A. (1992). Trapped in a guilt cage. *New Scientist*, *134*(1815), 33-35

Chapter 20

Animal Welfare Science as a Cross-Cultural Space

20.1 ANIMAL WELFARE SCIENCE AS A DISCIPLINE/PROFESSION

Animal welfare science as a discipline has been widely accepted since around 1980 and means many things to many people. An academic article might describe "animal welfare science" as nested within veterinary medicine or as a "new behavioral science". Elsewhere, animal welfare science is seen as having other disciplines nested within it, as suggested in the following quote:

> Animal welfare science and ethics are inextricably linked, and for brevity, we use the term animal welfare science to also include ethics.
>
> (Ward et al., 2018)

While it is important for each discipline engaged with animal welfare science to acknowledge its association, this does not mean that the older discipline or profession controls or encompasses animal welfare science—it simply engages with it.

The discipline a person claims in their personal identity (I am a …) is very important, but we often fill many roles. This book has illustrated those key concepts with some of the foundational history for context. The underlying themes of seeing welfare as having many domains that require collaborative efforts across disciplines have been set out, along with the benefits of remaining curious about what upsets the welfare of animals and how we might leverage scientific approaches to protect and promote the welfare of animals.

Keyword: interdisciplinary

DOI: 10.1201/9781003375456-25

20.2 CULTURAL CONTEXT

Around the year 2000, there was a tipping point in American culture where understanding things in their larger context became expected. This burst across our cultural awareness in waves expressed by the next generation of social movements, such as #MeToo, #BLM, and intersectionality. These changes presented serious challenges to movements that focus on the plight of non-human animals, as folding in human and environmental issues tends to bleed the focus away from animals. A backlash to these movements also arose, and the appropriation of some of its language to animal protection has been unfortunate. #All Lives Matter applied to farm animals is a clumsy usage that can easily be seen as anti-Black. Another characteristic of the times is the perception of tethered causes (e.g., vegan, leftish, queer) and their hypothetical links to a political identity, which amplifies division and erases the nuance of peoples' beliefs about animal rights and broader society.

Our new era requires a consideration of how animal welfare science operates within and across different cultural contexts. Within these, the norms and practices of smaller subcultures will exist. Take, for example, equestrians, which have smaller nested subcultures such a rodeo, which, in turn, would interact with factors like gender and nationality (for more on equestrian cultures, see Adelman and Thompson, 2017). This makes it hard for groups representing large sectors to do much of anything without greatly upsetting a significant proportion of their membership. Animal welfare science is not magically going to help all the animals in most need and is constrained by resources and human motivation; how it is pursued can also lead to accusations of neo-colonialism (Garcia & McGlone, 2022). Despite our many achievements, huge areas of need are still largely neglected. For example, there are limited resources to study the illegal smuggling of exotic animals in South America (Mota-Rojas et al., 2023).

Exactly how animal welfare science will function over the rest of this century is hard to say, but this section will examine how shifting the frame changes the way animal welfare science is or should be done (for best effect). The examples discussed below include:

- Studying welfare internationally
- Intersectional issues
- Affinity organizations
- DEI

Keyword: culture

20.3 STUDYING WELFARE INTERNATIONALLY

Animal welfare science, as we know it, began as a British concern. Elsewhere, there may be less awareness of animal welfare science, as described in this book, but relevant stakeholders will have broadly similar concerns. Many factors that are relevant to animal welfare operate across borders, for example, the international supply chain for agriculture (Bayvel et al., 2010). This global-scale operation interacts with an animal welfare tradition that is still strongly influenced by its British and generally Western roots. Indeed, Buller et al. (2018) state that "formal farm animal welfare science has by and large developed out of, and some would say remains still largely limited to, a Western industrial model of intensive, professionalized and technologized agricultural development". That said, animal welfare and closely related concepts are present and important in non-Western cultures, but to date, these knowledge traditions are not well integrated into animal welfare science as a discipline (e.g., China: Li, 2021; Yang et al, 2024).

As the field developed, more varied species and settings were accommodated to some extent, but some biases remain. For example, in an intensive system, one of the major options for remediation is to introduce more space and more contact with other animals and the environment. However, in pastoral systems, both the threats and the way they compromise welfare are expressed differently, and this varies further with differences in genetics, geography, and cultural knowledge systems. Any animal welfare assessment must be sensitive to the relevant environments and cultural sensibilities, and if necessary, reinvented from scratch. This is why, if you use the Three Circles, you must create that assessment from a blank page after observing the animals in their environment. In that fashion, you can easily focus on the important issues that might compromise welfare and why. As we move toward using big data, machine learning, and artificial intelligence (AI), we must also be mindful that AI is not allowed to shape an unfair future from having been trained on institutionally biased and for many, potentially irrelevant data sets. Until AI can sit peacefully on a hillside and watch animals, and until it can appreciate why animals do what they do, its understanding will be littered with misconceptions.

Keyword: international supply chain

20.4 INTERSECTIONALITY

The United States has a long and troubled history of politicizing citizenship and race, so much so that in the 1930s, a bemused fledgling Nazi party dispatched its lawyers to the United States to study its codified race laws. It was a revelation to them that the American legal system allowed "secret and twisting passageways of reasoning that led to whatever outcome was politically desirable" (Maddow, 2023). America, it seemed, was a nation of laws applied "any how" and often perversely (Whitman, 2017). For example, the US judicial system's refusal to appreciate that Black women might be multiply disadvantaged in Title VII discrimination cases gave rise to the term *intersectionality* (Crenshaw, 1989). As the experience of racism is clearly different for men versus women, descriptions were needed to break down the monolith of experience—being "Black", "female", "Queer", etc. Of course, intersectionality hasn't suddenly materialized. For example, the lived experience of being discriminated against for race and gender was eloquently described more than a century earlier by Sojourner Truth (1797–1883) in her 1851 speech, reportedly called "Ain't I a woman?"

Intersectionality also applies to domains such as science and animal welfare and challenges the idea that animal welfare science has a uniform core and shared identity. In this text, we have tried to present animal welfare science as a collection of past actions and exciting future opportunities. However, we are inevitably limited to parts of the field we have access to and can incorporate into our worldview. Entire bodies of work are inaccessible to us, such as the unique approaches of Francophone academics or the unrecorded efforts of animal sanctuaries in remote locations. Take this text as one window into a world that contains many wider opportunities.

Intersectionality is also apparent in conservation, practiced by the Global North and often imposed on the Global South. Unfair narratives that Black, Indigenous, and People of Color (BIPOC) were not interested in the outdoors, conservation, or veganism have been countered. The intersection of race and species is a problematic one (Alexis, 2020), as the human–animal divide has been used as the basis for white supremacy, with various minorities described as being too sensitive when they object to the comparisons of suffering (Ko & Ko, 2017). The Ko sisters (2017) argue that "we tend to be blind to the fact that in both the narrative of speciesism and the narrative of racism the members of the losing side both fall short of real human status and, as a result, their suffering and their deaths are mundane, normal, and expected".

A critical skill for animal welfare science is navigating how animal welfare research and its application are enmeshed in outcomes for other populations, including humans with varying characteristics. For example, a group that supports veganism with the goal of reducing animal suffering might acknowledge how their choices might increase the suffering of Latinx workers on produce farms. Rather than compare suffering, it is suggested that groups address issues directly and separately. For example, rather than conflate the violence suffered by female farm animals and women, address the violence that women face when working in agribusiness and separately, campaign for animal welfare by throwing a spotlight on the little-recognized issues of the violent handling of female animals while being inseminated and otherwise handled.

One group that takes this issue to heart is the Food Empowerment Project (https://foodispower.org/), a vegan-motivated mission to empower good food choices based on the treatment of farm workers, involvement of child labor in chocolate production, and supporting community access to healthy food. Having a less siloed approach can limit funding dedicated to only one issue or the other, but this should not be seen as an inconvenience or limitation. Like animal welfare, dealing with intersectionality is both a necessity and an asset. A group of this type can provide a new context and connections for existing research: for example, lactose intolerance being more prevalent in BIPOC communities, reflecting the establishment of dairy farming in Europe (with its associated lactase persistence mutation) while fermented milk products (that remove lactose) were used elsewhere (Segurel et al., 2020). In this way, intersectionality reveals fresh angles of research to create, communicate, and provoke change.

There will be those that strenuously reject the notion of intersectionality. While this position is currently being promoted within the federal government of the United States, it will inevitably clash with the reality of our demographics. Scholarship has repeatedly demonstrated that a diverse workforce is a more successful workforce (https://www.mckinsey.com/featured-insights/diversity-and-inclusion/diversity-matters-even-more-the-case-for-holistic-impact). In this respect, failing to recruit or to retain a diverse workforce is a pernicious form of self-sabotage that we can ill afford given the multitude of crises that are brewing (e.g. climate change, zoonoses, food insecurity, collapsing biodiversity).

Keywords: intersectionality, Sojourner Truth

20.5 AFFINITY ORGANIZATIONS

Being in a small minority within a profession puts you at a huge disadvantage and can be emotionally exhausting. One thing that can help is to find other people with similar experiences and benefit from support and collective action. One way to achieve this has been to create *affinity organizations* to advocate for and protect the interests of their members. Some Indigenous North American examples include The Native American Humane Society (NAHS), founded in 2014 by Diana Webster with a focus on free-roaming dogs (reservation dogs) and an approach that targets the connected human and animal community. Another area of focus has been veterinary medicine (where Native Americans are massively under-represented; Gelberg & Gelberg, 2007), which is supported by the Native American Veterinary Association (NAVA), founded in 2017. Lastly, Native American Veterinarian Services (NAVS) provides free veterinary care for Native Americans living on reservations.

While there is a national association of veterinarians, there are also smaller culture- and identity-specific associations such as the Association of Asian Veterinary Medical Professionals, the American Association of Veterinarians of Indian Origin, the Latinx Veterinary Medical Association, the National Association for Black Veterinarians, and Pride Veterinary Medical Community, as well as the Multicultural Veterinary Medical Association (founded in 2014).

One dramatic change in representation is the increased participation by women, even as veterinary leadership in many areas is still dominated by men. The Women's Veterinary Leadership Development Initiative (2013) was established to encourage and mentor women into leadership positions. Another new organization is Companions and Animals for Reform and Equity (CARE), founded in 2020. This is America's first BIPOC Animal Welfare organization with the aim of supporting human and animal welfare. CARE also hosts tailored Racial Equity Diversity and Inclusion training, and its own data show the bias within the animal welfare industry against Hispanics, African Americans, and most of all, poor people (Jenkins & Rudd, 2022).

The current shortage in veterinary medicine is fueled by many factors, including a large dropout rate for minorities during training and burnout through stress. Research funded by Merck (2021) (https://www.merck-animal-health-usa.com/about-us/veterinary-wellbeing-and-scholarships/veterinary-wellbeing-study-iii) revealed a third of veterinarians planning to leave their job. The non-profit "Not One More Vet" (nomv.org) was established after the suicide of a well-known animal behaviorist, Dr Sophia Yin, and has created a mental health certification (CLEAR) for the workplace to support veterinarians and support staff (https://nomv.org/clear-blueprint/).

Some of the stress experienced by veterinarians will be from angry clients upset with sky-high vet bills. Royal Canin research (2022, https://www.royal-canin.com/au/about-us/news/love-your-pet-love-your-vet-survey) conducted in Australia showed that clients commonly thought that veterinarians were better paid than they really are. (The US Bureau of Labor Statistics has an average wage of $119K/annum for 2023 in the United States). Hiked vet prices in the United Kingdom (up 50% since 2015, according to the Office for National Statistics) have coincided with corporate chains buying up independent practices. This situation of corporate capture is currently under investigation by the United Kingdom's Competition and Markets Authority (CMA).

Keyword: affinity organization

20.6 DIVERSITY, EQUITY, AND INCLUSION (DEI)

The DEI framework was the dominant approach to social justice in the context of American employment. DEI functioned as both a powerful protection and a set of principles that could easily be undermined when not implemented in good faith. Currently, DEI is under attack in the United States. DEI became unpopular because it shone a spotlight on inequality in a socially curated workforce riven with nepotism; questioned claims of "merit-" based employment that demonstrably were not and made the hegemonic (dominant) group members feel uncomfortable or annoyed at having to consider other social groupings. These factors provided obvious flashpoints for employees, especially those exposed to poor quality DEI training. It should be remembered that all of us will likely fall within a DEI protected group at some point in our lives, whether through disability, age, having a child, or as a veteran. DEI is not exclusively about gender, sexuality, or skin color; plenty of disadvantaged scholars, ranchers, and farmers have benefited greatly from federal DEI initiatives.

Diversity awareness and training arose around the time of the human rights movement in the 1960s, with greater uptake in the 1980s and 1990s by businesses dealing with, and protecting themselves from, civil rights suits. An outcome of this movement was the introduction of classes of people legally protected from discrimination and harassment, with accommodations for disability. The relevant laws include the Equal Pay Act of 1963, the Civil Rights Act of 1964 (and particularly title VII), the Age Discrimination in Employment Act 1967, Title I of the Americans with Disabilities Act of 1990, the Genetic Information Nondiscrimination Act of 2008, and the Pregnant Workers Fairness Act 2022. These resulted in our contemporary DEI framework, which is much more broadly applied. While there is continued debate about the best approach, and a reactive backlash, DEI represented a framework that was (and still is in many places) used in workplaces for the purposes of legal compliance. (Detailed DEI definitions can be found in many places, including from the American Medical Association's 2021 Advancing Health Equity: A guide to language, narrative and concepts.)

Academia has a troubled relationship with DEI, and minorities have traditionally faced barriers to participation, funding, and publication. More broadly, there have been calls to extend the implementation of DEI initiatives more commonly seen in the Global North to assist minorities disadvantaged in the Global South, such as Indigenous groups (Ocampo-Ariza et al., 2023). Universities have also been accused of tokenism, gaming DEI, or engaging in "performative allyship". Disciplines and social movements have a history of bias and discrimination (Jenkins & Rudd, 2022; Aiello et al., 2021) that persist to the present day, and it is part of our duty when participating in the activity of animal welfare science to call it out.

Keywords: Diversity, Equity, and Inclusion (DEI)

READINGS

Animal welfare science as a discipline/profession

Ward, S. J., Sherwen, S., & Clark, F. E. (2018). Advances in applied zoo animal welfare science. *Journal of Applied Animal Welfare Science, 21*(supl), 23–33. https://doi.org/10.1080/10888705.2018.1513842. PMID: 30325227.

Cultural context

Adelman, M., & Thompson, K. (2017). Introduction to equestrian cultures in global and local contexts. *Equestrian cultures in global and local contexts* (pp. 1–14). Springer International Publishing.

Garcia, A., & McGlone, J. J. (2022). Animal welfare and the acknowledgment of cultural differences. *Animals, 12*(4), 474.

Mota-Rojas, D., Strappini, A., Whittaker, A. L., Ghezzi, M., Gonçalves Titto, C., Calderón-Maldonado, N., & Orihuela, A. (2023). Controversial topics in animal welfare in Latin America: A focus on the legislation surrounding the human-companion animal relationship and animals used for recreational practices. *Animals, 13*(9), 1463.

Studying welfare internationally

Bayvel, A. D., & Cross, N. (2010). Animal welfare: A complex domestic and international public-policy issue—who are the key players? *Journal of Veterinary Medical Education, 37*(1), 3–12.

Buller, H., Blokhuis, H., Jensen, P., & Keeling, L. (2018). Towards farm animal welfare and sustainability. *Animals, 8*(6), 81.

Li, P. J. (2021). *Animal welfare in China: Culture, politics and crisis*. Sydney University Press.

Yang, Q., Dwyer, C. M., Vigors, B., Zhao, R., & Langford, F. M. (2024). Animal welfare with Chinese characteristics: Chinese poultry producers' perceptions of, and attitudes towards, animal welfare. *PLoS One, 19*(7), e0307061.

Intersectionality

Alexis, N. A. (2020). Beyond compare: Intersectionality and interspeciesism for co-liberation with other animals. In B. Fischer (Ed.), *The Routledge handbook of animal ethics* (p. 502). Routledge.

Crenshaw, K. (1989). Demarginalizing the intersection of race and sex: A black feminist critique of antidiscrimination doctrine, feminist theory and antiracist politics. *The University of Chicago Legal Forum, 140*, 139–167.

Ko, A., & Ko, S. (2017). *Aphro-ism: Essays on pop culture, feminism, and black veganism from two sisters*. Lantern Books.

Maddow, R. (2023). *Prequel: An American fight against fascism*. Crown.

Segurel, L., Guarino-Vignon, P., Marchi, N., Lafosse, S., Laurent, R., Bon, C., Fabre, A., Hegay, T., & Heyer, E. (2020). Why and when was lactase persistence selected for? Insights from Central Asian herders and ancient DNA. *PLoS Biology, 18*(6), e3000742. https://doi.org/10.1371/journal.pbio.3000742. PMID: 32511234; PMCID: PMC7302802.

Whitman, J. Q. (2017). *Hitler's American model: The United States and the making of Nazi Race Law*. Princeton University Press. https://doi.org/10.2307/j.ctvc774zw

Affinity organizations

Gelberg, S., & Gelberg, H. (2007). Native American tribal cultures: Implications for veterinary medical education. *Journal of Veterinary Medical Education, 34*(2), 60–73.

Jenkins Jr, J. L., & Rudd, M. L. (2022). Decolonizing animal welfare through a social justice framework. *Frontiers in Veterinary Science, 1509*. https://doi.org/10.3389/fvets.2021.787555

Diversity, equity, and inclusion (DEI)

Aiello, M., Bismar, D., Casanova, S., Manuel Casas, J., Chang, D., Chin, J.L., Comas-Diaz, L., Salvo Crane, L., Demir, Z., Garcia, M. A., Hita, L., Leverett, P., Mendez, K., Skawen Morse, G., O'Leary Sloan, M., Weil, M. C., & Blume, A. W. (2021). Protecting and defending our people: Nakni tushka Anowa (The Warrior's Path) Final Report. APA division 45 Warrior's Path Presidential Task Force (2020). *Journal of Indigenous Research, 9*, 8.

Jenkins Jr, J. L., & Rudd, M. L. (2022). Decolonizing animal welfare through a social justice framework. *Frontiers in Veterinary Science, 1509*(8), 787555. https://doi.org/10.3389/fvets.2021.787555

Ocampo-Ariza, C., Toledo-Hernández, M., Librán-Embid, F., Armenteras, D., Vansynghel, J., Raveloaritiana, E., Arimond, I., Angulo-Rubiano, A., Tscharntke, T., Ramírez-Castañeda, V., Wurz, A., Marcacci, G., Anders, M., Urbina-Cardona, J. N., de Vos, A., Devy, S., Westphal, C., Toomey, A., Sheherazade, C. Y., & Maas, B. (2023). Global South leadership towards inclusive tropical ecology and conservation. *Perspectives in Ecology and Conservation, 21*(1), 17–24. https://doi.org/10.1016/j.pecon.2023.01.002

Chapter 21

Themes

21.1 HOW WE GOT HERE

Animal welfare science coalesced at a very particular period in our history as we rejected *welfare* as a legitimate topic of investigation without its conversion into a science. This breakthrough happened in the wake of the cognitive revolution as "mind" took center stage in psychology. In North America, animal welfare scientists codified a multi-domain approach; being happy and free from suffering was now considered central to welfare, with the addition of other wholesome elements like health and naturalness. This framework could accommodate a broad church—those uncomfortable with notions of subjectivity with those who focused on it almost exclusively, often studying the exact same animal uses.

21.2 AN UNLEVEL PLAYING FIELD

We started this book with a title "The unavoidability of animal welfare" and Marian Dawkins' statement that animal welfare science "thinks big". So, let's circle back and consider what it is about our societies that threatens animals so much.

A space to live. In November, 2022, the eight billionth human was born. According to UN demographers, the Earth's human population will top out at approximately 10.3 billion in 2084 (Ritchie & Rodés-Guirao, 2024). A 2019 report from the Intergovernmental Science-Policy Platform on Biodiversity and Ecosystem Services noted that humans have already impacted three-quarters of the terrestrial surface of the planet, with areas of human–animal conflict projected to increase, especially across the African continent. This is of obvious concern for that continent's megafauna.

Politics. Animals cannot vote; they are not stakeholders. Their interests are only peripherally served when they align with those of the human population. As noted in the book *Zoopolis: A political theory of animal rights* (Donaldson & Kymlicka, 2011), human society is really a multi-species affair, but we rarely

DOI: 10.1201/9781003375456-26

see this acknowledged, and more often than not, animals are targeted by laws rather than assisted by them.

Economies. Governments in many developed countries, including the United States, are pegged to short-term election cycles and economies that must inexorably grow, putting pressure on animal populations either directly or indirectly. In the United States, the birth rate dipped below replacement over a decade ago, with immigration used to cover the shortfall. Western nations are unlikely to change their economic models any time soon, as to do so would cede short-term advantage to other countries. Nonetheless, economists like Herman Daly (Victor, 2022) have championed the need to transition to steady-state economies that are better matched to population trajectories, resource use, and environmental imperatives.

Diet. We need to eat. Meat eating has a direct impact on production animals, whereas plant monoculture pushes animals from the land—an indirect death sentence. The difference between the two is that land is still needed to generate animal feed. Therefore, eating less meat, especially beef, frees up land to feed ourselves directly and reduces our impact on climate change (Scarborough et al., 2023). Irrespective of debates about diet, the need to reform farming has never been more urgent. To that end, Oregon Congressman Earl Blumenauer, newly retired, spent nearly three decades trying to reform the US Farm Bill, noting that the present legislation "gives too much to the wrong people to grow the wrong food in the wrong places". Blumenauer's most recent attempt to introduce reform was endorsed by the ASPCA, Friends of the Earth, The Union of Concerned Scientists (UCS), and many others. Arrayed against him were giant agribusiness corporations and industry associations who parted with over $500 million to pay lobbyists over the last five years—this is pay-to-play on steroids (Omanjana & Stillerman, 2024). Powerful interests erode democracy, especially when the checks and balances that were designed to regulate their monopolies failed decades ago.

Monopolies. It should be no surprise to anyone living stateside that the US corporate scene is a "monopoly fest". How many power providers, telecom companies, or grocers do you have the option of using? US anti-monopoly laws (called antitrust) failed in the 1970s in the pursuit of an "economy of scale". The result was massive consolidation that has largely squeezed out mid- and small-sized operations. The good news was that the 2021 chair of the Federal Trade Commission (FTC), Lina Kahn, flagged this regulatory snafu in an article that went viral when she was just a law student (Khan, 2017). Later, Khan would leapfrog others to the top job at the FTC, where she filed multiple cases against the corporate giants accused of being "too big to care". Khan's tenure has now ended. If you want an idea of how dangerous consolidation is to workers and animal welfare, you just need to look at the meatpacking titans and their stranglehold over the industry.

21.3 MESSAGING

In a turbulent world, it can be difficult to attract attention to the plight of animal species. So, the messaging has to be right. We know that culturally, the *meme* (first coined by Richard Dawkins) can be an incredibly robust generator of emergent and sometimes destructive cultural movements. Could we interrogate our written cultural backstory for hints at the emergence of a culture of animal welfare?

Tools such as Google's NGRAM Viewer (https://books.google.com/ngram; Michel et al., 2011) can be used to track word use in books to give a broad-brush view of the ebb and flow of ideas. Obviously, it's an abstraction of the real picture, but it provides pointers as to cultural "pressure points". If we input welfare-related terms (Figure 21.1; top), we see that the idea of "humane education" (the little green bump) peaks around the time of the collective trauma experienced from World War I but has fallen away in modern times. "Vivisection" (purple line) generates two massive peaks, the first coinciding with UK activism (1870s), and the second after its spread to the United States (1890s). Searches with "animal liberation", "animal welfare", "animal ethics", and "animal rights" show the modern dominance of "animal welfare" (blue line), with Peter Singer's "animal liberation" (orange line) barely registering, despite being such a major influence for so many animal advocates. But even the modern-day peak of animal welfare has yet to surpass that of "vivisection" from the late 1800s. A clue as to why can be found in the *viral texts project*. Without the constraints of copyright, information exchange was quite extraordinary in the 1800s. A search of "vivisection" using the viral texts project database (https://viraltexts.org/) shows the UK newspaper stories that were copied, copied, and copied again (Cordell & Smith, 2024). If you do the search

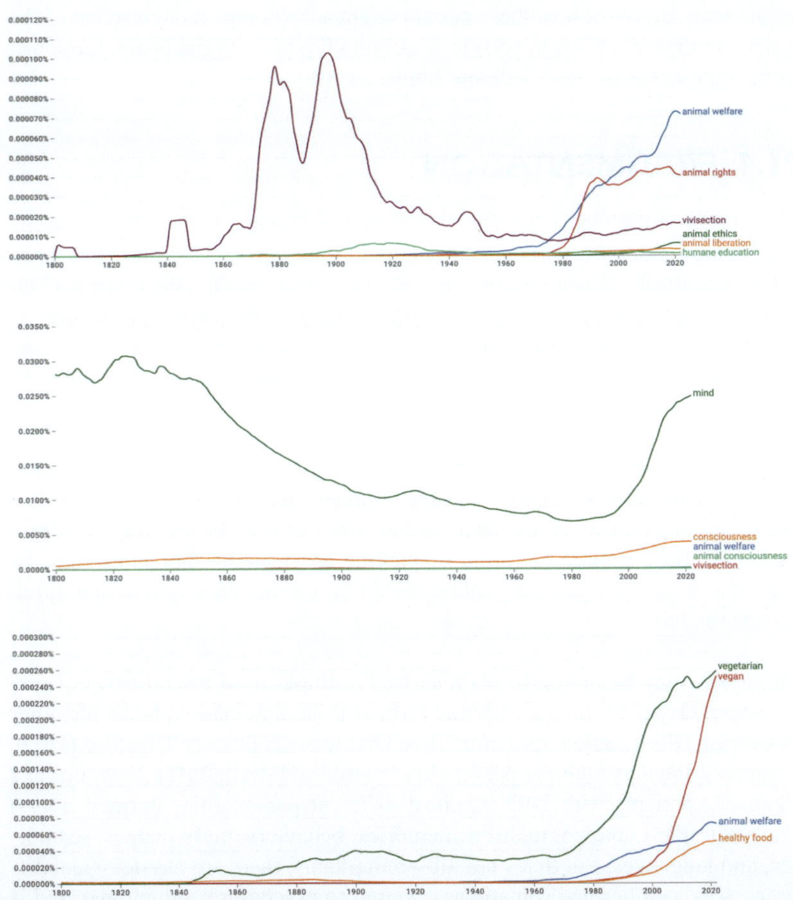

Figure 21.1 Google Books Ngram Viewer (https://books.google.com/ngrams) showing the frequency of word use since 1800 for three sets of word queries, with search terms shown at the right-hand side. The search term "animal welfare" was retained in each chart to facilitate a scale comparison.

yourself, you'll notice a large jump in hits in 1874. That date coincides with a visit from Éugène Magnan, a French physiologist and vivisector, who, in front of an audience of British medics, tortured a dog in the name of science. This shows the power of a single act to shock and provoke.

People are also fascinated by cognition and mind, and have been for a long time. So, perhaps not surprisingly, adding the term "mind" (Figure 21.1; middle) generated such a large hit that the other searches were flattened (the green line). Perhaps, tethering messaging about animal welfare to these interests might gain some traction. Lastly, mentions of "vegan" (red line) have massively increased since 2000, and are now comparable with "vegetarian" (green line), with both much greater than mentions of "healthy food" (Figure 21.1; bottom). According to the *2023 Plant-based foods state of the marketplace report,* 62% of US households now buy plant-based products, which have become mainstream with the "Beyond Meat" and "Impossible Foods" burgers and sausages.

When it comes to addressing the needs of animals who are suffering, this cultural backdrop can prove surprisingly irrelevant. Just as a person suffering from moderate depression is equally likely to be helped if they choose to receive counseling from a psychologist, psychoanalyst, or clergy person, a piglet screaming in pain after castration might be effectively assisted by a veterinarian with anesthetic, an immunologist developing immuno-castration, or an organic farm choosing to slaughter before sexual maturity. This assistance occurs irrespective of how these people might accept, reject, or conceive of the piglet's sentience (although outside of eliminating post-surgical pain, these different approaches do have different implications).

21.4 FRAGMENTATION

We still have little idea as to what the animal or human mind *is*. Our scientific approaches remain firmly pegged to physical functions and structures. Science has its guardrails. When researchers complain about methodologies or conclusions that they don't like, they may rush to apply a label of pseudoscience, as happened recently for the IIT theory of consciousness. But data are data—they are either methodologically and experimentally sound, or not. If you make the data up, then we call that fraud. In terms of research into the nature of consciousness, tremendous progress has been made in studying the brain's informational networks, processes, and pathologies. But we have yet to understand what consciousness is, because those experiments have yet to be designed. Just like imagining that we could decipher the power of sight by carving up an eye, we cannot fully explore the mind by picking apart brains. The mind must be studied as it is, and as we use it; not as we'd like it to be. We might have to think outside the box!

An answer may be hiding in plain sight. I'm thinking of the property of *dissociation*. Day-dreaming is a normal form of mild dissociation, but at the other end of the "dissociation spectrum" lies Dissociative Identity Disorder (DID), formerly known as multiple personality disorder (Howell, 2011). Now a robust diagnosis, patients with DID manifest different personalities (termed *alters*) along with their autobiographical memories, behaviors, body images, sexualities, and languages. But that's not all—remarkably, there are physiological differences as well (termed somatic as opposed to psychoform symptoms), and if these are triggered by alter switches, then we might have the beginnings of a methodological "handle" on the mind–body nexus. Alters can manifest differences in "dominant handedness, response to the same medication, allergic sensitivities, autonomic and endocrine function, EEG, VEP, and regional cerebral blood flow" (Birnbaum & Thomann, 1996). There are a multitude of anecdotal reports of patients needing different eyeglass prescriptions, and this variation has been published (Miller et al., 1991). One patient discovered, after 15 years of blindness, that one of her alters could see. As sight "spread" to her other alters, visual evoked potentials (VEPs) typical of sighted people would appear in EEG tests (Strasburger & Waldvogel, 2015). The unexplored biology of DID provides an opportunity to examine the mind–body connection and perhaps devise new therapies for these patients.

The problem is that Western science is itself remarkably prone to fragmentation, with disciplines calving like icebergs from older fields and then breaking up further. Experts two doors away from each other may not realize the skills and perspectives that they could bring to each other's work. Now, we find breakthroughs from reconnecting the parts, like new immunotherapies for cancer patients (uniting immunology and cancer). A fusion of ethology with ethnology also seems a natural solution to the study of mixed human–animal cultures, as suggested by the French philosopher Dominique Lestel (Chrulew et al., 2017).

Reconnecting disciplines has the potential to make our research so much more interesting. Take, for example, the following description of individual "X": a picky eater, xenophobic, and culturally conservative. Sounds familiar? X is a killer whale, and her culture is described in Whitehead and Rendells' 2014 publication, *The cultural lives of whales and dolphins.* Orca society is prone to fads, thrill killing, holidaying (off Brazil) in warm waters, and the rampant destruction of ecosystems. The rare phenomenon of menopause also occurs in Orca females. Are killer whales to the sea what human beings are to land, and are the differences largely down to environment? What does killer whale culture say about us?

Scientists can also trip themselves up in more obvious ways. The Harlow monkey experiments were indefensible, and Harlow made no attempt to hide his dislike of animals: "I really don't like animals. I despise cats, I hate dogs. How could you like a monkey?" (as quoted in Blum, 1994; Gluck, 1997). It is now suggested that Harlow's cruel experiments might have reflected a desire to find a cure for the depression that he himself had suffered (van Rosmalen et al., 2022). Irrespective of his motives, data recovered from a human under psychological torture are treated as suspect, as the "tortured" brain manifests disrupted brain function (Liddell et al., 2022). The same should hold for animals, and so publishing such material should be prohibited.

Scientists are not emotional blank slates, and lived experiences can create strange parallels with scientific lives. Are we really to imagine that as Turing developed his "Imitation Game" to flush out the machine intelligence masquerading as human (Part II), he never reflected on his own situation of having to hide his sexuality in a hostile wartime UK (https://www.turing.ac.uk/blog/what-alan-turing-means-us)? As Lynn Margulis (then Sagan) struggled to do the housework and look after the kids while Carl sat at his typewriter, did she pause to reflect on her theory of one organism captured by another and turned into its workhorse (Sagan, 1967)? Was Karl von Frisch's study of "fright substance" colored by his denunciation as a quarter Jewish to the Nazis, who were busy killing his students for their resistance activities (Zupanc & Wanninger, 2024)? And did the endless mouse universes of Calhoun resonate with his ancestry of Scots émigrés fleeing societal collapse (Calhoun, 1971)? And of course, the mind can be the superpower in your job, as it is for Temple Grandin in her work with animals.

Closer to home, animals don't have the luxury of meta-debates about the philosophy of science, and are caught up in more immediate crises. For our companion animals, we have the issue of pet-induced debt, alarming levels of pet surrender from Gen Zers, and a global unaffordability of medical care and pet food (https://www.lendingtree.com/debt-consolidation/pet-debt-report/). We don't have to have pets. So, if corporations keep turning the financial thumbscrews on their customers, will they create an unintended consequence of lowering pet ownership?

21.5 FIN

Let's end on a more positive note. The authors of *Zoopolis*, Sue Donaldson and Will Kymlicka, confessed to being Star Trek fans and imagined a first contact with a new species (Donaldson & Kymlicka, 2011). It's a sobering thought and puts our behavior into perspective. The two imagined how we might behave if we met, for example, an extraterrestrial parrot, and how we might strive to keep that lifeform alive, happy, and thriving. In short, we would treat our "guest" parrot so much better than an Earthly one. But what do we *really* know about the lives of our terrestrial animals?

You may not have come with us every step of the way through science, ethics, and philosophy. But if your work with animals inspires moments of insight or joy and an increasing sense of wonder, you're probably exactly where you need to be. There are animals on our planet with capabilities that we can't even begin to imagine. Sperm whale (*Physeter macrocephalus*) language is being decoded right now using machine learning and AI; it might be the first animal language to share a common structure (duality of patterning) with human language (Sharma et al., 2024). African elephants have unique names for each other (just like us and bottlenose dolphins) and make long-distance calls by chatting at frequencies so low that we can't hear them (Pardo et al., 2024). Sea slugs (*Elysia chlorotica*) become solar powered by sucking chloroplasts out of algae, so that they never have to eat again (Rumpho et al., 2011). Animals are spectacular, so let's do the best that we can for them.

READINGS

An unlevel playing field

Donaldson, S., & Kymlicka, W. (2011). *Zoopolis: A political theory of animal rights.* Oxford University Press.

Khan, L. M. (2017). Amazon's antitrust paradox. *The Yale Law Journal, 126*, 710.

Omanjana, G., & Stillerman, K. P. (2024). *Cultivating control: Corporate lobbying on the food and farm bill.* Union of Concerned Scientists. https://doi.org/10.47923/2024.15464

Ritchie, H., & Rodés-Guirao, L. (2024). Peak global population and other key findings from the 2024 UN World population prospects. Published online at OurWorldinData.org. 'https://ourworldindata.org/un-population-2024-revision' [Online Resource]

Scarborough, P., Clark, M., Cobiac, L., Papier, K., Knuppel, A., Lynch, J., Harrington, R., Key, T., & Springmann, M. (2023). Vegans, vegetarians, fish-eaters and meat-eaters in the UK show discrepant environmental impacts. *Nature Food, 4*(7), 565–574. https://doi.org/10.1038/s43016-023-00795-w

Victor, P. A. (2021). *Herman Daly's economics for a full world: His life and ideas.* Routledge.

Messaging

Cordell, R., & Smith, D. (2024). *Viral texts: Mapping* networks *of* reprinting *in 19th-century newspapers and magazines.* http://viraltexts.org

Michel, J. B., Shen, Y. K., Aiden, A. P., Veres, A., Gray, M. K., Google Books Team, Pickett, J. P., Hoiberg, D., Clancy, D., Norvig, P., Orwant, J., Pinker, S., Nowak, M. A., & Aiden, E. L. (2011). Quantitative analysis of culture using millions of digitized books. *Science (New York, N.Y.), 331*(6014), 176–182. https://doi.org/10.1126/science.1199644

Fragmentation

Birnbaum, M. H., & Thomann, K. (1996). Visual function in multiple personality disorder. *Journal of the American Optometric Association, 67*(6), 327–334.

Blum, D. (1994). *The monkey wars.* Oxford University Press.

Calhoun, J. B. (1971). Space and the strategy of life. In A. H. Esser (Ed.), *Behavior and environment* (pp. 329–387). Plenum Press.

Chrulew, M., Bussolini, J., & Buchanan, B (Eds.). (2017). *The philosophical ethology of Dominique Lestel.* Routledge.

Gluck, J. P. (1997). Harry F. Harlow and animal research: Reflection on the ethical paradox. *Ethical Behavior, 7*(2), 149–161. https://doi.org/10.1207/s15327019eb0702_6. PMID: 11655129.

Howell, E. (2011). *Understanding and treating: Dissociative identity disorder: A relational approach.* New York, Routledge.

Liddell, B. J., Das, P., Malhi, G. S., Felmingham, K. L., Outhred, T., Cheung, J., Den, M., Nickerson, A., Askovic, M., Aroche, J., Coello, M., & Bryant, R. A. (2022). Torture exposure and the functional brain: Investigating disruptions to intrinsic network connectivity using resting state fMRI. *Transl Psychiatry, 12*(1), 37.

Miller, S. D., Blackburn, T., Scholes, G., White, G. L., & Mamalis, N. (1991). Optical differences in multiple personality disorder. A second look. *The Journal of Nervous and Mental Disease, 179*(3), 132–135. https://doi.org/10.1097/00005053-199103000-00003

Sagan, L. (1967). On the origin of mitosing cells. *Journal of Theoretical Biology, 14*(3), 225–274. https://doi.org/10.1016/0022-5193(67)90079-3

Strasburger, H., & Waldvogel, B. (2015). Sight and blindness in the same person: Gating in the visual system. *PsyCh Journal, 4*(4), 178–185. https://doi.org/10.1002/pchj.109

van Rosmalen, L., Luijk, M. P. C. M., & van der Horst, F. C. P. (2022). Harry Harlow's pit of despair: Depression in monkeys and men. *Journal of the History of the Behavioral Sciences, 58*(2), 204–222. https://doi.org/10.1002/jhbs.22180. Epub 2022 Jan 18. PMID: 35040491; PMCID: PMC9305880.

Whitehead, H., & Rendell, L. (2015). *The cultural lives of Whales and Dolphins.* The University of Chicago Press.

Zupanc, G. K. H., & Wanninger, S. (2024). "Resistance leads to self-destruction": How an (a)political strategy helped Karl von Frisch succeed during the Nazi era. *Journal of Comparative Physiology. A, Neuroethology, Sensory, Neural, and Behavioral Physiology, 210*(2), 145–166. https://doi.org/10.1007/s00359-024-01697-3

Fin

Pardo, M. A., Fristrup, K., Lolchuragi, D. S., Poole, J. H., Granli, P., Moss, C., Douglas-Hamilton, I., & Wittemyer, G. (2024). African elephants address one another with individually specific name-like calls. *Nature Ecology & Evolution, 8*(7), 1353–1364. https://doi.org/10.1038/s41559-024-02420-w

Rumpho, M. E., Pelletreau, K. N., Moustafa, A., & Bhattacharya, D. (2011). The making of a photosynthetic animal. *The Journal of Experimental Biology, 214*(Pt 2), 303–311. https://doi.org/10.1242/jeb.046540. PMID: 21177950; PMCID: PMC3008634.

Sharma, P., Gero, S., Payne, R., Gruber, D. F., Rus, D., Torralba, A., & Andreas, J. (2024). Contextual and combinatorial structure in sperm whale vocalisations. *Nature Communications, 15*(1), 3617. https://doi.org/10.1038/s41467-024-47221-8

Glossary

A note on how long this glossary is: Animal Welfare Science is an interdisciplinary space that has captured terminology from many different fields and is somewhat prone to jargon. This glossary focuses on current use in the field but will vary from place to place and over time. Some of these terms are very useful, while others are just something we are in the habit of using, and you need to know what they mean to avoid unnecessary confusion.

A note on Latinate words: Many terms are based on Latin because that was historically the language of European scholarship. On the one hand, this is pointlessly obscure. On the other, these words are kind of cool and easy to understand if you know their constituent parts, which include *an* (without), *aisthēsis* (feeling), *allo* (same), *anthropoid* (human), *contra* (opposite), and *stasis* (stable or fixed).

Term	Definition	Most relevant book chapter and section
Ablation studies (neuroscience)	Cutting, removing, or injuring a part of the brain for research purposes.	6.7
Academia	The environment or community concerned with the pursuit of research, education, and scholarship.	19.2
Affective	Relating to feelings or emotion.	6.8
Affective neuroscience	The study of how the brain produces emotion.	6.8
Affinity organization	An educational organization formed around a shared goal or identity—including and representing individuals who share an aspect of identity.	20.5
Agonistic	Relating to aggression or conflict.	12.2
Allogrooming	To clean and maintain the appearance of a conspecific.	12.5
Allostasis	The systematic process of achieving a state of emotional and physiological stability—to re-establish the homeostatic set point.	3.1
Analgesia	A process that eliminates feelings of pain.	6.6
Ancient breed (dog)	An informal term used for dog breeds that arose before the formalization of pedigrees in the Victorian era.	17.4
Ancient contract	The conceptualization of domestication as an agreement between humans and animals in which both parties enter into a permanent relationship for their mutual benefit.	14.8
Anesthesia	A deliberate medical procedure that renders an individual unconscious.	1.1
Animal-assisted therapy	Providing an animal in an activity with structured goals to improve a person's ability to function (e.g., socially, emotionally, cognitively) —often provided by a licensed professional.	14.5

Term	Definition	Most relevant book chapter and section
Animal cruelty	Socially unacceptable behavior that intentionally causes unnecessary pain, suffering, or distress to and/or death of an animal.	15.3
Animal fads	Sudden and dramatic increases in pet ownership, use, or interest, fed by popular culture.	16.2
Animal Liberation Front (ALF)	A decentralized organization committed to animal rights, denoted a domestic terrorist organization by the US and UK because of threats made to scientists, bombings, and arson.	2.2
Animal rights	The position that animals are morally equivalent to human beings is based on being subject-of-a-life (rights) or capable of suffering (liberation), with the conclusion that exploiting animals for human purposes is not ethical.	1.1
Animal welfare	An ethically significant state of an animal ranging from negative to positive (defined more specifically within various models such as the 5 Freedoms and 3 Circles).	1.1
Animal welfare assessment	A structured method to determine or estimate the welfare of an animal according to categories or a scale.	9.1
Animal welfare assessment contests	A contest presents scenarios about animals under different conditions, and experts assess competitors on how well they determine which animals have better welfare and how they formulate and present their reasoning.	4.7
Animal welfare science	The use of scientific methods and theories to understand and advance animal welfare.	1.2
Animism	A sense that something/someone has a spiritual essence, identity, or presence beyond being simply dull matter.	8.9
Anna Karenina principle	If any of the many parts of a complex situation is bad, the whole situation is bad.	9.5
Anthropocentric	Regarding humankind as the central or most important element of existence.	1.1
Anthropodenialism	A blindness to the human-like characteristics of other animals.	8.13
Anthropogenic	Originating from human activities.	6.4
Anthropomorphism	The attribution, in error, of human qualities to nonhuman animals, objects, or phenomena. In relation to animals, this attribution may sometimes be valid, even if frowned-upon.	8.13
Anthrozoology	The study of interactions between humans and other animals.	4.1
Anxiety	Fear relating to something other than an immediate peril, for example, relating to an anticipated event.	3.2
APGAR	An assessment for babies within 1–5 minutes of birth, including—Appearance, Pulse, Grimace, Activity, and Respiration.	8.12

Term	Definition	Most relevant book chapter and section
Arcadian myth	An ideal of the good life that occurs in a pastoral/natural setting.	14.2
Artificial selection	Selectively breeding animals to perpetuate or develop desired traits.	6.10
Assessment	When a quality is measured or estimated concretely to support a decision or action.	9.1
Attachment	An association with an individual or object distinguished by staying near, protecting, feeling secure, and being distressed when separated.	16.5
Automata (re: Descartes)	A moving mechanical device that copies or mimics a human being or some other live organism.	1.1
Autonomy	The ability or right to self-govern or be independent.	3.3
Barbering (especially referencing mice)	Systematic plucking of another mouse's fur and whiskers.	12.2
Behavioral diversity	The richness of behavior as measured by the number of different behaviors performed and their relative frequency.	7.8
Behavioral genetics	The study of how genes and the environment influence behavior.	7.8
Behavioral sink	The behavioral collapse that results from overcrowding.	7.6
Berserk male syndrome	A proposed way in which hand-raised domestic animals, especially camelids, can become dangerous from mis-imprinting on people.	11.3
Bicultural	Having the cultural attitudes and customs of two nations, peoples, or ethnic groups.	4.4
Big Lick	An exaggerated gait of horses, most notably the Tennessee Walking Horse, where the front hooves are raised to reach chest high and forward.	6.9
Biomarker	A biological indicator of a normal or abnormal state, process, or disease.	4.7
Biophilia	An innate/genetically determined affinity of human beings with the natural world.	14.4
Biopsychism	The notion that all life forms are sentient.	8.4
Blind hens challenge	An ethically problematic scenario where one can disable animals ostensibly to improve their welfare—for example, blinding hens to prevent pecking and cannibalism.	7.3
Boarded	Specialty boarded refers to a veterinary professional's certification showing that they have advanced training and competency in a specialization.	1.3
Bodily integrity (of animals)	How well or richly an animal interacts with their environment. It may include a degree of bodily autonomy.	7.3

Term	Definition	Most relevant book chapter and section
Boycott	Decision to refuse to purchase a product or engage with an entity economically, with encouragement that others do the same.	18.8
Brahman Cattle (American Brahman)	A breed created in the mid-19th century from Zebu cattle imported from India and Brazil. Known for their hardiness, longevity, mothering skills, and lean beef production.	10.5
Breed	A stock of animals with distinguishing characteristics created by artificial selection.	6.10
Breed-specific legislation	Any law that limits or prohibits ownership of certain animals, typically breeds of dog. It is currently most used to limit or reduce the ownership of Pit Bull-type dogs.	14.7
Caponizing	Castration of a chicken (or other poultry).	12.4
Capture myopathy	When extreme stress causes severe muscle damage, often caused by the capture or restraint of wildlife. Can lead to organ failure and death.	11.1
Cartesian dualism	The problem of how to combine the physical body with the intangible mind, as discussed by René Descartes.	6.8
Charismatic megafauna	Large animal species with symbolic value or widespread popular appeal.	11.6
Chromatin	The complex of DNA and its bound proteins (histones).	4.7
Citizen science	Scientific research that includes the active participation of members of the public.	17.7
Codified	An arrangement of laws and rules.	18.1
Cognitive bias	A systematic distortion in perception or understanding of a phenomenon due to a flaw in information processing. For example, animals from chronically stressful environments tend to interpret ambiguous situations as being negative.	8.15
Cognitive dissonance	The strain caused by holding contradictory beliefs.	4.9
Cognitive revolution, the	A movement that started in the 1950s, and that united many fields of study to develop a scientific understanding of the mind and subjective experience, in addition to observable behavior.	1.2
Concentrated animal feed operation (CAFO)	CAFOs are intensive farming sites large enough that the USDA regulates their activities that might otherwise pollute waterways. Specifically, those containing over 1,000 "animal units" confined for over 45 days a year. (An animal unit is the equivalent of 1,000 pounds of "live" animal weight.)	6.4
Confucius foods	Foods that follow the recommendations of the highly influential Chinese philosopher Confucius (551–479 BCE).	17.5
Consciousness	Having awareness, for example, a sense of self or thinking about thinking. There are many definitions that are context dependent. You may find sentience and consciousness used interchangeably.	8.1

Term	Definition	Most relevant book chapter and section
Conservation	Protecting or restoring wildlife and their environments including the prevention of species extinction.	11.1
Conspecific	A member of the same species.	12.1
Conspecific proxy	Use of an animal of a different species to provide another (animal) with social support.	13.3
Contrafreeloading	Choosing to work for something when you could get the same reward for free.	7.1
Coping	Adapting to the environment without excessive strain.	3.1
Cortisol	A steroid hormone that is involved in the physiological stress response.	6.6
Co-sheltering	Accommodating animals in shelters provided for people.	16.4
CRISPR-Cas9	A method to precisely edit the genome of an organism.	6.11
Critical animal studies	An interdisciplinary approach to reveal and challenge power structures that harm animals, and to promote animal liberation.	14.2
Critical period (sensitive period)	A period in an individual's development when certain skills or characteristics are most easily acquired.	4.7
Cross-sucking	When members of a litter direct nursing or sucking behavior to each other, with the potential to cause serious injury.	12.2
Cube model	A model representing interacting factors as the dimensions of a cube.	4.3
Culling	1) Killing wild animals where their numbers have been deemed harmful or excessive. 2) Killing animals from a population of domesticated animals that are unneeded, deemed inferior, or not fit for purpose.	6.4
Cultural competencies	The ability to work effectively in cross-cultural situations.	17.3
Culture	A set of norms and other practices associated with a particular group. For example, religion or nation.	17.1
Dangerousness	Behavior/s that represents a threat to oneself or others' safety.	16.5
Dark Triad (a.k.a. MacDonald Triad)	There are two uses of this term. Referenced here is a historical theory that in male adolescents, there is an association between bed wetting, fire setting, and animal cruelty. It is also used to describe the association between narcissism, Machiavellianism, and psychopathy.	16.5
De-extinction	The use of scientific methods to recreate living individuals of an extinct species either as exact replicas (cloned) or proxies (edited from a template species).	11.4

Term	Definition	Most relevant book chapter and section
Delphi survey	A method for gathering a selection of diverse experts and facilitating their discussion until they reach a consensus on a task.	9.1
Deontology	An ethical approach to understanding our duties based on the rights accorded to others.	18.4
Depopulation	The killing of all members of a population (e.g., all members of a species within a geographical location).	6.4
Deviance	Conduct that does not conform to rules and norms, with implications of immorality.	18.1
Descriptive linguistics (vs. prescriptive)	Descriptive approaches describe actual language use, while prescriptive approaches specify how language should be used.	16.1
Diversity, Equity, and Inclusion (DEI)	A framework for avoiding discrimination while supporting a diverse workforce.	20.6
Dewatering (fish-killing)	Killing a fish by removing it from the water resulting in asphyxiation.	17.2
Dine'é	Dine'é refers to the Navajo Nation, with Diné referring to individuals.	6.4
Discipline	A branch of knowledge associated with an accumulated body of information and training methods, often at the tertiary level.	19.7
Dither fish	Small schooling fish who are added to an aquarium to promote calmer and less fearful behavior in large exhibit fish like cichlids.	13.1
Doctrine of mutual benefit, the	Based on legal doctrine, a person cannot continue to benefit from a deed without subscribing to its obligations.	14.5
Domesticated	A true-breeding stock of animals being exploited for a human purpose, and that has become more convenient to keep in captivity.	6.10
Dominance	The ability to secure limited resources in advance of other conspecifics.	12.3
Draize test	A method that assesses how irritating a substance is by applying it to the eye or skin of an animal and recording its effects.	2.1
Drive	An actual or potentially quantifiable expression of motivation.	3.4
Duty of care	An ethical position based on an intuitively recognized and moral value for caring relationships, especially of the vulnerable.	18.5
Dystocia	Difficult or abnormal labor and birth by any biological or medical cause.	12.4
Echinococcus granulosus	A tapeworm that causes the zoonotic disease referred to as hydatid disease.	6.3

Term	Definition	Most relevant book chapter and section
ECREE	The aphorism "extraordinary claims require extraordinary evidence", attributed to the astronomer, Carl Sagan.	8.3
Empathy	Mentally taking the place of another in an attempt to understand how they are feeling.	11.3
Endemic	Commonly occurring in an area.	6.3
End-of-life	A period that begins when a life-limiting condition is deemed not curable.	9.4
Environmental enrichment	Increasing the complexity of a captive animal's environment or experiences to improve their welfare (multiple more specific definitions exist).	7.6
Epigenome	The chemical modifications to the DNA and histone proteins of the genome.	4.7
Epistemology	The philosophical study of knowledge.	8.8
Ethical matrix, the	A conceptual tool designed to help decision-making.	18.7
Ethogram	A thorough description of all of the behaviors exhibited by members of a species.	7.4
Ethology	The scientific study of animal behavior.	3.3
Eudaimonia	A complete and spiritually fulfilled and fulfilling life, living life to its best—thriving.	18.6
Euthanasia, animal/ veterinary	Humane killing, specifically instantaneous death or unconsciousness, followed by death.	8.11
Falsifiable	Able to be proven incorrect through testing.	2.1
Feather pecking	Where poultry peck conspecifics—may escalate to feather pulling and cannibalism.	12.2
Fission / fusion society	A social group where the size and composition changes over time as animals move throughout the environment; the ability to live in a social or solitary fashion.	13.5
Five Freedoms, the	An animal welfare model that emphasizes freedom from hunger, discomfort, pain, fear, and behavioral restrictions.	2.7
Five Domains, the	An animal welfare assessment model that includes: 1 Nutrition, 2 Physical Environment, 3 Health, 4 Behavioral Interactions, and 5 Mental State.	9.2
Five Provisions, the	An animal welfare model that expands upon the five freedoms to include a range of positive factors rather than just the elimination of negative ones.	2.7
Flight zone	An area around an animal that when entered by a human handler will cause the animal to move away.	12.6
Fright substance (Schreckstoff)	A chemical cocktail released by certain fish taxa after injury. One chemical is a generalized alarm signal and the other is species-specific, so it might be interpreted as "alarm"—"for zebrafish".	10.2

Term	Definition	Most relevant book chapter and section
Gaze alternation	An inter-species method of getting aid or attention by repeatedly looking between a problem and an individual able to assist.	15.6
Gene drive	Introducing a gene to a population that causes it to rapidly spread through the population.	6.11
Global Neuronal Workspace Theory (GNWT)	A functional theory of consciousness.	8.6
Green criminology	The criminological study of harm and injustice to the environment, including nonhuman animals.	15.2
Grimace	Contortion and tensing of the face, associated with various forms of distress including pain and disgust, and can be coded according to severity scales created for various species.	8.12
Group processes	The behaviors of members in small working groups as they engage in decision-making and task performance.	14.3
H5N1	A subtype of the influenza A virus, which predominantly affects birds.	6.4
Hedonic state	A positive hedonic state is pleasurable, with hedonism the pursuit of pleasure. An individual with a low hedonic tone is less likely to feel pleasure. Anhedonia is the decreased ability to experience pleasure.	6.8
Hebb-Williams maze	A series of maze problems designed for rodents and used with many other species intended to assess intelligence/cognitive abilities.	3.2
Heterospecific	Between members of two different species.	12.1
Hierarchy of needs	A model showing that the basic requirements for life, such as food, must be in place before higher levels of well-being can be achieved for most individuals (from Abraham Maslow).	4.3
Hoarding (animal)	The accumulation of a large number of animals and failure to provide minimal standards of nutrition, sanitation, and veterinary care when a reasonable person would do so.	15.3
Homeostasis	The internal physical state that the animal returns to at rest. This can change with time, as is typical for adaptive responses that alter resilience.	3.1
Homunculus	A representation of a "small person" as part of a theoretical model. Suggesting an aspect of the process that is poorly understood.	3.3
Honest signaling	A communication that accurately reflects an animal's biological state—often because it requires effort to perform and so reflects an animal's health and vigor.	13.2

Term	Definition	Most relevant book chapter and section
Human–animal interactions	Human–animal interactions refer to the mutual and dynamic exchanges between humans and other animals.	15.1
Humane education	Is an "approach to teaching that draws connections between human rights, animal protection, and environmental sustainability". (per the Institute for Humane Education.)	14.6
Hunt saboteur	A person or group that deliberately disrupts a hunting activity, commonly associated with fox hunting using hounds.	2.2
Idiopathic	Without a known or identifiable medical cause.	13.4
Inbreeding depression	The tendency for biological function to decline with increased inbreeding.	6.10
Indigenous Traditional Ecological Knowledge (ITEK)	This is a holistic worldview (cosmology) based on a generationally acquired knowledge of place, living things, culture, and spirituality. There are multiple place and community specific ITEKs. (Other definitions might be used elsewhere.)	11.2
Infrahumanization	The tacitly held belief that members of a group you belong to are "more human" than people outside of the group.	14.3
Instinctive drift	An individual using an instinctive or automatic behavior in preference to one that is trained or learned (i.e. the tendency to revert to instinct).	19.6
Institutional Animal Care and Use Committee (IACUC)	A group that approves and oversees research at a facility according to scientific and ethical standards.	1.3
Inelastic demand	The willingness to dramatically increase expenditure for a resource, as required, indicating the resource is a need and not a luxury.	18.8
Infinite regression	An infinite series of entities governed by a recursive principle that determines how each entity in the series depends on, or is produced by, its predecessor.	3.3
Integrated Information theory (IIT)	A physicalist theory of consciousness.	8.6
Interdisciplinary	Interdisciplinary work involves connecting and integrating several academic schools of thought, professions, or technologies, along with their philosophies and cultures, in the pursuit of a common task.	19.7
International supply chain for animal products and live animals	The movement of live animals or animal products across international borders inclusive of production, storage, processing and packaging, distribution, and consumption.	20.3
Internship	Experiential learning, often unpaid, and often used as work experience.	19.1

Term	Definition	Most relevant book chapter and section
Intersectionality	How social categorizations, for example race, class, and gender, create overlapping and distinct kinds of discrimination against individuals and groups.	20.4
Judas goat	An animal trained to lead other herd animals calmly to a planned destination, usually slaughter.	12.6
Lab lore	The informal beliefs and practices passed down by members of a scientific laboratory.	17.1
Linear relation	When one variable is increased by a set amount, another variable is also increased by a uniform amount, such that the relation between the two can be plotted on a straight line.	2.2
Littermate syndrome (dogs)	A phenomenon reported in dogs raised from a young age with littermates where they become either aggressive to each other or overly dependent on each other.	13.4
Market regulation approach	Leaving consumers to decide acceptable production methods via labeling and choice rather than using government regulation.	18.9
Mātauranga Māori	The traditional knowledge of the Māori people of New Zealand.	4.4
Microbiome	The population of microorganisms in a location.	10.1
Mind–body problem	If the mind is intangible, how does it interact with the body?	6.8
Mirror test	An animal's body is marked while they are unconscious. They are provided with a mirror and if they direct attention to the mark this indicates some degree of self-awareness.	8.3
Mixed experience	A subjective experience with components of positive and negative valence.	4.8
Monism	A view of a single and fundamental underlying reality of matter *or* mind, with either mind subordinate to, and created by matter (matter is fundamental), or matter construed as a mental perception of mind (mind is fundamental). Dualism posits fundamental roles for matter and mind.	n/a
Motivation	A tendency to take a particular action.	3.4
Multi-domain model	A theoretical model that includes and groups together qualitatively different sources of evidence – for example, the Three Circles.	4.2
Multi-item survey	A tool to conduct an animal welfare assessment with preset questions to assess various domains, each with a rating scale. The tool may provide a score for each domain, and an overall score for welfare.	9.3
Multi-species ethnography	A more-than-human approach to ethnographic research and writing.	15.6

Term	Definition	Most relevant book chapter and section
Multi-species habitat	An artificial habitat that accommodates the needs of more than one species, with cohabiting species acting as enrichment for each other without compromising the welfare of either.	13.4
Musth	Male elephants in "Musth" are ready to mate, and this is associated with more aggressive or unpredictable behaviors.	12.1
Natural experiment	An observational study in which data are collected irrespective of variable natural conditions or other factors outside of the control of the investigators.	17.7
Natural history	Observations about the natural world, especially of plants and animals.	7.2
Navajo Churro	A breed of domestic sheep originating with the Spanish Churra sheep obtained by the Diné around the 16th century during the Spanish Conquest.	6.4
Needs, organic	Organic needs are resources and conditions required to remain alive.	3.4
Neural substrate	Often used to describe that part of the brain (its structure and function) that underlies a specific process or activity.	8.11
Neuroscience	The scientific study of the structure or function of the nervous system and brain.	6.7
Nociceptor	A sensory receptor for painful stimuli.	8.3
Nongovernmental organization (NGO)	A group that functions independently of any government with the objective of improving social conditions.	19.3
Non-use welfare	The economic value derived from good welfare.	19.6
Norms	Informal rules and expectations in a community or group of individuals.	17.1
Nutritional wisdom	When animals select their own food, they tend to select nutritionally suitable options.	3.6
Object play	Any play with a physical object.	13.2
One Health	An integrated, unifying approach that aims to sustainably balance and optimize the health of people, animals, and ecosystems.	6.4
Ontology	The philosophical study of our reality.	8.8
Open field	An enclosed open space used to assess fear in animals such as rodents who freeze when feeling at risk of predation.	3.2
Oxytocin	A hormone produced in the hypothalamus and released by the posterior pituitary that is associated with social bonding and positive mood.	14.2
Palliative care (hospice)	Care provided during the end-of-life period when life-limiting conditions impact quality of life.	9.4

Term	Definition	Most relevant book chapter and section
Panpsychism	The belief that consciousness is fundamental to the entire universe.	8.4
Paraprofessionals	People who are in a team in association with one or more licensed or credentialed practitioners.	19.5
Peer-review (refereed)	Peer review is the process of evaluating manuscripts by using a panel of people with expertise and qualifications in the subject of the academic journal.	19.2
Personality	A pattern of behavior (including internal behaviors like thinking and feeling) persisting over time.	16.3
Pest (species)	A species that is damaging to activities or resources valued by humans.	4.9
Physiological stress	Any condition that challenges an individual's physiological set-point (homeostasis).	3.2
Pillar model	A model representing factors as pillars, that require equivalent development to support a foundation or roof (which denotes the desired outcome, such as welfare).	4.3
Play	Exuberant and enjoyable interactions with others, objects, and the environment, theorized to function as skill development and practice for the unexpected.	4.8
Pluralism (re: work approach)	An approach based on consulting multiple opinions and sources of knowledge.	9.2
Polled	Lacking horns typical for the species, for genetic reasons.	6.9
Pound seizure law	A law enabling the legal seizure of animals from pounds for use in research testing or teaching.	17.2
Practical ethics	A framework for making ethical decisions in the real world.	18.7
Precautionary principle, the	A risk assessment in the face of uncertainty (i.e., a lack of actionable knowledge) that minimizes exposure to harm.	2.6
Precision livestock farming	The use of technology in farming to continuously monitor animals for the purposes of assessing health, welfare, productivity, or environmental impacts.	10.4
Predation	Killing an organism for food.	13.2
Predator avoidance training	Training animals to avoid predators, normally by the association of predator and predator-related cues with aversive stimuli.	13.2
Profession	Occupation that involves prolonged training and a formal qualification with ethical obligations.	19.7
PSE pork	A lower quality pork associated with stress (pink, soft, exudative).	7.7
Psycho-hydraulic model	A psychological model that represents behavioral drives as water pressure.	3.3

Term	Definition	Most relevant book chapter and section
Psychological scale	A survey that seeks to quantify a target's psychological quality with a rating from zero to some maximum number.	16.3
Psychopharmacology	The effect of pharmaceuticals on psychology (mind/behavior).	3.1
Qualia	The qualities of subjective experience e.g., how we experience color, taste etc.	8.2
Quality adjusted life years (QALY)	Usually a measure of a health outcome that takes into account life expectancy and quality of life.	4.7
Quality of life assessment	An assessment for how an animal is experiencing their life, based on scores from different domains such as psychological, physical, and social.	9.4
Qualitative behavioral assessment	A method for quantifying how people intuitively understand animal body language based on principal component analysis.	8.14
Qualitative research	A range of approaches to describe and understand individuals' lived experiences (the meaning people attach to their own beliefs, motivations, and actions).	16.3
Ractopamine	A livestock feed additive to promote muscle growth and sold in the US as Paylean, Optaflexx, and Topmax. Known for side effects on metabolism and behavior, especially when overdosed.	6.11
Rage syndrome (dogs)	Rage syndrome refers to an inherited condition of uncontrollable aggression in dogs.	12.4
Rat tickling	Human handlers using their hands to take part in "rough and tumble play" with rats.	15.4
Reasonable person test	A legal approach to understanding negligence by comparison with a hypothetical person acting with an appropriate or normative degree of care and attention.	15.3
Reintegrative shaming	A form of public shaming to motivate normative ethical behavior.	2.1
Representative sample	A subset of a group or population that accurately represents its characteristics.	10.2
Reputation	A widespread belief that an individual has specific positive and/or negative characteristics (especially amongst conspecifics).	12.5
Reverse zoonosis	A disease transmitted from humans to animals.	6.4
Rewilding (re: domesticated animals)	The study of domestic animals released into wild or wild-like environments.	7.5
Rinderpest	An RNA virus historically responsible for devastating "cattle plague" deaths.	5.2
Role model	One who serves as an exemplar for others.	12.1

Term	Definition	Most relevant book chapter and section
Rotation exhibit	A practice primarily in zoos of moving animals between exhibits so they can indirectly experience the presence of other individuals and other species without having direct interactions.	12.2
Satire	Literary form that ridicules the norms or conduct of a group of people.	17.6
Scientific materialism, functionalism, and physicalism	Philosophies of science that assert that our fundamental reality (including mind), is created by matter (materialism), or its functions (functionalism), or the physical world (physicalism). Also see "monism".	1.1
Scientific method	An approach to understanding and predicting the world, which may include features such as hypothesis testing, a systematic approach, and a shared body of knowledge.	1.2
Scientific theory	Some testable aspect of the world that can be proven to be right or wrong.	2.1
Sentience	Being able to experience feelings such as pain and pleasure.	8.1
Situated practice	Learning that relates to real-world applications.	n/a
Slaughter	Killing animals to harvest commodities such as meat or hides.	4.6
Social engineering	Attempts by a central authority to deliberately create, destroy, or modify social norms in a group or culture.	17.1
Social movement	A widespread effort to achieve a permanent change in how society operates.	17.2
Soring (gaited horse)	The practice of placing caustic substances on a horse's legs to create a high-stepping gait.	6.9
Spay/neuter	US terminology for the reproductive sterilization of a female or male dog. "Neuter" may also be used to describe the sterilization of either sex.	6.9
Speciesism	Inaccurate or unethical discrimination based on species.	11.6
Species specific behavior (species typical)	Behavior demonstrated by most members of a species in a similar manner.	7.8
Spectrum of care	A range of options in services such as medical care with acceptable outcomes that are all above the ethical minimum. These options allow providers to better accommodate what may be preferred, suitable, or affordable.	6.5
Spillover (infection)	Infection of a novel host population from an infected reservoir population. Subsequent infection between individuals of the novel host population may or may not occur.	6.4
Standard of care	Achieving an acceptable minimum standard of services (legal term).	6.5

Term	Definition	Most relevant book chapter and section
Stereotypy	A repeated behavior with no immediately obvious function.	8.16
Stunning	A method that renders an animal unconscious.	17.2
Suffering	A seriously negative or unpleasant experience of a sentient being.	3.5
Superfund site	A highly contaminated area that often poses an environmental hazard that is listed for a federally funded cleanup.	11.6
Telos	A set of criteria that encompass the nature of an animal and the elements needed to have a good life i.e., "pigness of the pig".	7.9
Theory of change	A living roadmap for achieving change that identifies the desired outcome(s), the starting point, and then constructs the "missing middle" using an inclusive consultative process.	16.4
Thought experiment	A hypothetical situation where the consequences are imagined.	18.2
Three Circles, the	A framework of animal welfare in which physical health, natural life, and subjective states are seen as overlapping concerns.	4.1
Three Rs, the	The goal to replace, reduce, and refine the use of animals in research, teaching, and testing.	2.4
Three Ss, the	Guiding principles for ethical animal research, specifically Good Science, Good Sense, and Good Sensibilities.	2.4
Tiered model	A model where some elements directly affect the outcome, but others only influence it indirectly.	4.3
Tradition	A belief or practice passed down from previous generations.	17.5
Two-Eyed-Seeing	A rubric for cross-cultural collaboration.	11.2
Umwelt	The philosophy that animals are inextricably connected with their environment as they understand and experience it.	3.3
Utilitarianism	A philosophical approach where risks and benefits are weighed when deciding whether to take or allow an action, and the goal is the greatest attainment of good for the greatest number of people (or individuals).	18.3
Valence	Re: emotion, being positive or negative in empirical value. For example, a pleasurable emotion is rated as positive.	6.8
Value-added products	Products that have been improved so that people will pay more than they would for the raw ingredients.	19.6
Vegan	Strictly speaking, a person who does not consume or use products from animals, especially food products.	2.5

Term	Definition	Most relevant book chapter and section
Veterinarians' oath	Graduating veterinarians swear this oath which includes the text: "I solemnly swear to use my scientific knowledge and skills for the benefit of society through the protection of animal health and welfare, the prevention and relief of animal suffering".	1.3
Veterinary care desert	A geographical area where veterinary care is unavailable or inaccessible.	6.5
Veterinary clinical trial/research	Research into spontaneous illness in client-owned animals to develop effective treatments. The animal is a patient rather than a test subject. Data may, or may not, be relevant to human medicine.	14.4
Visitor effect	How the presence of human observers impacts animals, such as zoo-exhibited animals. The effect may be positive, negative, or mixed.	15.5
Vivisection	Surgery without anesthesia, such that pain is consciously experienced.	1.1
Welfare-washing a.k.a. humane-washing	Misleading or false claims that misrepresent a product or practice as being better for animal welfare than it is.	18.9
Wild type	The phenotype (form) that is most typically found in nature.	11.5
Wilderness cure	A practice common in later 19th century America, to go to wild natural areas as a cure for diseases such as consumption (tuberculosis).	14.2
Wildlife harassment	An act that disregards the wild animal's welfare, disrupts normal behaviors, or risks injury.	14.5
Wind sucking	Abnormal behavior in horses where they forcefully gulp air into their stomach.	8.16
Yawn contagion	The tendency for one individual to yawn after they observe yawning by another in their social group or category.	15.6
Zooarchaeology	Within archeology, the study of animal remains, often bones.	17.4
Zoobiquity	The intersection of clinical human medicine, behavioral medicine, veterinary medicine, and evolutionary biology.	14.4
Zoonosis	Diseases that pass from animals to humans.	6.4

PEOPLE AND ANIMALS OF NOTE

Name	Contribution	Most relevant book chapter and section
Alex – parrot 1976–2007	An African grey parrot that demonstrated advanced spoken English language use, including identifying abstract categories.	8.9
Arnold Arluke, 1947–	A sociologist known for his enlightening work on the meaning of animals to people, best known for the classic book *Regarding Animals* (1996).	17.1
Jeremy Bentham, 1748–1832	English philosopher and founder of modern utilitarianism who stated that ethical concern is due to any being capable of experiencing suffering. Famously, Bentham supported a "greatest happiness principle" for all.	8.1
F.W. Rogers Brambell, 1901–1970	Zoologist and head of the committee that authored the "*Report of the Technical Committee to Inquire into the Welfare of Animals Kept Under Intensive Livestock Husbandry Systems*" which led to the development of the Five Freedoms.	2.6
Donald Broom, 1942–	Appointed to the first professorship in animal welfare (University of Cambridge) in 1986 with interests in ethics and animal cognition.	3.1
R "Rex" L, Burch, 1926–1996	Microbiologist and pioneer in laboratory animal welfare. Known for co-authoring "*The Principles of Humane Experimental Technique*".	2.4
Bruce Cattanach, 1932–2020	A renowned mouse geneticist also known for creating bob-tailed Boxer dogs able to be shown as pedigree dogs.	6.9
David Chalmers, 1966–	Leading Australian philosopher in the field of human cognition who coined the phrase, "hard problem" of consciousness. This is the problem of how the brain's physical structure and function can lead to subjective conscious experience.	6.7
Grahame Clark, 1907–1995	A British archeologist who drew attention to how plants and animals are an integral part of a site, providing context for any hominid remains or artifacts that might be found.	17.4
Marian Dawkins, 1945–	Oxford ethologist who has worked extensively on the use of preference assessment and the role of consciousness in animal welfare.	3.5
René Descartes, 1596–1650	Prominent French philosopher known for asserting that animals are not capable of experiencing suffering, as indicated by not expressing themselves through language.	1.1 & 6.8

Name	Contribution	Most relevant book chapter and section
Karl von Frisch, 1886–1982	German Austrian ethologist and Nobel Laureate. Frisch helped decipher the "bee dance", and chemical alarms called *Schreckstoff* (fright substance) released by injured minnows.	3.3
Temple Grandin, 1947–	An animal behaviorist who made revolutionary changes to how animal behavior was considered, especially in slaughterhouses.	10.3
Clever Hans – horse 1895–1916	Initially presented as capable of answering questions and doing math, Hans was later found to be reading human body language to know when to stop when tapping out answers.	8.9
Harry Harlow, 1905–1981	An American psychologist who conducted controversial experiments in studying attachment and social needs in infant rhesus monkeys.	6.7
Ruth Harrison, 1920–2000	A Quaker conscientious objector and animal welfare activist. Best known for her book *Animal Machines* which influenced the public and influential leaders including Peter Singer and Ingrid Newkirk.	2.6
Gwendolen Howard, 1894–1973	Reclusive British naturalist and musician who lived in *Bird Cottage* with a freely moving population of wild birds in the 1930s. Creator of animal narratives and author of *Birds as Individuals* (1952) and *Living with Birds* (1956) under the pseudonym Len Howard. Biography *Bird Cottage* written by Eva Meijer (2021).	7.2
Alicia Juarrero	Scholar, founder and president of VectorAnalytica Inc. Author of *Context Changes Everything: How Constraints Create Coherence* (2023). Expert in understanding top-down complex systems.	11.2
Immanuel Kant, 1724–1804	German philosopher of the European Enlightenment and hugely influential in developing rationalism and empiricism. Of relevance to consciousness research are his ideas of the whole and its parts.	1.1
Bernardo Kastrup, 1974 –	Leading voice for Analytical Idealism, a metaphysics that puts mind at the fore of its worldview.	8.8
Ronald "Ron" Kilgour, 1932–1988	A researcher of mainly livestock behavior who helped show the intelligence of these species and the value of studying them beyond what was needed to enhance productivity.	8.10
Christof Koch, 1956–	Koch is a physicist, neuroscientist, and ex-president of the Allen Institute for Brain Science. A leading voice in the field of consciousness research, Koch has helped to disseminate and explain complex ideas through popular books and online.	8.5 & 8.6

Name	Contribution	Most relevant book chapter and section
Konrad Lorenz, 1903–1989	An Austrian zoologist and Nobel Laureate. A founder of observational ethology and known for his studies on instinct, attachment, and imprinting.	3.3
Laurie Marker, 1954–	Zoologist and founder of the *Cheetah Conservation Fund*. A leading expert on Cheetah and a driving force in working to prevent their extinction.	13.3
Hal Markowitz, 1934–2012	Pioneer in environmental enrichment in zoos. Author of *Behavioral Enrichment in the Zoo* (1982).	13.2
Elder Albert D. Marshall, 1938–	An Elder of the Moose Clan of the Mi'kmaw Nation and leading advocate for the process of *Two-Eyed Seeing* (or *Etuaptmumk*), which can be used to help groups from different cultures to work together.	11.2
Lyle McNeal, 1942–	Utah State University professor who was instrumental in saving the Navajo Churro sheep; founder of the Navajo Sheep project (https://www.navajosheepproject.org/about).	6.4
David Mellor, 1942–	Physiologist turned animal welfare scientist who developed the Five Domains approach to animal welfare assessment.	9.2
Carol Newton, 1925–2014	After earning a PhD in physics and an M.D., Newton made advances in programming and the use of mathematics in biology and medicine. Newton taught and contributed to the NIH in these areas. Newton formulated the Three Ss (see above).	2.4
Jaak Panksepp, 1943–2017	Estonian born neuroscientist who pioneered *affective neuroscience*—the scientific study of emotion. Panksepp highlighted the similarities in human and animal emotions and described seven emotional systems.	6.8
Roger Penrose, 1931–	Mathematician, physicist, and Nobel Laureate. Devised the Orch OR theory of consciousness with Stuart Hameroff as an early and radical theory of consciousness involving quantum effects.	8.7
Irene Pepperberg, 1949–	A researcher best known for demonstrating advanced cognition in parrots by teaching them to use language.	8.9
Tom Regan, 1938–2017	American philosopher who proposed that because animals are the subject-of-a-life, they have the right not to be exploited by humans ("animal rights").	1.1
William Moy Stratton Russell, 1925–2006	A polymath—zoologist, psychoanalyst, lecturer in sociology, statistics, and cultural evolution. Best known for co-authoring *The Principles of Humane Experimental Technique* with R.L. Burch.	2.4

Name	Contribution	Most relevant book chapter and section
Hans Selye, 1907–1982	An endocrinologist who showed that the body responds in a similar (stereotypical) way to different stressors, creating the basis for understanding the physiological stress response.	3.2
Peter Singer, 1946–	An Australian moral philosopher who argued that because animals can suffer, their interests and the interests of humans must be given equal weight.	18.3
B.F. Skinner, 1904–1990	Psychologist and behaviorist who championed the study of observable behavior over the mind, and operant conditioning—i.e., behavior shaped by consequences.	6.7
Henry Spira, 1927–1998	Journalist and social activist known for his theory of reintegrative shaming.	2.1
Edward Taub, 1931–	A behavioral neuroscientist who researched neuroplasticity, including techniques for assisting stroke patients' recovery. His research became the subject of debate over the acceptable use of animals in research, a.k.a. "The Silver Springs Monkeys".	19.4
Henry David Thoreau, 1817–1862	An American philosopher and essayist who contributed to an idealized conception of the American lifestyle as self-reliance in connection with nature.	14.2
Nikolaas "Niko" Tinbergen, 1907–1988	Dutch biologist, ornithologist, and Nobel Laureate, who cofounded modern ethology. Known for his work on innate and learned behaviors, and in taking different perspectives when studying animal behavior.	3.3
Sojourner Truth, 1797–1883	Former slave who became a prominent activist for the rights of African Americans and for women.	20.4
Steven M. Wise, 1950–2024	An American legal scholar and theorist relating to the rights of animals.	7.3
D.G.M. Wood-Gush, 1922–1992	Ethologist-geneticist best known for studying the behavior of domesticated animals when released into large naturalistic areas (rewilding). Designed the first MSc in Applied Animal Behaviour and Welfare in 1990 at the University of Edinburgh.	7.5

Index